The Handbook of Environmental Chemistry

Edited by O. Hutzinger

Springer-Verlag Berlin Heidelberg GmbH

Volume 4 Part D

Airborne Particulate Matter

Volume Editors: T. Kouimtzis, C. Samara

With contributions by
S. Biegalski, I. Colbeck, W. Holländer,
T. Kouimtzis, P. Koutrakis, S. Landsberger,
J. M. Pacyna, S. N. Pandis, C. Pilinis,
C. Samara, E. Selin Lindgren, C. Sioutas,
A. I. Zouboulis

With 83 Figures and 50 Tables

Springer

Editor-in-Chief:
Professor Dr. Otto Hutzinger
University of Bayreuth
Chair of Ecological Chemistry and Geochemistry
P.O. Box 101251
D-95440 Bayreuth, Germany

Volume Editors:
Professor T. Kouimtzis
Dr. C. Samara
Aristotle University of Thessaloniki
Department of Chemistry
Environmental Pollution Control Laboratory
540 06 Thessaloniki, Greece

Library of Congress
The Handbook of environmental chemistry. Includes bibliographies and indexes. Contents: v. 1. The natural environment and the biogeochemical cycles / with contributions by P. Craig ... [et al] -- v. 2. Reactions and processes / with contributions by W.A. Bruggeman ... [et al] -- [etc.] -- v. 4. Air pollution / with contributions by H. van Dop ... [et al]
1. Environmental chemistry. I. Hutzinger, O.
QD31.H335 574.5'222 80-16607

ISBN 978-3-662-14881-5 ISBN 978-3-540-49145-3 (eBook)
DOI 10.1007/978-3-540-49145-3

Originally published by Springer-Verlag Berlin Heidelberg New York in 1995.
Softcover reprint of the hardcover 1st edition 1995

Typesetting: Macmillan India Ltd., Bangalore-25
SPIN: 10124961 52/3020 - 5 4 3 2 1 0 - Printed on acid-free paper

Preface

Environmental Chemistry is a relatively young science. Interest in this subject, however, is growing very rapidly and, although no agreement has been reached as yet about the exact content and limits of this interdisciplinary subject, there appears to be increasing interest in seeing environmental topics which are based on chemistry embodied in this subject. One of the first objectives of Environmental Chemistry must be the study of the environment and of natural chemical processes which occur in the environment. A major purpose of this series on Environmental Chemistry, therefore, is to present a reasonably uniform view of various aspects of the chemistry of the environment and chemical reactions occuring in the environment.

The industrial activities of man have given a new dimension to Environmental Chemistry. We have now synthesized and described over five million chemical compounds and chemical industry produces about one hundered and fifty million tons of synthetic chemicals anually. We ship billions of tons of oil per year and through mining operations and other geophysical modifications, large quantities of inorganic and organic materials are released from their natural deposits. Cities and metropolitan areas of up to 15 million inhabitants produce large quantities of waste in relatively small and confined areas. Much of the chemical products and waste products of modern society are relased into the environment either during production, storage, transport, use or ultimate disposal. These released materials participate in natural cycles and reactions and frequently lead to interference and disturbance of natural systems.

Environmental Chemistry is concerned with *reactions in the environment*. It is about distribution and equilibria between environmental compartments. It is about reactions, pathways, thermodynamics and kinetics. An important purpose of this Handbook is to aid understanding of the basic distribution and chemical reaction processes which occur in the environment.

Laws regulating toxic substances in various countries are designed to assess and control risk of chemicals to man and his environment. Science can contribute in two areas to this assessment: firstly in the area of toxicology and secondly in the area of chemical exposure. The available concentration ("environmental exposure concentration") depends on the fate of chemical compounds in the environment and thus their distribution and reaction behaviour in the environment. One very important contribution of Environmental Chemistry to the above mentioned toxic substances laws is to develop laboratory test methods, or mathematical correlations and models that predict the environmental fate of new chemical compounds. The third purpose of this Handbook is to help in the basic understanding and development of such test methods and models.

The last explicit purpose of the handbook is to present, in a concise form, the most important properties relating to environmental chemistry and hazard assessment for the most important series of chemical compounds.

At the moment three volumes of the Handbook are planned. Volume 1 deals with the natural environment and the biogeochemical cycles therein, including some background information such as energetics and ecology. Volume 2 is concerned with reactions and processes in the environment and deals with physical factors such as transport and adsorption, and chemical, photochemical and biochemical reactions in the environment, as well as some aspects of pharmacokinetics and metabolism within organisms. Volume 3 deals with anthropogenic compounds, their chemical backgrounds, production methods and information about their use, their environmental behaviour, analytical methodology and some important aspects of their toxic effects. The material for volumes 1, 2 and 3 was more than could easily be fitted into a single volume, and for this reason, as well as for the purpose of rapid publication of available manuscripts, all three volumes are published as a volume series (e.g. Vol. 1; A, B, C). Publisher and editor hope to keep the material of the volumes 1 to 3 up to date and to extend coverage in the subject areas by publishing further parts in the future. Readers are encouraged to offer suggestions and advice as to future editions of "The Handbook of Environmental Chemistry".

Most chapters in the Handbook are written to a fairly advanced level and should be of interest to the graduate student and practising scientist. I also hope that the subject matter treated will be of interest to people outside chemistry and to scientists in industry as well as government and regulatory bodies. It would be very satisfying for me to see the books used as a basis for developing graduate courses on Environmental Chemistry.

Due to the breadth of the subject matter, it was not easy to edit this Handbook. Specialists had to be found in quite different areas of science who were willing to contribute a chapter within the prescribed schedule. It is with great satisfaction that I thank all authors for their understanding and for devoting their time to this effort. Special thanks are due to the Springer publishing house and finally I would like to thank my family, students and colleagues for being so patient with me during several critical phases of preparation for the Handbook, and also to some colleagues and the secretaries for their technical help.

I consider it a privilege to see my chosen subject grow. My interest in Environmental Chemistry dates back to my early college days in Vienna. I received significant impulses during my postdoctoral period at the University of California and my interest slowly developed during my time with the National Research Council of Canada, before I was able to devote my full time to Environmental Chemistry in Amsterdam. I hope this Handbook will help deepen the interest of other scientists in this subject.

This preface was written in 1980. Since then publisher and editor have agreed to expand the Handbook by two new open-ended volume series: Air Pollution and Water Pollution. These broad topics could not be fitted easily into the headings of the first three volumes.

All five volume series will be integrated through the choice of topics covered and by a system of cross referencing.

The outline of the Handbook is thus as follows:

1. The Natural Environment and the Biogeochemical Cycles,
2. Reactions and Processes,
3. Anthropogenic Compounds,
4. Air Pollution,
5. Water Pollution.

Bayreuth, June 1991 Otto Hutzinger

Fifteen years have passed since the appearance of the first volumes of the solid scientific information in Environmental Chemistry has been well received, and with the help of many authors and volume-editors we have published a total of 24 books.

Although recent emphasis on chemical contaminants and industrial processes has broadened to include toxicological evaluation, risk assessment, life cycle analysis and similar approaches there is still a need for presentation of chemical and related facts pertaining to the environment. The publisher and editor therefore decided to continue our five volume series.

Bayreuth, March 1995 Otto Hutzinger

Contents

Introduction XI

Particle Emission from Outdoor and Indoor Sources
I. Colbeck 1

In Situ Particle Formation/Reaction Mechanisms
S. N. Pandis, C. Pilinis 35

Sources, Particle Size Distribution and Transport of Aeorsols
J. M. Pacyna 69

Physical, Chemical and Optical Properties of Atmospheric Aerosols
C. Pilinis, S. N. Pandis 99

Chemical Mass Balance
J. M. Pacyna, E. Selin Lindgren 125

Sampling of Airborne Particulate Matter
W. Holländer 143

Analysis of Inorganic Particulate Pollutants by Nuclear Methods
S. Landsberger, S. Biegalski 175

Methods for Measuring Atmospheric Acidic Particles and Gases
C. Sioutas, P. Koutrakis 201

Analysis of Organic Particulate Matter
C. Samara 233

Particle Counting and Particle Size Analysis
W. Holländer 253

Particulate Emission Control
T. Kouimtzis, A. I. Zouboulis 279

Subject Index 337

Introduction

Airborne particulate matter, the suspended material of the atmosphere, is one of the most important parameters dealt with in air pollution studies due to its adverse health and environmental impacts. Although global estimates show that the production of atmospheric particulates is dominated by natural sources, under local urban conditions the ambient aerosol may be totally dominated by human activities.

The volume *Airborne Particulate Matter* is part of the Volume 4 series: *Air Pollution* of *The Handbook of Environmental Chemistry*.

The volume provides a compilation Preface of today's knowledge regarding the emission, behavior, impacts, physical characterization, analysis and emission control of airborne particulates. This multidiscipline subject has been covered by leading experts with practical experience in their fields. Each author (or group of authors) was permitted to employ, within reasonable limits, his or her personal style in organizing and presenting a particular subject area and consequently it has been difficult to treat all the subject material in a homogeneous manner. Moreover, owing to space limitations some topics could not be treated in great detail or were totally omitted. Instead, the authors have provided a rich list of references for the benefit of the interested reader.

The volume is designed to serve as a basis text, as well as a comprehensive reference book. We hope and expect that it will provide a valuable source not only for scientists, but also for environmental managers and decision makers. We also hope that the volume will prove to have an equally high value for advanced undergraduate or graduate students.

The editors wish to thank the contributing authors for their time and effort to contribute a chapter within a prescribed schedule. They are also pleased to acknowledge the support received from Mr. P. Enders of Springer for his advice throughout the stages of preparation of the volume. Finally, they are grateful to the editor-in-chief of *The Handbook of Environmental Chemistry*, Prof. Dr. O. Hutzinger, for giving them the opportunity to become this volume's editors.

Th.Kouimtzis
C. Samara

Particle Emission from Outdoor and Indoor Sources

I. Colbeck

Institute for Environmental Research, University of Essex, Colchester, CO4 3SQ

Introduction 2
Mineral Dust 7
Sea Salt Particles 10
Extraterrestrial Sources 10
Volcanic Aerosol 11
Forest Fires/Slash Burning 11
Organic Particles 12
Industrial Sources 12
Gas-to-Particle Formation 20
Bioaerosols 22
Indoor Aerosols 23
Airborne Particulate Elemental Carbon (PEC) 24
Concluding Remarks 29
References 30

Summary

Aerosols, by their nature, are somewhat unstable in the sense that concentration and particle properties change with time. The sources of aerosols may be either natural or anthropogenic. However any source may be described as either primary (particles emitted directly into the atmosphere) or secondary (formed by chemical reactions of gaseous components). With industrialisation and urbanisation increasing, man emits substantial quantities of particulate material although, globally, natural contributions far exceed emissions from man's activities.

Introduction

Microscopic particles are ubiquitous in our environment. They are produced by natural processes and by man, both intentionally and unintentionally. Wind blown dust, smoke from forest fires and volcanic material are examples of naturally produced particles. Automobile exhaust and smoke from power generation are examples of anthropogenic particles. Many such particles have the ability to affect not only climate but also our health and quality of life. These particles, which may be manufactured or naturally produced, are called aerosols. The term aerosol, meaning a suspension of solid or liquid particles in a gaseous medium, originates from military research during World War I. The name is associated with Donnon [1], although the first publication of the term was due to Schmauss [2] where it was used as an analogy to hydrosol, a stable suspension of solid or liquid droplets in a gas or vapour.

The particle size of interest in aerosol behaviour ranges from molecular clusters of 0.001 μm to fog droplets and dust particles as large as 100 μm. Figure 1 illustrates the typical size range of various aerosols. There are various types of aerosol which are classified according to physical form and method of generation. The commonly used terms are [3]:

dust – a solid particle formed by mechanical disintegration of a parent material, such as in crushing, grinding and blasting;
fume – solids produced by physicochemical reactions such as combustion, sublimation or distillation;
smoke – visible aerosol produced by some sort of oxidation process such as burning;

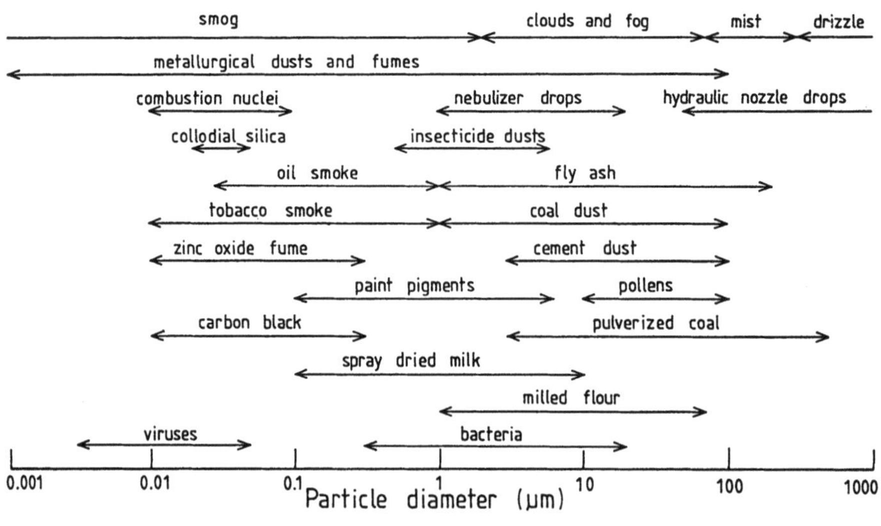

Fig. 1. Particle size range for aerosols

fog and mist – liquid particle aerosol produced by the disintegration of liquid or combustion of vapour;
smog – the term is derived from a combination of smoke and fog. It consists of solid and liquid particles created, at least in part, by photochemical reactions.

In the UK the following terminology is used:

grit – large particles (> 76 μm diameter) which will rapidly settle out of the air due to gravity and are just visible to the naked eye;
fume – very small particles (< 1 μm diameter) which can remain suspended in the air for long periods and which are visible only by electron microscopy;
dust – the intermediate size range (1 μm < diameter < 76 μm) which can be seen under an ordinary optical microscope.

Emissions of grit, dust and fume from various processes are controlled under the Clear Air Acts and the Environmental Protection Act 1990.

A number of terms describe the shape and origin of particles in an aerosol and include:

agglomerate – a group of particles held together by van de Waals forces or surface tension;
aggregate – a heterogeneous particle in which the various components are not easily broken up.

All the above terms have arisen from popular usage and in the general scientific literature the words are used loosely and interchangeably.

Aerosol sources can be classified as primary or secondary. Primary aerosols are those emitted directly in particulate form from sources such as power stations, motor vehicles etc. Secondary aerosols refer to particles formed within the atmosphere from condensation of vapours, or as a result of chemical reaction species.

Once aerosols are in the atmosphere, their size, number and chemical compositions are changed by several mechanisms until they are ultimately removed by natural processes. Measurements of particle size distributions normally show up to three groups of particles (Fig. 2). These are as follows.

Nucleation mode, < 0.2 μm diameter. In this mode particles have recently been either emitted from processes involving condensation of hot vapours, or freshly formed within the atmosphere by gas to particle conversion. Such particles account for the preponderance of particles by number, but because of their small size account for little of the total mass of airborne particles.

Accumulation mode, 0.2 – 2 μm diameter. These particles have grown from the nucleation mode by coagulation or condensation of vapours. They generally account for most of the aerosol surface area and a substantial part of the aerosol mass.

Coarse mode, > 2 μm diameter. These particles are mainly formed by mechanical attrition processes, and hence soil dust, sea spray and many industrial dusts fall within this mode.

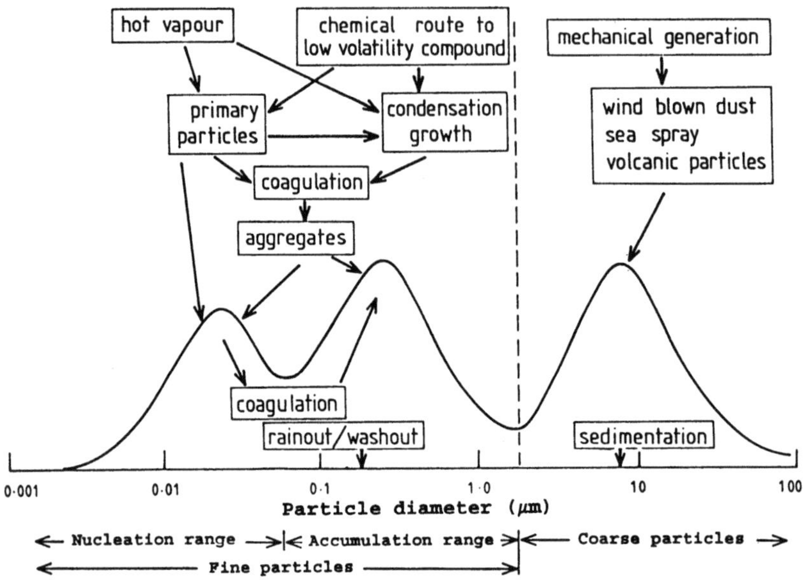

Fig. 2. Schematic diagram of a typical size distribution and formation mechanisms for atmospheric particles

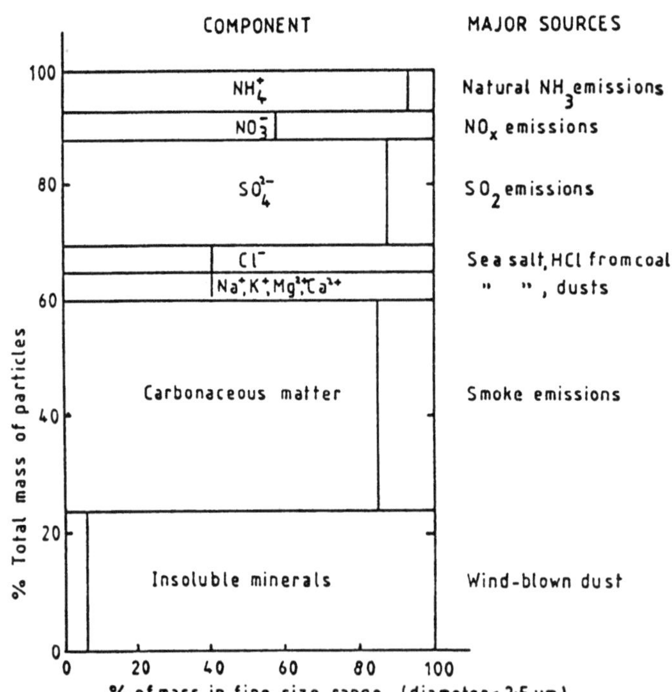

Fig. 3. Composition of atmospheric particles. Fine particles < 2.5 μm, coarse 2.5–15 μm. Data from Leeds 1982/83. Reproduced with permission from Clarke AG (1992) The atmosphere. In: Harrison RM (ed) Understanding our environment. Royal Society of Chemistry

Studies in which the chemical composition has been determined as a function of particle size have shown a division between the fine and coarse modes. The major components of the fine fraction of atmospheric aerosols are sulphate, nitrate, lead, carbonaceous material and condensed organic matter. The coarse fraction consists mainly of crustal material. Figure 3 shows a typical breakdown of the components of the total suspended particles for an urban area [4]. It is evident that the fine particles are dominated by ammonium sulphate and nitrate plus carbonaceous material. Roughly one third of this is elemental carbon and the other two thirds organic carbon which together comprise smoke. Wind blown dusts and sea salt dominate the coarse particles.

In addition to the distinction between primary and secondary aerosols a distinction may be made between natural and anthropogenic sources. Some sources

Table 1. Source strength of atmospheric aerosol particles in Tg per year for radii smaller than 100 μm (after d'Almeida et al. [91])

Aerosol Component	Source Strength	Reference
Natural Aerosol Sources		
○ Extraterrestrial dust	10	SCEP [84]
	16–18	Cameron [85]
○ Sea-salt	1000	SMIC [86], Junge [87]
	10000	Blanchard and Woodcock [22]
○ Desert dust	500	Peterson and Junge [88]
	1800–2000	d'Almeida [89]
○ Volcanic debris	25–150	SMIC [86]
	250	SCEP [84]
○ Biogenic/biologic	80	Bach [90]
○ Gas to particle conversion (sulfate, organics, nitrate)	570	SCEP [84]
	345–1100	SMIC [86]
Subtotal	3266–13448	
Anthropogenic Aerosol Sources		
○ Direct particulate emission	10–90	SMIC [86]
○ Gas to particle conversion (carbonaceous substances, organics, sulfate, nitrate)	275	SCEP [84]
	175–325	SMIC [86]
○ Forest fires and slash-burning debris[a]	3–150	SMIC [86]
Subtotal	188–565	
Total	3454–14013	

[a] Indicates unknown amount of indirect natural contributions

Table 2. Global emission estimates for major aerosol types in the 1980s. Sulfates and nitrogen are assumed to occur as ammonium salts (after Andreae [92])
Unit: Tg per year (dry mass)

Source	Estimated flux low	high	"best"	Particle size Category
Natural				
Primary				
Soil dust (mineral aerosol)	1000	3000	1500	mainly coarse
Seasalt	1000	10000	1300	coarse
Volcanic dust	4	10000	30	coarse
Primary organic aerosols	26	80	50	coarse
Secondary				
Sulfates from biogenic gases	60	110	90	fine
Sulfates from volcanic SO_2	10	30	20	fine
Organic matter from biogenic VOC	40	200	50	fine
Nitrates from NO_x	10	40	20	mainly coarse
Sum Natural	2150	23460	3070	
Anthropogenic				
Primary				
Industrial dust etc. (except soot)	40	130	100	fine and coarse
Soot	10	30	20	mainly fine
Biomass burning	50	190	80	fine
Secondary				
Sulfates from SO_2	120	180	140	fine
Nitrates from NO_x	20	50	40	mainly coarse
Organics from anthropogenic VOC	5	25	10	fine
Sum anthropogenic	245	555	290	
Total	2400	24000	3460	

are clearly natural (e.g. volcanic emissions) and some anthropogenic (e.g. industrial emissions). However there are also some sources where the distinction between natural and anthropogenic is more difficult to make (e.g. biomass burning and soil dust emissions). Tables 1, 2 and 3 summarize the source strengths of atmospheric aerosol. It is evident that there is a great uncertainty and variability in the estimates. On a global scale natural sources will dominate due to their emission from large area sources such as the oceans which cover over 70% of the globe. In contrast anthropogenic emissions over relatively smaller areas, such as cities, are likely to exceed the contributions from natural sources.

Table 3. Global aerosol production in Tg per year (after Preining [93])

	Jaenicke [94]	Hantel [95] estimates by Bach [96]	Monin [97]
Natural			
Direct			
Sea salt	0.2–2.0	0.5	1.0
Mineral dust	0.1–1.8	0.1	0.2
Volcanic dust	0.0–0.1	0.2	–
Forest fires	0.0–0.2	–	–
Subtotal	0.3–4.1	0.8	1.2
Conversions			
Sulphates	0.3–2.4	0.5	0.2
Nitrates	0.1–0.7	0.1	0.4
Hydrocarbons	0.1–1.1	0.1	0.2
Subtotal	0.5–4.2	0.7	0.8
Man made			
Direct	0.0–0.1	0.3	0.1
Conversions	0.2–0.3	0.4	0.2
Subtotal	0.2–0.4	0.7	0.3
Grand total	1.0–8.7	2.2	2.3

Mineral Dust

Continental sources of coarse aerosol arise mainly from the action of wind on loose soil and is primarily of importance due to the potential source areas that cover about one third of the total land mass. Dust sources, which include deserts and arid regions, vary in their geographical locations and are reflected in their chemical and physical properties. Regions in which dusts are major constituents of the atmospheric aerosol include the Sahara [5, 6], the Saudi Arabian peninsula [7, 8], the US Southwestern deserts [9, 10] and the Soviet Central Asian desert of Tadzhikistan [11]. The observation of mineral aerosol in these areas is fraught with difficulties due to the sporadic nature of seasonal dust storms and other events producing large amounts of suspended dust. Table 4 lists the major locations of dust sources which range from semi-arid to hyperarid conditions.

Atmospheric burdens vary from 500 Tg per year to 3000 Tg per year, while mass concentrations vary in the range 30–9500 $\mu g\, m^{-3}$ with an aerosol number concentrations between 250 and 2600 cm^{-3}. Substantial differences in source strengths are seen between the two hemispheres; this is related to the arid land area involved.

Arid weathering produces debris with a high coarse particle content where the fine dust is either blown away by the wind of washed into the sub-surface by runoff. Suspension of the soil-derived dust into the atmosphere occurs when the drag velocity of the wind exceeds the threshold required to entrain erodible surface particles. Different terrains have differing threshold velocities depending on a number of factors in which the soil particle size distribution is not necessarily

Table 4. Major global dust source locations with dust-storm frequency as days per year (after Middleton [98])

Source	Dust-storm Day Frequency
Africa	
Bodele Depression	22.5
S. Mauritania, N. Mali	27.4
Central Algeria	
Libya, Egypt	17.8
Namib Desert	Not Available
Middle East	
N. Saudi, Jordon, Syria	14.9
Lower Mesopotamia	27
S.W. Asia	
Makran Coast	27.3
Seistan Basin	80.7
Afghan Turkestan	46.7
Upper Indus Plain	18.9
Thar Desert	17.8
Former USSR	
Turkmenistan	65.5
Kara Kum	60
Rostov	23.3
Altay	25.1
Alma Ata	47.7
Kazakhstan	45.9
China	
Taklimakan Desert	32.9
Gansu Corridor	37.3
Mongolia	34.4
Americas	
SW USA	12.0[a]
Atacama Desert	Not Available
Australia	
Australian Desert	Not Available

[a] Days on which visibility < 11.3 km.

the controlling parameter. This effect is largely due to soil moisture and surface crusting by salts or organic matter [12]. Measured values of drag wind velocity range from 0.17 $m\,s^{-1}$ for uncrusted river channel deposits and dune sands to 2.5 $m\,s^{-1}$ for crusted or armoured alluvial fan sediments. These velocities correspond to mean wind velocities of 5–15 $m\,s^{-1}$ at a height of 10 m above ground level. A list of threshold velocities vs soil type is given in Table 5.

Generation of airborne dust also depends on the type of soil where the total clay and sand/silt plus clay content is important. Field tunnel tests have indicated that sediments classified as sand, loamy sand, and sandy loam are potentially significant sources of aerosol, as highly or moderately erodible material.

Table 5. Wind speed for initiation of dust production vs undisturbed or disturbed soil type (after Gillette [99])

Approximate range of friction velocity (cm/sec)	Soil type
20–40	Disturbed soils having less than 50% clay and less than 20% pebble (< 1 cm diameter) cover
25–40	Tilled bare sand soils
40–65	Disturbed pebbly soils
45–70	Bare clay soils that have been disaggregated by natural forces
40–150	Disturbed soils having a high salt content or more than 50% clay
140–200	Undisturbed sandy soils having a crust and soils covered with fine gravel
> 150	Undisturbed soils haivng more than 50% clay and surface crusts and salt-crusted soils
> 180	Soils covered by coarse (>5 cm) pebbles

The suspension of soil dust produces a particle size distribution which reflects the distribution of soils and rock weathered during the course of erosion. However, gravitational sedimentation will rapidly deplete particles larger than about 10μm. An analysis of Saharan dust indicates the presence of sub- 0.1 μm particles, of which the size was unexpected as only secondary mechanisms were thought to produce such particles. The exact mechanisms were attributed by Lerman [13] to weathering and not to break-up alone. For instance, water-soluble soil material may be dissolved to decrease particle sizes. Most distributions are similar except for those from alluvial land (Goundam) which contain a sub-micron fraction, and those from sand-dunes (Dar Albeida) exhibiting sizes below a sharp cut-off at about 12 μm.

During sandstorms particles larger than 100 μm may be suspended for short periods [14] while long-range transport involves sizes in the range 0.1–5 μm, observed up to 5000 km away [15]. As another example, mineral dust with $d < 2$ μm of Australian origin has also been found in Antarctic aerosol [16].

An examination of particle morphology may indicate whether dusts are proximal or distal in their origin. Coarse particles (10–20 μm) are generally characterized by an unpolished and porous structure which suggest recent mobilization, while particles below this size are more spherical. Measurement of particle morphology may be characterized by the ratio d_{max}/d_{min} or elongation. An investigation by Andronova et al. [17] for Tadzhikistan dust gave the following values of elongation as a function of overall size: $d_{max}/d_{min} \approx 1$ for $d < 10$ μm; 1.3 for $d <$ 10–20 μm; 1.5 for $d <$ 20–30 μm; 1.6 for $d <$ 30–50 μm; 1.7 for $d <$ 50–60 μm; 1.8 for $d <$ 60–70 μm; 1.5 for $d <$ 70–80 μm; 1.8 for $d <$ 80–100 μm and 2.0 for $d <$ 100–120 μm.

The chemical composition of aerosol dominated by wind-blown dust shows a close relation to material from the earth's crust. Abundances of elements may be described by an enrichment factor defined as (element X/$Fe_{AEROSOL}$)/(element X/$Fe_{CRUSTAL}$) where the ratio is normalised to the iron concentration. Other common crustal elements used to calculate enrichment factors or to trace particles back to their sources include aluminium, silicon, iron, titanium and calcium. If a Saharan dust is considered then the majority of elements have enrichments close to unity, in other words they are in the proportions of average crustal rock. Comparison to a global average indicates enrichment factors of over ten for more than half the elements indicating the dependence of the aerosol composition on local geological surface materials.

Mean values are more representative as the aerosols tend to be derived from a number of source areas and hence of fairly homogeneous composition [18]. As a general rule almost all analyzed elements are found to be enriched in the aerosol except for silicon and this is attributed to the greater sedimentation velocity of quartz relative to clay. Further details on the physical and chemical properties of crustal derived aerosol are given in articles by Schutz and Jaenicke [19], Patterson and Gillette [20], d'Almeida [21], and d'Almeida and Schutz [14].

Sea Salt Particles

The generation of aerosols by the oceans was thoroughly investigated in the 1950s and since that time a clearer picture of the maritime source has emerged. There are four mechanisms of bubble building and a subsequent jet drop: raindrops, snowflakes, supersaturation of sea water by temperature change, and white caps [22]. The latter is the major one. White caps cover about 2% of the sea surface and the flux of bubbles to the surface is about $2 \times 10^6\ m^{-2}s^{-1}$. Once the seawater droplets formed by the bursting bubbles are in the air they can evaporate, leading to concentrated saline droplets or pure sea salt particles. A bubble of radius 1 mm is likely to produce droplets of 3–15 μm radius and containing 4–5 pg salt. The source region for sea salt particles is a very thin layer at the air-sea interface, the sea surface microlayer, together with the material, particulate or dissolved, scavenged by rising bubbles. The salt particle size distribution follows a power law distribution roughly proportional to $r^{-2.5}$ above 1 μm radius. The number of particles of different sizes increases nearly uniformly with windspeed. This suggests no strong differentiation in size of spray droplets with wind action on the waters. As expected from the particle generation mechanism involved, the particle volume distribution of marine aerosols has a characteristic dominant mode in the coarse particle range.

Extraterrestrial Sources

Possibly the most fundamental natural background level for atmospheric aerosol consists of extraterrestrial dust. The overall contribution to atmospheric aerosols is small although this material undoubtedly makes up a much larger share of particles found above 30–40 km altitude. The origins and composition of extraterrestrial dust have been reviewed by Millman [23]. The origin of this dust is thought to be associated predominantly with debris from comets and meteorites.

Comet and meteorite debris represents another source of material yielding particles ranging from > 1 μm to some millimetres in radius. The flux of material with a radius > 100 μm is about 10^{-9} $m^{-2}s^{-1}$ and that of material with a radius > 1 μm is about 10^{-4} $m^{-2}s^{-1}$. Hidy [24] has pointed out that the large particles are fluffy and compacted aggregates, while the submicrometer sized ones are rather dense. Extraterrestrial particles, occur, to a great extent, in the polar regions. Their residence time ranges from some months to years. This source is small in comparison to the others. However, just occasionally there is a very large meteorite impact such as in Tunguska, Siberia in 1908 that is capable of mechanically injecting large quantities of dust into the atmosphere.

Volcanic Aerosol

Volcanic emissions consist of solid particles (ash) and gases (mainly SO_2, CO_2 and H_2O) in variable concentrations. Ash is believed to stem from breakup or disaggregation with disintegration of the mountain. Source strength and size distribution of ash are functions of the intensity of the volcano and the state of the crustal disruption. In addition to ash, sulphuric acid aerosols are formed by SO_2 gas-to-particle conversion. Assuming that the ratio of ash to magma is unity and that 50% of the ash from an eruption is suspended as a dust cloud, an annual eruption of 1.1×10^7 m^3 would yield an equivalent aerosol rate of $\sim 10^4$ ton/day [24].

The size distribution of volcanic aerosol shows two modes. Particles in the size range 0.001 μm to 1 μm are mainly H_2SO_4 droplets while particles larger than 1 μm are likely to be ash. The El Chichon eruption produced a cloud with a bimodal size distribution with modal radii at 0.4 μm and 1.2 μm [25, 26]. Although volcanic ash forms a major part of the erupted component from volcanoes, it is not as important as H_2SO_4 aerosols due to particle coagulation and thus a short residence time of volcanic ash in the atmosphere.

Forest Fires/Slash Burning

In large areas of the world, fires are set by man either accidently or deliberately. Such fires are now an integral part of the ecological evolution in areas with rapidly growing vegetation. Forest fires can produce large quantities of submicron particles. In the United States it has been estimated that forest fires emit approximately 3×10^5 tons of aerosol particles per day. The composition of smoke from forest fires varies tremendously. It consists predominantly of volatile and nonvolatile carbon material, with significant amounts of sulphate, nitrate and chlorine.

Organic Particles

Natural emissions of organic carbon material are produced by marine and continental sources. The ocean is a dominant source of organic carbon due to the injection of naturally derived marine surfactants from bubble bursting processes. Terrestrial sources of primary organic particulate carbon include natural products emitted from vegetation such as terpenes, plant waxes and macromolecular plant fragments [27, 28]. Particles are also produced by the disintegration of biogenic bulk material and occur mostly in tropical forests and forested savannah. Pollen, spores, fungi, algae etc. may be produced where the biosphere experiences a seasonal cycle. In tropical continents where emissions of isoprene and other hydrocarbons are large, the contribution of natural non-methane hydrocarbons to the aerosol mass and number concentration is likely to be significant. The global emission of natural non-methane hydrocarbons is estimated to be approximately 700 Tg per year out of which about 10% may be converted to aerosols.

Industrial Sources

In many parts of the world anthropogenic sources of particulate matter dominate. Even with the introduction of emission control efforts anthropogenic sources are still important. Any incombustible material present as inorganic impurities in fuel, which passes through the combustion process, will produce aerosol particles. Coal produces more aerosol material than oil. Emissions from modern coal fired power stations are an order of magnitude lower than those of a few decades ago as a result of more efficient particle removal systems. In many third world countries the old technology is still in use and some 3% of the ash emitted there is from coal firing. Emission of ash from current power stations in the UK is limited to 115 $mg\,m^{-3}$ and this requires the removal of over 99% of the ash originally present in the fuel gases. The limit for new stations has been reduced to 50 $mg\,m^{-3}$ in line with European requirements.

Table 6 summarizes the total US emission estimates for 1940–1989 [29]. Tables 7 and 8 speciate the emissions into five major categories: transportation, stationary source fuel combustion, industrial processes, solid waste disposal and miscellaneous sources. It can be seen that emissions from fuel combustion and industrial processes have been substantially reduced since 1970 as a result of the installation of air pollution control equipment (Tables 9 and 10).

Table 6. Particulate emissions in the United States for 1940 to 1989 (10^{12} g/year)

1940	1950	1960	1970	1975	1980	1981	1982	1983	1984	1985	1986	1987	1988	1989
23.1	24.9	21.6	18.5	10.6	8.5	8.0	7.1	7.1	7.4	7.4	6.8	7.0	7.5	7.2

Table 7. 1940–1970 summary of estimated emissions of particulate (PM/TSP) (teragrams/year)

Source Category	1940	1950	1960	1970
Transportation Highway Vehicles	0.2	0.3	0.6	0.9
Aircraft	0.0	0.0	0.0	0.1
Railroads	2.4	1.7	0.1	0.1
Vessels	0.1	0.1	0.0	0.0
Other Off-Highway	0.0	0.0	0.0	0.1
Transportation Total	2.7	2.1	0.7	1.2
Stationary Source Fuel Combustion				
Electric Utilities	1.3	2.0	2.8	2.3
Industrial	3.3	2.8	1.8	1.6
Commercial-Institutional	0.4	0.5	0.1	0.1
Residential	2.5	1.7	1.0	0.6
Fuel Combustion Total	7.5	7.0	5.7	4.6
Industries-Processes				
Iron and Steel Mills	3.0	3.5	1.7	1.2
Primary Metal Smelting	0.6	0.6	0.5	0.6
Secondary Metals	0.2	0.3	0.2	0.2
Mineral Products	2.0	2.9	3.8	2.9
Chemical	0.3	0.4	0.3	0.2
Petroleum Refining	0.0	0.0	0.1	0.1
Wood Products	0.5	0.8	0.9	0.7
Food and Agriculture	0.8	0.8	0.9	0.8
Mining Operations	1.3	3.4	4.1	3.9
Industrial Processes Total	8.7	12.7	12.5	10.5
Solid Waste Disposal				
Incineration	0.3	0.3	0.4	0.4
Open Burning	0.2	0.3	0.5	0.7
Solid Waste Total	0.5	0.6	0.9	1.1
Miscellaneous				
Forest Fires	2.9	1.7	1.0	0.7
Other Burning	0.8	0.8	0.8	0.4
Misc. Total	3.7	2.5	1.8	1.1
Total of All Sources	23.1	24.9	21.6	18.5

Table 8. Estimates of national emissions of particulate (PM/TSP) (teragrams/year)

Source Category	1970	1975	1979	1980	1981	1982	1983	1984	1985	1986	1987	1988	1989
Transportation													
Highway Vehicles	0.9	1.0	1.1	1.1	1.1	1.1	1.0	1.1	1.1	1.1	1.1	1.2	1.3
Aircraft	0.1	0.1	0.1	0.1	0.1	0.1	0.1	0.1	0.1	0.1	0.1	0.1	0.1
Railroads	0.1	0.1	0.1	0.1	0.1	0.1	0.0	0.0	0.0	0.0	0.0	0.0	0.0
Vessels	0.0	0.0	0.0	0.0	0.0	0.0	0.0	0.0	0.0	0.0	0.0	0.0	0.0
Other Off-Highway	0.1	0.1	0.1	0.1	0.1	0.1	0.1	0.1	0.1	0.1	0.1	0.1	0.1
Transportation Total	1.2	1.3	1.4	1.3	1.3	1.3	1.3	1.3	1.4	1.4	1.4	1.5	1.5
Stationary Source Fuel Combustion													
Electric Utilities	2.3	1.5	1.0	0.8	0.7	0.6	0.6	0.6	0.4	0.4	0.4	0.4	0.4
Industrial	1.6	0.6	0.5	0.5	0.5	0.4	0.3	0.3	0.3	0.3	0.3	0.2	0.2
Commercial-Institutional	0.1	0.1	0.1	0.1	0.1	0.1	0.0	0.0	0.0	0.0	0.0	0.0	0.0
Residential	0.6	0.5	0.9	1.0	1.0	1.1	1.1	1.2	1.0	1.0	1.0	1.0	1.1
Fuel Combustion Total	4.6	2.8	2.5	2.4	2.3	2.2	2.0	2.1	1.8	1.8	1.8	1.7	1.8
Industrial Processes	10.5	5.2	3.8	3.3	3.0	2.6	2.4	2.8	2.8	2.6	2.5	2.7	2.7
Solid Waste Disposal													
Incineration	0.4	0.3	0.2	0.2	0.2	0.1	0.1	0.1	0.1	0.1	0.1	0.1	0.1
Open Burning	0.7	0.3	0.2	0.2	0.2	0.2	0.2	0.2	0.2	0.2	0.2	0.2	0.2
Solid Waste Total	1.1	0.6	0.4	0.4	0.4	0.3	0.3	0.3	0.3	0.3	0.3	0.3	0.3
Miscellaneous													
Forest Fires	0.7	0.6	0.8	1.0	0.8	0.6	1.0	0.8	1.0	0.8	0.9	1.2	0.9
Other Burning	0.4	0.1	0.1	0.1	0.1	0.1	0.1	0.1	0.1	0.1	0.1	0.1	0.1
Misc. Organic Solvent	0.0	0.0	0.0	0.0	0.0	0.0	0.0	0.0	0.0	0.0	0.0	0.0	0.0
Misc. Total	1.1	0.7	0.9	1.1	0.9	0.7	1.1	0.9	1.1	0.9	1.0	1.3	1.0
Total of All Sources	18.5	10.6	8.9	8.5	8.0	7.1	7.1	7.4	7.3	6.8	7.0	7.5	7.2

Table 9. Emissions of particulate (TM/TSP) from fuel combustion (gigagrams/year)

Source Category	1970	1975	1979	1980	1981	1982	1983	1984	1985	1986	1987	1988	1989
Coal													
Electric Utilities	2220	1420	860	720	640	490	510	550	392	387	370	350	354
Industrial	1300	360	250	250	280	220	110	110	120	93	80	80	84
Commercial-Institutional	40	40	30	30	30	40	20	20	19	19	20	20	14
Residential	80	20	10	10	10	20	20	20	15	15	20	20	15
Coal Total	3640	1840	1150	1010	960	770	660	700	547	513	490	470	466
Fuel Oil													
Electric Utilities	110	120	120	100	90	70	60	50	38	55		50	60
Industrial	80	70	70	60	50	50	30	40	36	39		30	33
Commercial-Institutional	60	40	30	40	30	30	20	20	17	20	20	20	20
Residential	20	10	10	10	10	10	10	10	9	9	10	10	12
Fuel Oil Total	270	240	230	210	180	160	120	120	100	123	30	110	125
Natural Gas													
Electric Utilities	6	5	6	6	6	5	5	5	5	4	5	4	5
Industrial	20	20	20	20	20	20	20	20	16	15	10	20	18
Commercial-Institutional	3	3	4	4	3	4	3	3	3	3	3	4	4
Residential	7	7	7	6	6	6	6	6	6	6	6	6	6
Natural Gas Total	36	35	37	36	35	35	34	34	30	28	24	34	33
Wood													
Industrial	180	120	130	130	120	110	100	100	96	96	100	90	94
Residential	460	490	870	990	1020	1110	1110	1120	993	1002	1010	960	1026
Wood Total	640	610	1000	1120	1140	1220	1210	1220	1090	1098	1110	1050	1121
Other Fuels													
Industrial	40	40	30	30	20	20	20	20	21	19	20	20	19
Residential	4	3	3	2	2	2	2	2	2	2	2	2	2
Other Fuel Total	44	43	33	32	22	22	22	22	23	21	22	22	21
Fuel Combustion Total	4630	2768	2450	2408	2337	2207	2046	2096	1788	1783	1766	1686	1765

Table 10. Emissions of particulate (PM/TSP) from industrial processes (gigagrams/year)

Source Category	1970	1975	1979	1980	1981	1982	1983	1984	1985	1986	1987	1988	1989
Cattle Feed Lots	20	20	20	20	20	20	20	20	21	22	20	20	19
Cotton Ginning	20	20	20	20	30	20	10	20	24	17	30	30	21
Metallic Ore Mining	530	320	210	180	200	110	110	130	129	114	130	130	153
Coal Mining	350	250	290	310	310	320	300	350	314	313	310	320	336
Crushed Stone	1350	760	570	450	380	340	370	400	421	482	480	580	574
Sand and Gravel	50	40	50	40	40	30	30	40	42	46	50	45	46
Clays	1610	290	150	130	70	60	70	80	79	64	60	50	46
Potash/Phosphate Rock	40	30	30	30	10	10	10	10	12	6	5	7	9
Feed and Grain Milling	70	60	50	40	50	50	30	50	49	58	40	40	46
Lumber and Plywood	80	70	80	70	70	60	70	80	82	92	100	100	96
Pulp Mills	620	220	120	140	90	100	100	120	117	96	100	109	86
Chemicals	220	120	140	140	120	100	110	130	118	90	100	100	101
Petroleum Refining	60	70	50	50	40	40	30	30	21	20	20	20	18
Asphalt Paving and Roofing	560	320	130	110	90	90	110	140	118	124	130	130	125
Glass	40	40	30	30	30	30	30	30	28	25	20	20	25
Cement	1580	640	520	380	290	220	240	270	266	213	200	200	215
Brick and Tile	40	30	20		10	7	10	10	14	12	10	10	10
Concrete, Lime, Gypsum	580	290	140	130	100	80	80	90	90	89	90	90	92
Clay Sintering	100	40	10	10	10	10	10	10	10	10	10	6	5
Iron and Steel	1190	570	400	310	300	200	180	180	160	140	140	160	155
Ferroalloys	160	90	40	30	30	20	20	20	20	20	20	20	20
Iron and Steel Foundries	150	70	60	50	40	40	30	30	40	30	40	40	50
Primary Nonferrous Smelters	390	200	100	90	90	60	70	70	70	50	50	54	55
Secondary Nonferrous Smelters	60	50	50	40	40	30	30	40	35	30	30	34	40
Grain Elevators	670	590	550	490	550	510	280	430	490	390	350	350	350
Total	10540	5200	3830	3300	3010	2557	2350	2780	2768	2552	2535	2665	2693

Table 11. Emissions of particulate (PM/TSP) from transportation (gigagrams/year)

Source Category	1970	1975	1979	1980	1981	1982	1983	1984	1985	1986	1987	1988	1989
Highway Vehicles													
Gasoline-powered													
Passenger cars	610	680	620	570	540	560	550	550	540	540	540	590	620
Light trucks-1	80	100	90	90	90	80	90	90	100	110	120	130	130
Light trucks-2	20	30	70	70	70	70	70	70	70	70	60	70	080
Heavy duty vehicles	60	50	60	60	60	50	50	50	45	40	40	40	050
Motorcycles	4	8	8	7	5	4	4	4	3	3	4	4	004
Total-Gasoline	774	868	848	797	765	764	764	764	758	763	764	834	884
Diesel-powered													
Passenger cars	0	1	5	9	10	20	20	20	20	20	20	10	010
Light trucks	0	0	1	3	5	5	5	6	4	4	3	4	004
Heavy duty vehicles	130	180	230	250	280	270	250	270	340	310	320	360	370
Total-Diesel	130	181	236	262	295	295	275	296	364	334	343	374	384
Highway Vehicle Total	904	1049	1084	1059	1060	1059	1039	1060	1122	1097	1107	1208	1268
Aircraft	100	80	70	70	70	70	80	80	90	85	80	80	080
Railroads	60	50	60	50	50	50	40	40	40	35	40	40	035
Vessels	40	30	30	30	30	30	30	30	30	30	30	30	030
Farm Machinery	40	50	70	60	60	60	60	60	70	70	70	70	070
Construction Machinery	10	10	20	20	20	20	20	20	20	20	20	20	020
Industrial Machinery	20	20	30	20	20	20	20	20	10	10	10	10	015
Other Off-Highway Vehicles	4	5	5	5	5	5	5	5	5	5	5	7	005
Transportation Total	1178	1294	1369	1314	1315	1314	1294	1315	1387	1352	1362	1465	1523

Emissions from transportation decreased substantially between 1940 and 1960 as the result of the obsolescence of coal-burning railroad locomotives. Due to increased motor usage, and in particular diesel powered vehicles, particulate emissions from transportation has risen since 1960. This increase is demonstrated in Table 11.

Coal combustion by electric utilities has increased from approximately 321 million tons in 1970 to 7648 million tons in 1989 while particulate emissions over the same period decreased by 83% as a result of emission control equipment. It is estimated that if the 1970 level of control had remained in effect in 1989, electric utility emissions would have more than doubled. Emissions from certain sources, such as residential coal use, have declined simply because of reduced use, whilst those from many industrial processes reflect changes in production levels along with an increase in control levels. It must be remembered that fugitive particulate emissions (emissions from uncontrolled sources) are incompletely accounted for in the emission totals. Such emissions include unpaved roads, paved road resuspension, wind erosion, agricultural tilling and construction activities. Generic categories of fugitive dust sources are given in Table 12. A major portion of the fugitive emissions are relatively large particles that settle out of the atmosphere rapidly and are not readily captured by particulate air quality monitors. Techniques for emission quantification and available models for emission estimation and control performance estimation for fugitive dust sources have been described by Cowherd [30] and Kinsey and Cowherd [31]. Particulate control measures have been most effective in reducing emissions of large and intermediate size particles. It is unlikely that the emission of small particles has been reduced to the extent that total particulate emissions have been reduced.

In the UK similar trends are observed [32]. Of current concern are the particulate emissions that arise from diesel vehicles and contain a mixture of soot,

Table 12. Generic categories of open dust sources (Cowherd et al. [100])

UNPAVED TRAVEL SURFACES
roads
parking lots and staging areas
storage piles
PAVED TRAVEL SURFACES
streets and highways
parking lots and staging areas
EXPOSED AREAS (WIND EROSION)
storage piles
bare ground areas
MATERIALS HANDLING
batch drop (dumping)
continuous drop (conveyor transfer, stacking)
pushing (grading, scraping)
tilling

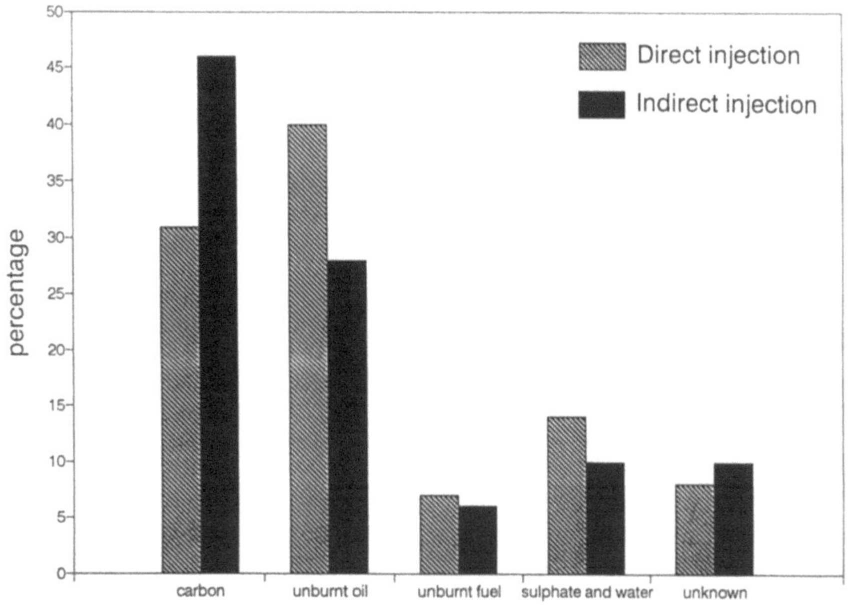

Fig. 4. Typical composition of diesel exhaust particulate matter for direct injection and indirect injection engines

Table 13. Source contributions to total UK emissions, 1991 (%)

ROAD VEHICLES		PM_{10}	Black smoke
Cars	petrol	9	3
	diesel	1	3
	total	10	6
LGV	petrol	1	<1
	diesel	1	3
	total	2	3
HGV	large	5	2
	small	7	16
	total	12	28
Buses		2	6
Motorcycles		<1	<1
Exhaust emissions		27	42
Evaporation			
Total road vehicles		27	42
OTHER SOURCES			
Domestic		47	36
Commercial etc.		1	1
Power stations		7	5
Refineries		<1	1
Agriculture		<1	<1
Other industry		18	15
Civil aircraft		<1	<1
Shipping		<1	1

unburned fuel and hydrocarbons. as shown in Fig. 4. The contribution of different source sectors to total UK emissions of PM_{10} and black smoke is given in Table 13. Estimated emissions of particles in the former Soviet Union from industrial sources are available from Berlyand [33]. Estimates of industrial emissions of particles for individual cities as well as for broad economic regions are quoted. The inventory includes emissions of specific particles including lead, organic and inorganic dust, cement dust and others.

Gas-to-Particle Formation

The main source of secondary particles is the atmospheric oxidation of SO_2 to sulphuric acid, and of NO_2 to nitric acid. Sulphur and nitrogen bearing gases are generated both naturally and anthropogenically. Oxidation of these precursors occurs both in the gas phase and in liquid droplets. Non-volatile products are formed which either condense onto pre-existing aerosol particles or form new particles. Virtually all the aerosol mass produced by this process is in the fine size range.

The natural production of sulphur containing gases include emissions from marine flora, the terrestrial biota and via volcanic eruptions. A large part of the sulphate from biogenic gases is derived from dimethyl sulphide (DMS) emitted from the oceans. Recent estimates of 10–50 Tg S per year are significantly lower than earlier estimates. Anthropogenic emissions stem mostly from combustion processes such as coal burning, oil burning, etc. Nowadays industrial SO_2 emissions surpass natural emissions. Industrial sulphur aerosol loadings have doubled since 1950, following a slow increase over the preceding 100 years. The current anthropogenic emission is between 70 and 80 Tg per year, roughly half of which is oxidised to aerosol sulphate before being deposited [34]. Roughly 90% of the emissions are from industrial regions in the northern hemisphere, and little sulphate is transported to the southern hemisphere. Details of both sulphur dioxide and reduced sulphur emission inventories are given by Graedel et al. [35]. Table 14 summarizes the emissions of key short-lived sulphur gases.

Hidy [24] has identified the following precursor gases of sulphate:

sulphur dioxide – produced mainly by fossil fuel combustion and volcanic eruption;
hydrogen sulphide – generated biogenically and through volcanic production;
carbon disulphide – biogenic production;
carbonylsulphide – biogenic production;
dimethyl sulphide and dimethylate sulphide – produced by bacteria and freshwater green and blue algae.

The main chemical routes leading to sulphate formation are summarized in Table 15. The gas phase oxidation of SO_2 is dominated by the reaction of SO_2 with hydroxyl radicals (OH). The major aqueous phase oxidation routes for SO_2

Table 14. Estimated sources of short lived sulphur gases (Tg S per year)

Anthropogenic emissions (mainly SO_2)	70–80
Biomass burning (SO_2)	0.8–2.5
Oceans (DMS)	10–50
Soils and plants (DMS and H_2S)	0.2–4
Volcanic emissions (mainly SO_2)	7–10

Table 15. Major chemical pathways of atmospheric sulphate and nitrate formation

Gas phase	
	$SO_2 + OH(+ H_2O + O_2) \rightarrow H_2SO_4 + HO_2$
	$NO_2 + OH \rightarrow HNO_3$
	$NO_2 + O_3 \rightarrow NO_3 + O_2$
	$NO_3 + RH \rightarrow HNO_3 + R$
	$NO_3 + NO_2 \rightarrow N_2O_5$
	$N_2O_5 + H_2O \rightarrow 2HNO_3$
Aqueous Phase	
	$S(IV) + H_2O_2 \rightarrow S(VI) + H_2O$
	$S(IV) + O_3 \rightarrow S(VI) + O_2$
	$S(IV)(+ 1/2O_2) \xrightarrow{Mn^{2+}} S(VI)$
	$S(VI)(+ 1/2O_2) \xrightarrow{Fe^{3+}} S(VI)$
	$HSO_3^- + OH(+ HSO_3^- + O_2) \rightarrow SO_4^{2-} + SO_4^- + H_2O + H^+$
	$SO_4^- + H_2O_2 \rightarrow HSO_4^- + HO_2$
	$S(IV) + HO_2 \rightarrow S(VI) + OH$
	$HSO_3^- + Cl_2^-(+ HSO_3^- + O_2) \rightarrow SO_4^{2-} + SO_4^- + 2H^+ + 2Cl^-$
	$NO_3 + HSO_3^-(+ HSO_3^- + O_2) \rightarrow NO_3^- + 2H^+ + SO_4^- + SO_4^{2-}$
	$NO_3 + H_2O_2 \rightarrow NO_3^- + H^+ + HO_2$

include reaction with hydrogen peroxide (H_2O_2), ozone (O_3), oxygen (O_2), hydroxyl radicals and hydroperoxyl radicals (HO_2) [36–38]. The reaction involving H_2O_2 is very fast and its rate is not very sensitive to pH because the aqueous kinetics and S(IV) solubility vary in opposite directions as the pH varies. However the aqueous phase oxidation of SO_2 by O_3 is influenced by pH and takes place at a significant rate only in droplets with pH above 4. While the extent of H_2O_2 oxidation may be limited by the amount of H_2O_2 present, the amount of O_3 is rarely limiting to the extent of oxidation by the species. The reaction of SO_2 and O_2 requires the presence of metal catalysts such as iron and manganese. Again this reaction occurs at a faster rate when the pH is high.

In addition to sulphate, nitrate is another major constituent in atmospheric aerosol and is the end product of a wide variety of reactions involving trace gases. Biological processes in soil lead to the release of all of the common nitrogen oxides, NO, NO_2 and N_2O. The amounts involved are very uncertain but for NO and NO_2 are of the order of 10 Tg N per year. The most important oxides of nitrogen forming nitrates are NO_2, NO, N_2O_5 and NH_3. Nitrogen dioxide is

Table 16. Emissions of nitrogen compounds (Tg N per year)

Lightning	8
NH_3 oxidation	1–10
From stratosphere	0.5
Biogenic production	8
Biomass burning	12
Natural total	33
Anthropogenic	21

produced both naturally and anthropogenically through a variety of mechanisms including oxidation of ammonia, lightning, combustion processes and oxidation of NO. It is estimated that half of the NO_x is oxidised to nitric acid which is present in both the gas and the condensed phase (in aerosols and droplets). As can be seen from Table 16, natural emissions of nitrogen-containing gases are greater than anthropogenic emissions.

The gas phase chemistry of NO_2 is governed by OH radical chemistry during daytime and by nitrate (NO_3) radical chemistry during nighttime. The aqueous phase chemistry of NO_2 is not significant for nitrate formation.

Both nitric acid and hydrochloric acid vapour, the latter arising mainly from refuse incineration and coal combustion, react reversibly with ammonia gas, largely arising as a product of decomposition of animal urine, to form ammonium salts [39]:

$$NH_4NO_3(\text{aerosol} \Leftrightarrow NH_3(\text{gas}) + HNO_3(\text{gas})$$

$$NH_4Cl(\text{aerosol}) \Leftrightarrow NH_3(\text{gas}) + HCl(\text{gas}) \ .$$

The salts are appreciably volatile and may release their parent acids under conditions of low atmospheric ammonia, high temperature, or reduced humidity. Particles formed by gaseous reactions are initially very small (< 0.2 µm) but grow rapidly by either particle-particle coagulation or by surface accumulation of material from the gas, as illustrated in Fig. 2.

Bioaerosols

Bioaerosols are particles of variable biological origin, e.g. pollen, fungal spores, bacterial cells, viruses, protoza, excreta or fragments of insects, skin scales or hair of mammals, or other components, residues or products of organisms, such as bacterial saccharides [40]. Outdoors airborne fungal spores originate from soil, vegetation and decaying plants and animals [41–45]. Differences in quantity and quality of sources and meteorological conditions cause a strong temporal and spatial variation in airborne concentrations [46]. For most genera, the highest numbers in the outdoor air are found during summer and autumn. During these

Table 17. Mean values and ranges of airborne fungal spore levels in different types of houses (after Pasanen and Kalliokoski [46])

House type	Airborne spore counts (colony forming units m^{-3}) Arithmetic mean	Geometric mean	Range
Urban/suburban	150–1220	40– 410	< 14000
Rural/farm	–	80–6320	< 21180
Damp/mouldy	340–3060	80– 820	< 449800

seasons the outdoor air is the main source of fungi in the indoor air. Recent studies indicated that the outdoor air spora influences the presence of fungi in indoor environments, but the indoor air spora is not a simple reflection of the presence of fungi in outdoor air [47, 48].

In addition to the outdoor air, there are numerous indoor sources of biological particulate matter such as food stuffs [49–51], house plants and flower pots [52–55], house dust [56–61], pets and their bedding material [62–64], textiles, carpets, wood material and furniture stuffing, from which spores of Alternaria, Aspergillus, Botrytis, Claclosporium, Penicillium, Scopulariopsis and yeast cells are occasionally released into the air. Values of fungal spore levels in various indoor environments are summarised in Table 17. In nonindustrial indoor environments, the most important source of airborne bacteria is the presence of humans. As a result, the concentration in air of normal human skin bacteria is often used as an indicator of indoor air quality. In industrial environments strong sources of bioaerosols may exist when organic material is handled, such as plants, hay, straw, wood chips, cereal grain and tobacco.

All natural and anthropogenic waters contain large numbers of microorganisms. Hence water or liquid droplets resulting from rain, splashes or bubbling processes may contain bioaerosols. These may remain airborne after the water evaporates. In nonindustrial localities bioaerosol sources can develop due to microbial growth in a building's ventilation system or in the structure itself. The prerequisite for microbial growth is excessive and accessible moisture. Standing water is a good reservoir for microbial growth and a potential source of microbial aerosols when disturbed.

Indoor Aerosols

Indoor aerosol concentrations are associated with both indoor and outdoor sources. The identification of sources and the assessment of their relative contributions can be a complicated process due to the presence of a number of indoor sources, which can vary from building to building. There are also uncertainties associated with estimating the impact of outdoor sources on the indoor environment [65, 66]. Many indoors activities generate pollutants. These include heating, cooking, cleaning, smoking, the use of a wide variety of consumer products, and the simple act of moving about and stirring up particles. Particles studied indoors have predominantly been those in the fine mode (< 2 μm) resulting from cigarette smoking or as emissions from combustion appliances. Coarse mode particles from reentrainment of fibres, dust, animal and human dander and mould spores constitute the second most common form of indoor pollution [67]. Only limited data are available for particulate emissions from indoor sources, except for tobacco smoke [68, 69].

Koutrakis et al. [66] apportioned aerosol emissions to cigarette smoke, gas stores, wood burning stores and kerosene heaters. Of these, cigarette smoking was found to be the most important with a particle emission rate per cigarette of 12.7 mg m^{-3}. Gas stoves did not contribute to the observed indoor aerosol concentrations. However, concentrations of respirable suspended particulate between 5 and 1000 μg m^{-3} have been reported elsewhere [70]. Unknown indoor sources were found to be significant contributors. He suggested that these unknown sources were resuspension of aerosols from indoor surfaces, generation of biological aerosols for humans, and other indoor activities. Kerosene heaters and woodburning stoves were identified as significant sources of particulate. This is in agreement with the findings of other workers [71, 72].

The use of wood as a residential house heating fuel is widespread in the US. In localities where wood is the predominant house heating fuel, wood stoves have been shown to contribute as much as 80% of the ambient PM_{10} concentration during winter months. The US Environmental Protection Agency has now introduced emission standards for woodburning stoves. Prototypes of each model stove must pass an emission test performed in a laboratory for that model line to be certified for manufacture and sale. The results of field studies to establish the emission rates of typical, uncontrolled conventional technology stoves and the degree of emission control achieved by newer stoves designed to reduce emissions are summarised by McCrillis and Jaasma [72]. Their findings indicate the stoves subject to Phase I particulate emission limits (stoves manufactured after July 1, 1988) and to Phase II (manufactured after July 1, 1990) had 26% and 44%, lower emissions, respectively, than those stoves not subject to emission control. Most of this reduction resulted from increased combustion efficiency and hence a lower fuel consumption.

Airborne Particulate Elemental Carbon (PEC)

Carbonaceous material is a principal component of airborne particulate matter. In most ambient atmospheric situation, it contributes 10–20% of the total aerosol mass. The main forms in which carbon occurs in aerosols are organic carbon which normally contributes 60–80% of total carbon mass, and PEC, also known as black carbon and graphitic carbon. PEC is produced during the incomplete combustion of various fuels. PEC emissions may be calculated from annual combustion of fuels and their PEC emission factors. The results for London in 1980 are shown in Table 18. It is clear from this and Table 19 that the vast majority of the emissions come from diesels. It can also be seen that PEC emissions have increased at an average rate of 2×10^5 kg per annum and this increase is due entirely to the diesel emission.

No direct measurements exist for the emissions of soot particles from biomass burning. Penner et al. [73] have produced a model-related global emissions inventory for soot particles and estimate a global biomass burning flux of approximately 7 Tg C per year. Penner et al. [74] have shown that the estimated emissions depend to a great extent on the method used. By using estimated emission factors and published fuel production and use statistics for various fuels they derived a PEC global emission of 12.6 Tg C per year. They then showed that PEC and sulphur dioxide concentrations are well correlated and within one region the ratio of PEC to SO_2 appears to be relatively constant. They then constructed a global inventory of PEC emissions from published inventories for the emissions of sulphur. By this method the total emissions were approximately 24 Tg C per year. This figure is considerably higher than previous estimates of PEC based on the use of emission factors and applied to data on the worldwide use of fossil fuels. Table 20 shows the estimated emissions of PEC by region for the two methods. The difference between the methods results from regions where the inventories are inadequate or non-existing. In regions where the PEC/S ratios and sulphur emissions are well defined the difference between the two techniques is 50% whereas in areas of great uncertainty the difference is approximately 300%. Clearly, extensive field measurements are required. Whichever method is used, the global anthropogenic soot flux is several times that due to biomass burning. It should be noted that most of the anthropogenic source is in the northern hemisphere and much of the biomass burning is in the tropic or southern mid-latitudes.

Emissions of organic aerosols in the greater Los Angeles area have been subject of much research [75, 76]. As with any emission inventory various techniques of measuring particulate emissions have been used. Recently a series of papers have been published containing results of emission source tests conducted by a

Table 18. Particulate elemental carbon emissions, London 1980 (after Hamilton and Mansfield [101])

Fuel type	Amount of fuel consumed (10^3 tonnes)	PEC emission factor (% by mass)	Amount of PEC emitted (tonnes)	% of PEC total by fuel type
Coal: domestic	50	0.0001	0.05	0.004%
Coal: industrial	250	0.0001	0.25	0.02%
Solid smokeless fuel	350	0.0001	0.35	0.03%
Gas oil	1000	0.01	100	8%
Fuel oil	600	0.002	12	1%
Petrol	2200	0.002	44	4%
Diesel	520	0.2	1040	87%

Table 19. Particulate elemental carbon emissions in Western Europe (after Hamilton and Mansfield [101])

	Area (m^2)	Total PEC emissions ($\times 10^6$ kg per year)	% Diesel	Areal emission rate ($\mu g\ m^{-2} S^{-1}$)	Total PEC emissions ($\times 10^6$ kg per year)	% Diesel	Areal emission rate ($\mu g\ m^{-2} S^{-1}$)
Austria	8.4×10^{10}	2.2	82	8.3×10^{-4}	3.5	92	1.3×10^{-3}
Belgium	3.0×10^{10}	4.5	76	4.8×10^{-3}	6.8	89	7.2×10^{-3}
Denmark	4.3×10^{10}	2.8	71	2.1×10^{-3}	3.6	86	2.6×10^{-3}
Finland	3.6×10^{11}	2.6	81	2.3×10^{-4}	3.5	89	3.1×10^{-4}
France	5.5×10^{11}	23	78	1.3×10^{-2}	29.5	80	1.7×10^{-3}
F.R.G.	2.5×10^{11}	28	75	3.6×10^{-3}	34	84	4.3×10^{-3}
Greece	1.3×10^{11}	2.6	81	6.3×10^{-4}	3.6	87	8.8×10^{-4}
Iceland	1.0×10^{11}	0.4	90	1.3×10^{-4}	0.5	97	1.6×10^{-4}
Ireland	6.9×10^{10}	0.9	80	4.1×10^{-4}	1.2	86	5.5×10^{-4}
Italy	3.0×10^{11}	19	84	2.0×10^{-3}	30	92	3.2×10^{-3}
Luxembourg	2.6×10^{9}	0.3	63	3.7×10^{-3}	0.6	89	7.3×10^{-3}
Netherlands	3.6×10^{10}	6.6	86	5.8×10^{-3}	6.9	94	6.1×10^{-3}
Norway	3.2×10^{11}	2.9	90	2.9×10^{-4}	3.1	91	3.1×10^{-4}
Portugal	9.2×10^{10}	2.5	92	8.6×10^{-4}	3.0	93	1.0×10^{-3}
Spain	5.0×10^{11}	13.0	85	8.2×10^{-4}	15.0	93	9.5×10^{-4}
Sweden	4.5×10^{11}	4.4	75	3.1×10^{-4}	4.1	85	2.9×10^{-4}
Switzerland	4.1×10^{11}	1.9	53	1.5×10^{-3}	2.2	63	1.7×10^{-3}
Turkey	7.8×10^{11}	5.0	92	2.0×10^{-4}	10.6	96	4.3×10^{-4}
UK	2.4×10^{11}	19.0	84	2.5×10^{-3}	22.3	90	2.9×10^{-3}

Table 20. Particulate elemental carbon emission (Tg C per year) by region (after Penner et al. [74])

Region	Ratio method	Emission factor method
U.S.A	1.10	0.40
Rest of North America	0.18	0.19
Mexico	1.03	0.08
Central America	0.06	0.10
Brazil	0.44	0.71
Rest of South America	1.20	0.33
China	3.66	2.68
Japan and South Korea	0.82	0.43
Rest of Asia (except U.S.S.R.)	4.14	2.17
U.K. and West Germany	0.61	0.28
Spain	0.39	0.04
Rest of Western Europe	1.46	0.27
Eastern Europe (except Poland, U.S.S.R.)	0.46	1.10
Poland	0.88	0.45
former U.S.S.R.	5.60	1.93
Oceania	0.24	0.12
Africa	1.56	1.30
Total	23.84	12.61

single method [76–82]. Cass and co-workers selected various source types for testing that contribute approximately 80% of the organic aerosol emitted in Los Angeles ranging from paved road dust to cigarettes. Figure 5 illustrates the average chemical composition of fine (d < 2 μm) particulate emissions from various sources. It can be seen that the contribution of the organic mass emitted to the total particulate mass varies enormously. For natural gas appliances ≥ 99% of the total particulate mass can be attributed to organic material whereas for oil fired boilers this figure is only 6%. The results of the study, shown in Table 21, indicate that fine organic aerosol emissions in the western portion of the Los Angeles basin are 29.8 tonnes per day. Over 40% of these organic aerosol emissions are from anthropogenic pollution sources. Meat frying and charboiling together account for approximately 21% of all fine organic particulate matter released into the atmosphere. A study of particulate matter emitted by vehicles (both petrol and diesel engines) measured more than 100 compounds quantitatively [78]. Not only does traffic contribute particulate emissions from their exhausts but also particles worn from tyres and brake linings and by resuspension of road dust. Rogge et al. [79] estimated that vehicular exhaust emissions and tyre debris contribute 32.7 and 4.3% of the organic mass in road dust respectively. Road dust may be considered to be chemically similar in many respects to the primary portion of the atmospheric aerosol and indeed bears a dynamic relationship to the atmospheric aerosol through resuspension into and redeposition from the atmosphere.

Table 21. Revised estimate of the major sources of fine aerosol organic carbon (OC) emissions within an 80×80 km heavily urbanized area surrounding Los Angeles for 1982 (after Hildemann et al. [76])

Source type		OC emitted,[a] Kg/day	% of OC emissions
(1)	Meat-cooking operations		
	charcoal broiling	4938	16.6
	frying	1393	4.7
(2)	Paved road dust	4728	15.9
(3)	Fireplaces	4173	14.0
(4)	Noncatalyst gasoline vehicles		
	automobiles	2088	7.0
	other vehicles	1372	4.6
(5)	Diesel vehicles		
	heavy-duty trucks	1242	4.2
	other vehicles	617	2.1
(6)	Surface coating	1433	4.8
(7)	Forest fires	877	2.9
(8)	Cigarettes	802	2.7
(9)	Catalyst-equipped gasoline vehicles		
	automobiles	780	2.6
	other vehicles	79	0.3
(10)	Organic chemical processes	692	2.3
(11)	Brake lining	690	2.3
(12)	Roofing tar pots	556	1.9
(13)	Tire wear	414	1.4
(14)	Misc industrial point sources	393	1.3
(15)	Natural gas combustion		
	residential/commercial	30	0.1
	other sources	262	0.9
(16)	Misc petroleum industry processes	278	0.9
(17)	Primary metallurgical processes	228	0.8
(18)	Railroad (diesel oil)	211	0.7
(19)	Residual oil stationary sources	206	0.7
(20)	Refinery gas combustion	195	0.7
(21)	Secondary metallurgical processes	167	0.6
(22)	Mineral industrial processes	158	0.5
(23)	Other organic solvent use	106	0.4
(24)	Jet aircraft	92	0.3
(25)	Asphalt roofing manufacturing	81	0.3
(26)	Coal burning	76	0.3
(27)	Wood processing	74	0.2
(28)	Residual oil-fired ships	66	0.2
(29)	Structural fires	63	0.2
(30)	Distillate oil stationary sources		
	industrial	13	0.04
	other	23	0.08
Other sources		226	0.8
Total		29822	

[a] Annual average emissions stated at a daily rate

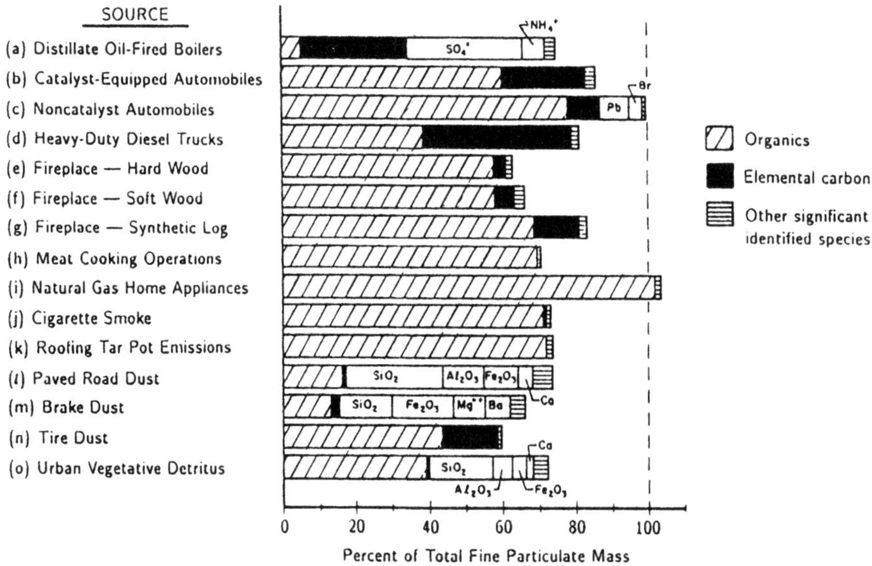

Fig. 5. Average chemical composition of fine particulate emissions from various source types. Reprinted with permission from Environ Sci Technol 25:753, 1991

Concluding Remarks

The particle size of interest in aerosol behaviour ranges from molecular clusters of 0.001 µm to fog droplets and dust particles as large as 100 µm: a variation of 10^5 in size. In nature, aerosol particles are found over a wide range of at least four decades in size from tens of angstroms to more than ten micrometers. Three size modes can generally be distinguished: the nucleation mode (< 0.2 µm), the accumulation mode (0.2–2.0 µm) and the coarse mode (> 2.0 µm). Although forming a small part of the mass of the atmosphere, aerosols have the potential to influence, significantly, both the radiative transfer through the atmosphere and the atmospheric water cycle. However, at the present time, the climatic effects of aerosols can only be estimated with considerable uncertainty [83] due to lack of sufficient detailed observations and emission data. An inventory of emissions is required and this may be related to the physical and chemical properties of the aerosol, e.g. the production source reflects the chemical composition. Residence time and source strength also have a significant influence on aerosol climatology.

Primary sources of aerosols are mainly of natural origin, whereas secondary sources (e.g. gas to particle conversion) are predominantly anthropogenic. The origins of aerosols are quite diverse and the sources vary both spatially and temporally. Emissions of aerosols from natural sources have altered, in strength, only slightly during the past centuries. However, anthropogenic emissions have grown dramatically especially during the 20th century. It is expected that the globally averaged aerosol concentration will stabilise due to emission controls, though their geographic distribution will change.

References

1. Whytlaw-Gray R, Speakman JB, Campbell JHP (1923) Smokes: part I – A study of their behaviour and a method of determining the number of particles they contain. Proc Roy Soc 102:600–615
2. Schmauss A (1920) The chemistry of fog, cloud and rain. Umschau 24:61–63
3. Willeke K, Baron PA (1993) Aerosol measurement: principles, techniques and applications. Van Nostrand Reinhold, New York
4. Clarke AG, Willison MJ, Zek EM (1984) A comparison of urban and rural aerosol composition using dichotomous samplers. Atmos Environ 18:1767–1775
5. Carlson TN, Caverly RS (1977) Radiative characteristics of Saharan dust at solar wavelengths. J Geophys Res 82:3141–3152
6. Fouquart Y, Bonnel BM, Caoui Roquai M, Santer R (1987) Observations of Saharan aerosols: results of the ECLATS field experiment. J Clim Appl Met 26:28–52
7. Levin ZL, Joseph JH, Mekler Y (1980) Properties of Sharav (Khamsin) dust – comparison of optical and direct sampling data. J Atmos Sci 37:882–891
8. Ackerman SA, Cox SK (1982) The Saudi Arabian heat low: aerosol distribution and thermodynamic structure. J Geophys Res 97:8991–9002
9. DeLuisi JJ, Furakava PM, Gillette DP, Schuster BG (1976) Results of a comprehensive atmospheric aerosol-radiation experiment in the southwestern United States. J Appl Met 15:441–470
10. Brazel AJ (1989) Dust climate in the American Southwest. In palaeoclimatology and palaeometeorology: modern and past patterns of Global atmospheric transport, Leinen M and Sarntheim M (eds), Kluwer Academic, Dordrecht
11. Golitsyn G, Gillette DA (1993) Introduction: A joint Soviet-American experiment for the study of Asian desert dust and its impact on local meteorological conditions and climate. Atmos Environ 27A:2467–2470
12. Gillette DA, Adams J, Muhs D, Kihl R (1982) Threshold friction velocities required and rupture moduli for crushed desert soils for input of soil particles into air. J Geophys Res 87:9003–9015
13. Lerman A (1979) Geochemical processes: water and sediment environments. John Wiley, New York
14. d'Almeida GA, Schutz L (1983) Number, mass and volume distribution of mineral aerosol and soils of the Sahara. J Clim Appl Meteor 22:233–243
15. Prospero JM, Glaccum RA, Nees RT (1981) Atmospheric transport of soil dust from Africa to South America. Nature 289:570–572
16. Parungo F, Bodhaine BA, Bortniak J (1981) Seasonal variation in Antarctic aerosol. J Aerosol Sci 17:491–504
17. Andronova AV, Gomess L, Smirnov VV, Ivanov AV, Shukorova LM (1993) Physico-chemical characteristics of dust aerosols deposited during the Soviet-American experiment (Tadzhikistan, 1989). Atmos Environ 27A:2487–2493
18. Schutz L, Sebert M (1978) Mineral aerosols and source identification. J Aerosol Sci 18:1–10
19. Schutz L, Jaenicke R (1974) Particle number and mass distribution above 10^{-4} cm radius in sand and aerosol of the Sahara desert. J Appl Meteor 13:863–870
20. Patterson EM, Gillette DA (1977) Commonalities in measured size distributions for aerosols having a soil-derived component. J Geophys Res 82:2074–2082
21. d'Almeida GA (1989) Desert aerosol: characteristics and effects on climate. In palaeoclimatology and palaeometeorology: modern and past patterns of Global atmospheric transport. Leinen M and Sarntheim M (eds), Kluwer Academic, Dordrecht.
22. Blanchard DC, Woodcock AE (1980) The production, concentration and vertical distribution of the sea salt aerosol. Ann NY Acad Sci 338:330–347
23. Millman PM (1979) Interplanetary dust. Naturwissenschaften 66:134–139
24. Hidy GM (1984) Aerosols: an industrial and environmental science. Academic Press, Orlando
25. Knollenberg RG, Huffman D (1983) Measurements of aerosol size distribution in El Chichon cloud. Geophys Res Letters 10:1025–1028

26. Oberbeck VR, Danielsen EF, Snetsinger KG, Ferry GV (1983) Effects of the eruption of El Chichon on stratospheric aerosol size and composition. Geophys Res Letters 10:1021–1024
27. Graedel TE (1979) Turpenoids in the atmosphere. Rev Geophys Space Phys 17:937–947
28. Simoneit BRT, Mazurek MA (1981) Air pollution: the organic constituents. Crit Rev Envir Control 11:219–276
29. United States Environmental Protection Agency (1991) National air pollutant emission estimates, 1940–1989. EPA/450/4–91/004. Environmental Protection Agency, Research Triangle Park
30. Cowherd C (1993) Fugitive dust emissions. In aerosol measurement: principles, techniques and applications. Willeke K, Baron PA (eds), Van Nostrand Reinhold, New York
31. Kinsey JS, Cowherd C (1992) Fugitive emissions. In air pollution engineering manual Bounicore AJ, Davis WT (eds), Van Nostrand Reinhold, New York
32. QUARG (1993) Diesel vehicle emissions and urban air quality. Second report of the quality of urban air review group, Department of the Environment
33. Berlyand ME (1990) Emissions of harmful substances. Annual report on the state of air pollution and anthropogenic emissions in cities and industrial centres of the Soviet Union. State Comm on Hydrometeorol, Leningrad
34. Langner J, Rodhe H (1991) A global three-dimensional model of the tropospheric sulphur cycle. J Atmos Chem 13:255–263
35. Graedel TE, Bates TS, Bouwman AF, Cunnold D, Dignon J, Fung I, Jacob DJ, Lamb BK, Logan JA, Marland G, Middleton P, Pacyna JM, Placet M, Veldt C (1993) A compilation of inventories of emissions to the atmosphere. Global Biogeochem Cycles 7:1–26
36. Faust BC (1994) Photochemistry of clouds, fogs and aerosols. Environ Sci Technol 28:217A–221A
37. Pandis SN, Seinfeld JH (1989) Sensitivity analysis of chemical mechanisms for aqueous-phase atmospheric chemistry. J Geophys Res 94:1105–1126
38. Martin LR (1994) Aqueous sulfur (IV) oxidation revisited. In Environmental Oxidants, Nriagu JO, Simmons MS (eds), John Wiley and Sons, New York
39. Allen AG, Harrison RM, Erisman JW (1989) Field measurements of the dissociation of ammonium nitrate and ammonium chloride aerosols. Atmos Environ 23:1591–1599
40. Nevalainen A, Willeke K, Liebhaber F, Pastuszka J, Burge H, Henningson E (1993) Bioaerosol sampling. In aerosol measurement: principles, techniques and applications. Willeke K and Baron PA (eds), Van Nostrand Reinhold, New York
41. Hamilton ED (1959) Studies on the air spora. Acta allergologica 13:143–175
42. Bhati HS, Gaur RD (1979) Studies on aerobiology - atmospheric fungal spores. New Phytologist 82:519–527
43. Larsen LS (1981) A three-year survey of micro fungi in the outdoor air of Copenhagen 1977–79. Grana 20:197–198
44. Beaumont F, Kauffman HF, deMonchy JGR, Sluiter HJ and deVries K (1985) Volumetric aerobiological survey of conidial fungi in the north-east Netherlands. Allergy 40:181–186
45. Burge HA (1986) Some comments on the aerobiology of fungus spores. Grana 25:143–146
46. Pasanen AL, Kalliokoski P (1991) Airborne fungal spores in different home environments. In quality of the indoor environment. Lester JN, Perry R and Reynolds GL (eds), Selper Publications, London
47. Fradkin A, Tobin RS, Tarlo SM, Tucuuc-Porretta M, Malloch D (1987) Species identification of airborne moulds and its significance for the detection of indoor pollution. J Air Pollut Control Assoc 35:51–53
48. Verhoeff AP, van Wijnen JH, Brunekreef B, Fischer P, van Reenen-Hoekstra ES, Samson RA (1992) The presence of viable mould propagules in indoor air in relation to home dampness and outdoor air. Allergy 47:83–91
49. Hirsch SR, Sosman JA (1976) A one-year survey of mould growth inside twelve homes. Ann Allergy 36:30–38
50. Lumpkins ED, Corbit SL, Tiedeman GM (1973) Airborne fungi survey 1. Culture-plate survey of the home environment. Ann Allergy 31:361–370
51. Deak T (1991) Foodborne yeasts. Adv Appl Microbiol 36:179–278
52. Mullins J, Harvey R, Seaton A (1976) Sources and incidence of airborne Aspergillus fumigatus (Fres.). Clin Allergy 6:209–217
53. Staib F, Tompak B, Thiel D, Blisse A (1978) Aspergillus fumigatus and Aspergillus niger in two potted ornamental plants, cactus (Epiphyllum truncutum) and clivia (Clivia miniata). Biological and epidemiological aspects. Mycopathologia 66:27–30
54. Green AR (1979) An updated assessment of the critical environmental factors involved in the prevention of allergic disease. Ann Allergy 42:372–383

55. Burge HA, Solomon WR, Muilenberg ML (1980). Aeroallergen exposure risks imposed by indoor plant growth. J Allergy Clin Immunol 65:210
56. Miller JD, Laflame AM, Sobol Y, Lafontaine P, Greenhalgh R (1988) Fungi and fungal products in some Canadian houses. Int. Biodeterioration 24:103–120
57. Swaelby MA, Christensen CM (1952) Molds in house dust, furniture stuffing and in the air within house. J Allergy 23:370–374
58. Schaffer N, Seidmon EE, Brunskin S (1953) The clinical evaluation of airborne and house dust fungi in New Jersey. J Allergy 24:348–354
59. Hyde HA (1973) Atmospheric pollen grains and spores in relation to allergy II. Clin Allergy 3:109–126
60. Gravesen S (1978) Identification and prevalence of culturable mesophilic microfungi in house dust from 100 Danish houses. Allergy 33:268–272
61. Calvo A, Drendo A, Castello R (1982) Fungal spores in house dust. Ann Allergy 49:213–219
62. Solomon WR (1976) A volumetric study of winter fungus prevalence in the air of midwestern homes. J Allergy Clin Immunol 57:46–55
63. Lowenstein H, Gravesen S, Larsen L, Lind P, Schwartz B (1986) Indoor allergens. J Allergy Clin Immunol 78:1035–1039
64. Luczynska CM, Li Y, Chapman MD, Platts-Mills TAE (1990) Airborne concentrations and particle size distribution of allergen derived from domestic cats (Felis domesticus). Measurements using cascade impactor, liquid impinger and a two-site monoclonal antibody assay for FEL-d-1. Am Rev Respir Dis 141:361–367
65. Yocom JE (1982), Indoor-outdoor air quality relationships. J Air Pollut Control Assoc 32:500–520
66. Koutrakis P, Briggs SLK, Leaderer BP (1992) Source apportionment of indoor aerosols in Suffolk and Onondaga Countries, New York. Environ Sci Technol 26:521–527
67. United States Environmental Protection Agency (1987) Compendium of methods for the determination of air pollutants in indoor air. EPA/600/8–87/014. Environmental Protection Agency, Research Triangle Park
68. Katzenstein AW (1992) Environmental tobacco smoke and lung cancer risk: epidemiology in relation to confounding factors. Environ Int 18:341–345
69. US Department of Health and Human Services (1989) Reducing the health consequences of smoking: 25 years of progress. A report to the surgeon general. DHHS Publication (CDC) 89–8411, Washington DC
70. Lebowitz MD, Holberg CJ, O'Rourk MK, Corman G, Dodge R (1983) Gas Store usage, CO, TSP and respiratory effects. Proceedings 76th Annual Meeting, APCA, paper No. 83–91
71. National Research Council (1981) Indoor pollutants, National Academy Press, Washington DC
72. McCrillis RC, Jaasma DR (1993) Woodstove emission measurement methods: comparison and emission factors update. Environ Monitor Assess 24:1–12
73. Penner JE, Ghan SJ, Walton JJ, Dignon J (1990) The global budget and cycle of carbonaceous soot aerosols. Paper presented at 7th International Symposium of Comm on Atmos Chem and Global Pollution Chamrousse, France
74. Penner JE, Eddleman H, Novakov T (1993) Towards the development of a global inventory for black carbon emissions. Atmos Environ 27A:1277–1295
75. Cass GR, Boone PM, Macias ES (1982) In particulate carbon: atmospheric life cycle. Wolff GT, Klimisch RL (eds) Plenum Press, New York
76. Hildemann LM, Markowski GR, Cass GR (1991) Chemical composition of emissions from urban sources of fine organic aerosol. Environ Sci Technol 25:744–759
77. Rogge WF, Hildemann LM, Mazurek MA, Cass GR, Simoneit BRT (1991) Sources of fine organic aerosol. 1. Charbroilers and meat cooking operations. Environ Sci Technol 25:1112–1125
78. Rogge WF, Hildemann LM, Mazurek MA, Cass GR, Simoneit BRT (1993) Sources of fine organic aerosol. 2. Noncatalyst and catalyst-equipped automobiles and heavy-duty diesel trucks. Environ Sci Technol 27:636–651
79. Rogge WF, Hildemann LM, Mazurek MA, Cass GR, Simoneit BRT (1993) Sources of fine organic aerosol. 3. Road dust, tire debris, and organometallic brake lining dust: roads as sources and sinks. Environ Sci Technol 27:1892–1904
80. Rogge WF, Hildemann LM, Mazurek MA, Cass GR, Simoneit BRT (1993) Sources of fine organic aerosol. 5. Natural gas home appliances. Environ Sci Technol 27:2736–2744

81. Rogge WF, Hildemann LM, Mazurek MA, Cass GR, Simoneit BRT (1993) Sources of fine organic aerosol. 4. Particulate abrasion products from leaf surfaces of urban plants. Environ Sci Technol 27:2700–2711
82. Rogge WF, Hildemann LM, Mazurek MA, Cass GR, Simoneit BRT (1994) Sources of fine organic aerosol. 6. Cigarette smoke in the urban atmosphere. Environ Sci Technol 28:1375–1388
83. Hansen JE, Rossow W, Fung J (1990) The missing data on global climate change. Issues in science and technology, 62–69
84. SCEP (1970) Man's impact on the global environment. Report on the study of critical environmental problems, MIT Press Cambridge, Mass
85. Cameron AGW (1981) Elementary and nuclidic abundances in the solar system. In essays in nuclear astrophysics, Barnes CA, Clayton DD, Schramm DN (eds), CUP, Cambridge
86. SMIC (1971) Inadvertent climate modification. Report on the study of man's impact on climate. Mathews AW, Kellog WW, Robinson AD (eds), MIT Press Cambridge, Mass
87. Junge CE (1979) The importance of mineral dust as an atmospheric constituent. In Saharan dust, mobilization, transport and deposition. Morales C (ed), John Wiley and Sons, Chichester
88. Peterson JT, Junge CE (1971) Sources of particulate matter in the atmosphere. In man's impact on climate. Matthews AW, Kellog WW, Robinson AD (eds), MIT Press Cambridge, Mass
89. d'Almeida GA (1986) A model for Saharan dust transport. J Clim Appl Meteor 25:903–916
90. Bach W (1976) Global air pollution and climatic change. Rev Geophys Space Phys 14:429–474
91. d'Almeida GA, Koepke P, Shettle EP (1991) Atmospheric aerosols: global climatology and radiative characteristics. A Deepak Publishing, Hampton, Virginia
92. Andreae MO (1994) Climate effects of changing atmospheric aerosol levels. In world survey of climatology, Vol XX Future climate of the world. Henderson-Sellers A (ed)
93. Preining O (1992) Global climate change due to aerosols. In global atmospheric chemical change. Hewitt CN, Sturges WT (eds), Elsevier Applied Science, London
94. Jaenicke R (1988) Aerosol physics and chemistry. In Landolt-Bornstein, New series, Hellwege KE, Madelung O (eds), Springler, Berlin
95. Hantel M (1989) Climate modeling. In Landolt-Bornstein, New series, Hellwege KE, Madelung O (eds), Springler, Berlin
96. Bach W (1984) Our threatened climate. Ways of averting the CO_2 problem through rational energy use, D. Reidel, Dordrecht
97. Monin AS (1986) An introduction to the theory of climate, D. Reidel, Dordrecht
98. Middleton NJ (1989) Climatic controls on the frequency, magnitude and dust storms: examples from India/Pakistan, Mauritania and Mongolia. In palaeoclimatology and palaeometeorology: modern and past patterns of global atmospheric transport, Leinen M, Sarntheim M (eds), Kluwer Academic, Dordrecht
99. Gillette DA (1980) Major contributions of natural primary continental aerosols: source mechanisms. Ann NY Acad Sci 338:348–358
100. Cowherd C, Kinsey JS, Wallance DD, Grelinger MA, Cuscino TA, Neulicht RM (1986) Identification, assessment and control of fugitive particulate emissions. EPA/600/8–86/023
101. Hamilton RS, Mansfield TA (1991) Airborne particulate elemental carbon: its sources, transport and contribution to dark smoke and soiling. Atmos Environ 25A:715–723

In Situ Particle Formation/Reaction Mechanisms

Spyros N. Pandis[1] *and Christodoulos Pilinis*[2]

[1] Departments of Chemical Engineering and Engineering & Public Policy, Carnegie Mellon University, Pittsburgh, PA 15213
[2] Rosenstiel School of Marine and Atmospheric Science, University of Miami, Miami, FL 33149

Introduction 37
Thermodynamic Principles 38
  Conditions for Equilibrium 39
  Ideal Gas 39
  Ideal Gas Mixture 40
  Ideal Solutions 40
  Non-ideal Solutions 41
  Pure Solid Compounds 41
  Solutions of Electrolytes 41
Chemistry of Gaseous Aerosol Precursors 42
  Sulfuric Acid (H_2SO_4) 42
  Nitric Acid (HNO_3) 44
  Ammonia (NH_3) 45
  Organic Gases 45
Aerosol Nucleation 46
  Homogeneous Nucleation 46
  Homogeneous-Heteromolecular Nucleation 50
Condensation/Evaporation 52
  The Continuum Regime 52
  Free Molecule Regime 53
  Transition Regime 54
  The Accommodation Coefficient 54
  Mass Transfer Timescales 55
Inorganic Aerosol Material Formation 56
  Atmospheric Gas-Aerosol Equilibrium Models 56
Organic Aerosol Material Formation 63
  Biogenic Aerosol 63
  Anthropogenic Aerosol 63
The Equilibrium Assumption 64
  Timescales for Equilibrium 65
  Equilibrium Assumption and Aerosol Distribution 65
References 66

Summary

A significant fraction of the atmospheric particulate matter is not emitted directly as particles, but it is formed in-situ by gas-to-particle conversion. This section provides an overview of the fundamental and applied aspects of the processes leading either to the formation of new particles or to the transfer of material from the gas phase onto preexisting aerosols. The thermodynamic principles that govern the partitioning of species between the gas and aerosol phases are discussed first. An overview of the gas-phase reaction pathways leading to the formation of the main gaseous aerosol precursors (H_2SO_4, HNO_3, NH_3 and condensible organic gases) is presented next. The basic theory for the formation of new particles either from a single vapor or from two gaseous species is presented including brief discussion of the observed discrepancies with laboratory results. The next paragraph is a basic treatment of the mass transport of material from the gas to the particulate phase, with emphasis on the interaction of the gas molecules with the particle surface. The framework established in the above paragraphs is then used for the discussion of inorganic and organic aerosol formation mechanisms and the discussion of the gas-aerosol phase equilibrium assumption.

Introduction

The atmospheric aerosol is a highly complex entity, representing a mixture of primary emissions such as sea-salt and soot, and secondary species formed during the photooxidation of various pollutants. Primary particles, once emitted, undergo chemical and physical transformations that result in a continuously changing chemical composition and particle size distribution. The relative importance of primary and secondary aerosol depends on the geographical location with its particular emissions and on the corresponding atmospheric chemistry. The major primary aerosol constituents are crustal elements (Si, Al, Ca, Fe, K, Ti, Mn, Sr, etc.) and their compounds, sea-salt, soot (also called elemental carbon, EC), and a series of organic compounds [1]. The secondary material is inorganic (mainly sulfates, nitrates, chloride, ammonium and water) or carbonaceous material. The various pathways for the aerosol formation in the atmosphere are depicted in Fig. 1. Particles of diameter less than 2 μm are largely responsible for the severe visibility deterioration observed in urban areas. These particles are also respirable and may contain toxic or mutagenic compounds. Control of this fine aerosol component of the urban aerosol is an important problem facing today's air pollution control agencies. Secondary aerosol usually dominates this fine aerosol fraction.

The focus of this chapter will be on secondary aerosol matter. This includes formation of new particles in the atmosphere by nucleation, but also in-situ formation of condensable compounds and their mass transport onto the already existing

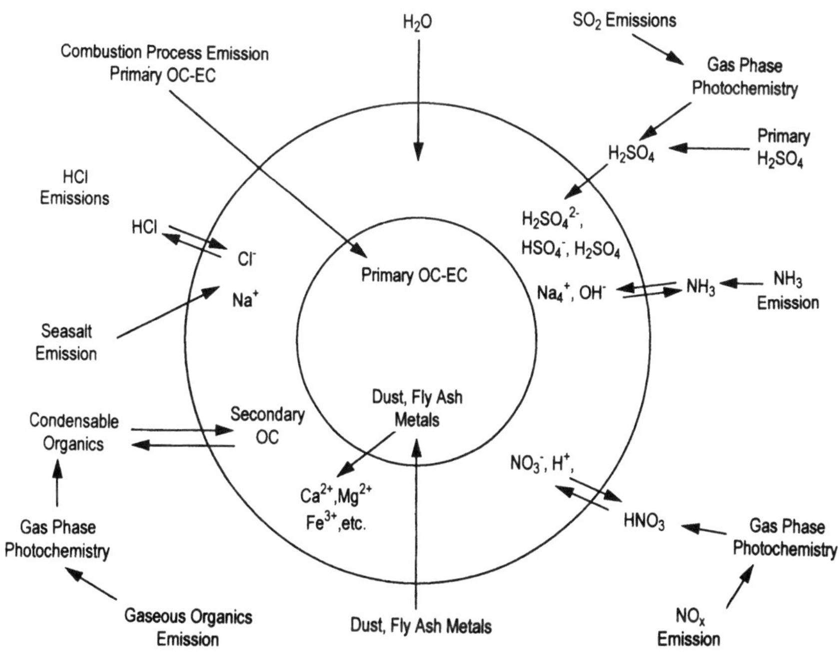

Fig. 1. Schematic description of the main pathways for production of atmospheric aerosol

aerosol (primary or secondary). The main thermodynamic principles that govern the multiphase aerosol/gas-phase system will be presented initially. The reaction mechanisms by which the gaseous precursors of aerosol are formed in the atmosphere will be then discussed. The classical nucleation theory will be developed and the in-situ formation of new particles will be examined. Next, the theoretical relationships that describe the transport of material between the gas and aerosol phases will be derived. The thermodynamic and mass transport principles will be applied to the atmospheric formation of inorganic and organic aerosol matter. Finally, the validity of the equilibrium assumption between the gas and aerosol phases in the atmosphere will be investigated.

Thermodynamic Principles

An atmospheric air parcel can be thermodynamically viewed as an open, homogeneous system which may exchange heat, pressure, work, and mass with its surroundings. Let us assume that an air parcel contains k chemical species, has a temperature T, pressure p, volume V, and entropy S. According to the first and second laws of thermodynamics, the incremental change in the internal energy U of this system for reversible processes is [2]

$$\mathrm{d}U = T\mathrm{d}S - p\mathrm{d}V + \sum_{i=1}^{k} \mu_i \,\mathrm{d}n_i \tag{1}$$

where n_i is the number of moles and μ_i is the chemical potential of species i. The study of several atmospheric processes can be facilitated by introducing other thermodynamic variables in addition to the internal energy U. One of the most useful is the Gibbs free energy, G, defined as

$$G = U + pV - TS. \tag{2}$$

Combining these two equations one finds that

$$\mathrm{d}G = -S\mathrm{d}T + V\mathrm{d}p + \sum_{i=1}^{k} \mu_i \,\mathrm{d}n_i \tag{3}$$

and the chemical potential of species i can be now formally defined as

$$\mu_i = \left(\frac{\partial G}{\partial n_i}\right)_{T,P,n_j}. \tag{4}$$

The chemical potential has an important function in the system's thermodynamic behavior analogous to pressure or temperature. A temperature difference determines the tendency of heat to pass from one body to another and a pressure difference determines the tendency for bodily movement. A difference in chemical potential can be viewed as the cause for chemical reaction or for mass transfer from one phase to the other.

Conditions for Equilibrium

The condition for chemical equilibrium in a reactive system at constant temperature and pressure is obtained by minimizing the Gibbs free energy of the system. It can be shown [2] that this is equivalent to setting

$$\sum_{i=1}^{k} v_{ij}\mu_i = 0 \text{ for every } j \tag{5}$$

where v_{ij} is the stoichiometric coefficient of species i in reaction j. This is the most general condition of equilibrium for a reacting system and is applicable whether the reactants and products are solids, liquids or gases. For example, for an air parcel where the following reactions take place,

$$2NH_3\,(g) + H_2SO_4(g) \rightleftarrows (NH_4)_2SO_4\,(s)$$
$$NH_3(g) + HNO_3(g) \rightleftarrows NH_4NO_3\,(s)$$

at equilibrium the chemical potentials of gas-phase NH_3, HNO_3, and H_2SO_4 and of the solids $(NH_4)_2SO_4$ and NH_4NO_3 should satisfy

$$\begin{aligned}\mu_{(NH_4)_2SO_4} &= 2\mu_{NH_3} + \mu_{H_2SO_4}\\ \mu_{NH_4NO_3} &= \mu_{NH_3} + \mu_{HNO_3}.\end{aligned} \tag{6}$$

Determination of the equilibrium composition of this multiphase system, therefore requires determination of the chemical potentials of all species of interest in the gas, solid and aqueous phases.

Ideal Gas

An ideal gas can be defined as a gas whose chemical potential $\mu(T, p)$ at temperature T and pressure p is

$$\mu(T, p) = \mu^o(T, 1 \text{ atm}) + \text{RT} \ln p \tag{7}$$

where μ^o is the standard chemical potential defined at a pressure of 1 atm and therefore is a function of temperature only. R is the ideal gas constant. Pressure p actually stands for the ratio $(p/1 \text{ atm})$ and is dimensionless. One can prove that, for such a gas, $pV = nRT$[2]. Thus, the above definiton of the ideal gas in terms of its chemical potential entails the ideal gas law.

Deviations from ideal gas behavior are customarily expressed in terms of the compressibility factor $C = pV/nRT$. Both dry air and water vapor have compressibility factors C, $0.998 < C < 1$, for the pressure and temperature ranges of atmospheric interest [3]. Hence, both dry air and water vapor can be treated as ideal gases with an error of less than 0.2% throughout the range of meteorological interest.

Ideal Gas Mixture

A gaseous mixture is defined as ideal if the chemical potential of its i-th component satisfies the equation

$$\mu_i = \mu_i^o(T) + RT \ln p + RT \ln y_i \tag{8}$$

where $\mu_i^o(T)$ is independent of composition and depends only on temperature, p is the pressure of the mixture and y_i is the mole fraction of compound i. For $y_i = 1$, this equation is simplified to Eq. (7), and μ_i^o is therefore precisely the same in both equations, that is the Gibbs free energy per mole of the gas in its pure state at unit pressure. One can define the partial pressure, p_i, of compound i as $p_i = y_i\ p$ and Eq. (8) can be rewritten in a more compact form

$$\mu_i = \mu_i^o(T) + RT \ln p_i. \tag{9}$$

One can also prove that Eq. (9) is equivalent to the more traditional ideal gas mixture definition, $p_i V = n_i RT$. The atmosphere can be treated as an ideal gas mixture [4].

Ideal Solutions

Atmospheric aerosols in high relative humidities are liquid, forming aqueous solutions of ammonium, nitrate, sulfate, chloride, sodium, etc. A solution is defined as ideal if the chemical potential of every component is a linear function of the logarithm of its aqueous mole fraction x_i according to the relation

$$\mu_i = \mu_i^*(T,P) + RT \ln x_i. \tag{10}$$

A multicomponent solution is ideal only if Eq. (10) is satisfied by every component. A solution approaches ideality as it becomes more and more dilute in all but one of its components (the solvent).

Assuming that an ideal solution is in equilibrium with an ideal gas mixture over it, then for the i-th component the condition of equilibrium is $\mu_i^{soln.} = \mu_i^{vap}$ and from Eqs. (9) and (10) one obtains

$$p_i = \exp\left(\frac{\mu_i^* - \mu_i^o}{RT}\right) x_i = K_i(T,p) x_i. \tag{11}$$

The chemical potentials μ_i^* and μ_i^o depend only on temperature and pressure and therefore K_i is independent of the solution composition. Setting x_i equal to one, one concludes that K_i is equal to the vapor pressure p_i of the pure component i, and Eq. (11) is equivalent to Raoult's law.

Non-ideal Solutions

Atmospheric aerosols are usually concentrated aqueous solutions that deviate significantly from ideality. The degree of deviation from ideal behavior is usually described by defining the activity coefficient γ_i such that the equation

$$\mu_i = \mu_i^*(T,P) + RT\ln(\gamma_i x_i) \tag{12}$$

is satisfied, however large the deviation from ideality. The activity coefficient γ_i is in general a function of pressure and temperature together with the mole fractions of all substances in solution. The chemical potential μ_i^* for the solutes is defined as their chemical potential at infinite dilution when their activity coefficient approaches unity [2]. The product of the mole fraction x_i of a solution component and its activity coefficient γ_i is defined as the activity of the component, $\alpha_i : \alpha_i = \gamma_i x_i$, and the chemical potential of species i is given by

$$\mu_i = \mu_i^*(T,P) + RT\ln\alpha_i. \tag{13}$$

Pure Solid Compounds

The chemical potential of a pure solid compound i can easily be derived from Eq. (10) setting x_i equal to one and is simply

$$\mu_i = \mu_i^*(T, P). \tag{14}$$

Solutions of Electrolytes

Most of the inorganic aerosol components dissociate upon dissolution, i.e. NH_4NO_3 dissociates in forming NH_4^+ and NO_3^-. The concentration of each ion in the aqueous solution is traditionally expressed in the molality scale (moles of solute per kg of water) and the chemical potential of each ion for a NH_4NO_3 solution is

$$\begin{aligned} \mu_{NH_4^+} &= \mu^o_{NH_4^+} + RT\ln(\gamma_{NH_4^+} m_{NH_4^+}) \\ \mu_{NO_3^-} &= \mu^o_{NO_3^-} + RT\ln(\gamma_{NO_3^-} m_{NO_3^-}) \end{aligned} \tag{15}$$

where $m_{NH_4^+}$ and $m_{NO_3^-}$ are the ion molalities and $\gamma_{NH_4^+}$ and $\gamma_{NO_3^-}$ the corresponding activity coefficients. The dissociation reaction $NH_4NO_3 \leftrightarrows NH_4^+ + NO_3^-$ is at equilibrium and therefore the chemical potential of NH_4NO_3 satisfies (applying Eq. 5):

$$\mu_{NH_4NO_3} = \mu_{NH_4^+} + \mu_{NO_3^-} = \mu^o_{NH_4NO_3} + RT\ln(\gamma_{NH_4^+}\gamma_{NO_3^-}m_{NH_4^+}m_{NO_3^-}). \quad (16)$$

Because of the form of these expressions, the only quantity that can be measured experimentally is the product of the activity coefficients $\gamma_{NH_4NO_3} = \gamma_{NH_4^+}\gamma_{NO_3^-}$. These activities are measured in solutions of various ionic strengths I,

$$I = \frac{1}{2}\sum_i z_i^2 m_i \quad (17)$$

where z_i is the charge of ion i, and the results are fitted in semi-empirical expressions as functions of I [5].

In multicomponent non-ideal electrolyte solutions the chemical potential of the binary electrolytes is influenced by the existence of the other ions due to interactions of ion pairs, ion triplets, etc. A rigorous theoretical treatment of these solutions is impractical due to the large number of unknown parameters involved. Thus, one generally resorts to semi-empirical approaches that do not involve as many unknown parameters as the rigorous theory, for example the mixing rule of Kusik and Meissner [6].

Chemistry of Gaseous Aerosol Precursors

The major gaseous aerosol precursors in the atmosphere are sulfuric acid (H_2SO_4), nitric acid (HNO_3), hydrochloric acid (HCl), ammonia (NH_3) and a series of organic compounds with usually more than five carbon atoms per molecule. The atmospheric reactions leading to the production of these precursors are reviewed in the subsequent paragraphs.

Sulfuric Acid (H_2SO_4)

Sulfuric acid is produced in the atmosphere during the oxidation of sulfur dioxide (SO_2). Sulfur dioxide is generated during combustion of any sulfur-containing fuel and is therefore a byproduct of industrial processes that consume sulfur-containing raw materials. The rate of SO_2 emission can be estimated by knowing the rate of fuel consumption, the percentage sulfur in the fuel, and the fact that more than 90% of the sulfur is emitted as SO_2. Cullis and Hirschler [7] estimated the global anthropogenic emissions of SO_2 for 1976 to be 104 Tg S y^{-1}. Because of the relatively high sulfur content of bituminous coals burned in great quantities in several countries, coal combustion accounts for roughly 60% of all SO_2 emitted. Petroleum refining and combustion of other fuel contribute around 30% and smelting of metallic ores and other industrial processes around 10% of the global anthropogenic SO_2 emissions. The uncertainty involved in these estimates is illustrated by comparing the 104 Tg S y^{-1} estimated by Cullis and Hirschler [7] with the 72 Tg S y^{-1} estimated for the same year by Möller [8].

Once emitted, the gaseous SO_2 is initially oxidized inside the plume and later in the ambient troposphere to sulfuric acid by reactions occurring in the gas, aqueous, and aerosol phases. Several pathways have been proposed for the homogeneous oxidation of SO_2 to H_2SO_4 [9], but all available evidence suggests that the only gas phase process that is fast enough and efficient enough is the reaction of SO_2 with the OH radical

$$SO_2 + OH \xrightarrow{M} HOSO_2$$

The effective bimolecular constant for this reaction is 9×10^{-13} cm^3 $molecule^{-1}$ s^{-1} at 298 K with an uncertainty of $\pm 50\%$ [10]. The $HOSO_2$ adduct then reacts rapidly with O_2 to produce sulfur trioxide (SO_3) [11]

$$HOSO_2 + O_2 \rightarrow HO_2 + SO_3.$$

The SO_3 finally reacts rapidly with water vapor to form sulfuric acid vapor

$$SO_3 + H_2O \rightarrow H_2SO_4.$$

The overall reaction is usually expressed as

$$SO_2 + OH(+O_2, H_2O) \rightarrow H_2SO_4 + HO_2.$$

Sulfur compounds are emitted into the atmosphere from a variety of natural sources, including volcanic eruptions and biological activity in vegetation, soils, and water ecosystems. However, in contrast to anthropogenic emissions which are entirely in the form of SO_2, these natural emissions are primarily in the form of reduced sulfur compounds. Naturally emitted sulfur compounds include hydrogen sulfide (H_2S), dimethyl sulfide (CH_3SCH_3), carbon disulfide (CS_2), dimethyl disulfide (CH_3SSCH_3), carbonyl sulfide (COS) and methyl mercaptan (CH_3SH). Natural sulfur emissions were estimated to be around 100 Tg S y^{-1} by Cullis and Hirschler [7] and 50–100 Tg S y^{-1} by Möller [12]. These sulfur compounds are oxidized in the atmosphere and it has often been assumed that they are first converted to SO_2 and subsequently to sulfate. Still relatively little is known about the oxidation of most reduced sulfur compounds in the atmosphere.

A lot of attention has recently been focused on dimethyl sulfide (DMS), a metabolic by-product of marine phytoplankton emitted by the ocean waters. Charlson et al. [13] have postulated the existence of a feedback mechanism by which global climate perturbations caused by anthropogenic greenhouse gas emissions may either be dampened or accelerated by changes in the DMS ocean flux.

DMS reacts in the remote marine atmosphere mainly with the OH radical. The DMS-OH reaction constant is 8×10^{-12} cm^3 $molecule^{-1}$ s^{-1} at 298 K [14]. Two pathways have been proposed for the OH attack on DMS, a hydrogen abstraction and a OH addition [15]. The abstraction pathway leads mainly to the formation of SO_2 while the addition of OH to the sulfur atom leads to the formation of methanesulfonic acid (MSA). The relative importance of hydrogen atom abstraction vs addition to the sulfur atom is uncertain because of its complex dependence on both temperature and NO_x availability. Use of the available kinetic data suggests that, under typical atmospheric conditions, hydrogen abstraction accounts for around 70% of the overall reaction. Field measurements have suggested that MSA yields range from 6% in tropical areas [16], to about 20% in mid latitudes [17], to around 30% near the polar regions [18]. Berresheim et al. [19] recently suggested that the dominant oxidant product of DMS in the marine atmosphere may not be SO_2 or MSA but dimethylsulfone ($DMSO_2$). Moreover, Bandy et al. [20] proposed that the major pathway for oxidation of DMS to H_2SO_4 is via SO_3 and not SO_2. The importance of DMS and its oxidation products as a climate feedback mechanism has recently been reviewed by Lawrence [21].

Nitric Acid (HNO_3)

Nitric acid is formed in the atmosphere during the NO_2 oxidation by OH, reaction of the NO_3 radical with organics and N_2O_5 hydrolysis. The attack of OH on NO_2

$$NO_2 + OH \xrightarrow{M} HNO_3$$

is the most important HNO_3 source in the atmosphere and the recommended effective bimolecular reaction constant at 300 K is 1.1×10^{-11} cm^3 $molecule^{-1}$ s^{-1} [10]. The nitrate radical NO_3 is formed during the reaction of NO_2 with O_3:

$$NO_2 + O_3 \rightarrow NO_3 + O_2$$

but during daytime it photolyses rapidly

$$\begin{aligned} NO_3 + h\nu &\rightarrow NO + O_2 \\ &\rightarrow NO_2 + O^* \end{aligned}$$

and its daytime concentration is negligible [22, 23]. However, the nighttime concentration of NO_3 can be significant and its contribution to atmospheric chemistry important. NO_3 has been shown to react rapidly with organic gases, including alkenes, cresols and biogenics [24]. The NO_3 may also react with NO_2 to generate N_2O_5

$$NO_3 + NO_2 \rightarrow N_2O_5$$

which in turn can react with water vapor to form nitric acid [23]

$$N_2O_5 + H_2O \rightarrow 2HNO_3.$$

An upper limit of 1.3×10^{-21} cm^3 molecule^{-1} s^{-1} has been established for this last reaction [25] and the reaction appears to be slow. However Russell and co-workers [23] showed that, despite the slow nature of this reaction, it can still produce significant quantities of nitric acid under suitable conditions (e.g. low NO concentration). The heterogeneous conversion of N_2O_5 in liquid aerosols or water droplets is a subject of current research and there are indications that this reaction can be significantly faster than the gas-phase reaction.

Ammonia (NH_3)

NH_3 plays a significant role in atmospheric chemistry because it is the only common highly soluble base present in the atmosphere. It is mainly produced by biological processes and hence exists in both clean and polluted environments. It reacts very slowly with OH (reaction constant 1.6×10^{-13} cm^3 molecule^{-1} s^{-1} at 298 K) and its atmospheric lifetime with respect to this reaction is of the order of 70 days [9]. Hence NH_3 is usually assumed to be nonreactive in the gas-phase.

Organic Gases

Several organic gases emitted directly to the atmosphere have been recognized as organic aerosol precursors. During the atmospheric oxidation of these gases by OH, O_3, and NO_3, low vapor pressure products are formed that can then condense to the aerosol phase as secondary organic aerosol material. Grosjean [26] reviewed the organic aerosol-forming potential of several hydrocarbons. Most paraffins do not generate aerosol, even at high concentrations. Some aerosol can be formed from the more reactive branched paraffins having more than six carbon atoms, e.g. isooctane. Acetylenics do not form aerosol. Alkenes with fewer than six carbon atoms do not form aerosol; those with six or more carbon atoms form aerosol when they yield, after rupture of the double bond, a fragment with at least five carbon atoms. For example, 1-heptene forms much more aerosol than 3-heptene, and 2,4,4-trimethyl-1-pentene (isooctene) forms more aerosol than its isomer trans-4-octene. Cyclic olefins and diolefins form more aerosol than 1-alkenes that have the same number of carbon atoms. Heavier unsaturated cyclic compounds such as indene and terpenes form even more aerosol. Carbonyl compounds (ketones, C_{1-7} aldehydes, dialdehydes) do not generate aerosol. Significant aerosol quantities are formed during the photooxidation of aromatic hydrocarbons. Organic aerosol yields for the atmospheric oxidation of several organics are given in Table 1.

Aerosol Nucleation

Homogeneous Nucleation

Homogeneous nucleation (self-nucleation) is the formation of new particles from a supersaturated vapor without the assistance of condensation nuclei or ions. Even when a vapor is below saturation, the attractive forces between the molecules, such as Van der Waals forces, lead to the formation of molecular clusters. These clusters are formed continuously but are unstable and continuously disintegrate. A snapshot of a mixture of a vapor A in air would show that all molecules of A exist independently or in clusters of two, three, or possibly four molecules.

Let us consider a closed system under constant pressure and temperature containing a spherical liquid drop of radius r containing g molecules of species A surrounded by air containing N_v A molecules. The total Gibbs free energy of the system is

$$G = N_v \mu_v + g\mu_l + 4\pi r^2 \sigma \tag{18}$$

where μ_v and μ_l are the chemical potentials per molecule of A in the gas and liquid phase respectively and σ is the surface tension of the bulk liquid. The Gibbs free energy of the system prior to the drop formation was

$$G_0 = (N_v + g)\mu_v \tag{19}$$

and therefore the Gibbs free energy change required for the drop formation is

$$\Delta G = G - G_0 = g(\mu_l - \mu_v) + 4\pi r^2 \sigma = g(\mu_l - \mu_v) + \xi g^{2/3} \tag{20}$$

where $\xi = (4\pi)^{1/3}\sigma(3v_m)^{2/3}$ with v_m equal to the molecular volume of A. The above change in the Gibbs free energy corresponds to the reversible work required to form the drop. Assuming that the vapor phase is ideal and it is in thermodynamic equilibrium with the droplet, one obtains

$$\mu_l - \mu_v = -kT \ln \frac{p_A}{p_A^o} = -kT \ln S \tag{21}$$

where p_A is the equilibrium partial vapor pressure of A over the liquid droplet and p_A^o is the equilibrium vapor pressure over a flat surface. The ratio p_A over p_A^o is defined as the supersaturation S. If $\mu_l > \mu_v$, then the vapor is subsaturated and it is the more stable phase and ΔG increases continuously as the number of molecules of A in the droplet, g, increases. If $\mu_l = \mu_v$, the gas-phase is saturated

Table 1. Measured and estimated aerosol yields

Species	Aerosol yield ($\mu g\ m^{-3}\ ppm^{-1}$)	Comments	Reference
Alkanes	0	estimated	[27]
Propane	0	estimated	[27]
n-Butane	0	estimated	[27]
Isobutane	0	estimated	[27]
n-Pentane	0	estimated	[27]
Cyclopentane	0	estimated	[27]
Pentane isomers	0	estimated	[27]
n-Hexane	0	estimated	[27]
Methylcyclopentane	17	estimated	[27]
Cyclohexane	17	estimated	[27]
Hexane isomers	0	estimated	[27]
n-Heptane	7	estimated	[27]
Methylcyclohexane	120	estimated	[27]
C7-Cycloparaffins	120	estimated	[27]
Heptane isomers	0	estimated	[27]
n-Octane	98	estimated	[27]
Ethylcyclohexane	275	estimated	[27]
C8-Cycloparaffins	275	estimated	[27]
Octane isomers	98	estimated	[27]
n-Nonane	236	estimated	[27]
C9-Cycloparaffins	466	estimated	[27]
Nonane isomers	236	estimated	[27]
n-Decane	348	estimated	[27]
C10-Cycloparaffins	687	estimated	[27]
Decane isomers	348	estimated	[27]
n-Undecane	479	estimated	[27]
C11-Cycloparaffins	945	estimated	[27]
Undecane isomers	479	estimated	[27]
n-Dodecane	626	estimated	[27]
C12-Cycloparaffins	1100	estimated	[27]
Dodecane isomers	626	estimated	[27]
n-Tridecane	865	estimated	[27]
C13-Cycloparaffins	1265	estimated	[27]
Tridecane isomers	865	estimated	[27]
n-Tetradecane	972	estimated	[27]
C14-Cycloparaffins	1443	estimated	[27]
Tetradecane isomers	972	estimated	[27]
n-Pentadecane	1300	estimated	[27]
C15-Cycloparaffins	1717	estimated	[27]
Pentadecane isomers	1300	estimated	[27]
Alkenes			
Ethene	0	estimated	[27]
Propene	0	estimated	[27]
1-Butene	0	estimated	[27]
1-Butene isomers	0	estimated	[27]
2-Butene	0	estimated	[27]
2-Butene isomers	0	estimated	[27]
1,3 Butadiene	0	estimated	[27]
1-Pentene	0	estimated	[27]
1-Pentene isomers	0	estimated	[27]
2-Pentene isomers	0	estimated	[27]
Cyclopentene	132	measured	[28]
1-Hexene	0	estimated	[27]
1-Hexene isomers	0	estimated	[27]
2-Hexene isomers	0	estimated	[27]
Cyclohexene	278	measured	[28]
1-Heptene isomers	78	measured	[29]
2-Heptene isomers	78	estimated	[27]
1-Octene isomers	227	measured	[30]

Table 1. (continued)

Species	Aerosol yield (μg m^{-3} ppm^{-1})	Comments	Reference
2-Octene isomers	227	estimated	[31]
1-Nonene isomers	304	estimated	[27]
2-Nonene isomers	304	estimated	[27]
Decene isomers	470	estimated	[27]
Undecene isomers	743	estimated	[27]
Dodecene isomers	878	estimated	[27]
Tridecene isomers	1024	estimated	[27]
Tetradecene isomers	1180	estimated	[27]
Pentadecene isomers	1350	estimated	[27]
Bioalkenes			
Isoprene	0	measured	[32]
α-Pinene	762	measured	[33]
β-Pinene	720	measured	[32]
Monoterpenes	762	estimated	[31]
Aromatics			
Benzene	0	measured	[34]
Toluene	424	measured	[34]
Ethylbenzene	440	measured	[34]
Isopropylbenzene	334	measured	[34]
n-Propylbenzene	138	measured	[34]
Isobutylbenzene	334	estimated	[27]
m-Xylene	419	measured	[34]
o-Xylene	428	measured	[34]
p-Xylene	180	measured	[34]
1,3,5-Trimethylbenzene	577	measured	[34]
1,2,3-Trimethylbenzene	496	measured	[34]
1,2,4-Trimethylbenzene	251	measured	[34]
Naphthalene	400	estimated	[31]
2,3-Dimethylnaphthalene	600	estimated	[31]
Methylnaphthalenes	500	estimated	[31]
Tetralin	400	estimated	[31]
Other			
Methyl ethyl ketone	0	estimated	[27]
Alkyl phenols	220	estimated	[27]
Glyoxal	0	estimated	[27]
Methyl glyoxal	0	estimated	[27]
Higher aldehydes	0	estimated	[27]
Acetone	0	estimated	[27]
Higher PAN analogues	0	estimated	[27]
Aromatic aldehydes	5	estimated	[27]
Peroxy benzoyl nitrate	5	estimated	[27]
Phenols	192	estimated	[27]
Nitrophenols	285	estimated	[27]

in A, but ΔG is still increasing continuously as g increases. In both these cases a liquid drop will not be formed because of the additional work needed to create the droplet interface. Finally, if $\mu_l < \mu_v$ (supersaturated vapor), ΔG increases with increasing g for small g values, exhibits a maximum where the negative volume term starts dominating the positive surface term and starts decreasing for higher g values. The number of molecules in the droplet at this maximum is denoted as g^* (critical size nucleus) and the corresponding Gibbs free energy as ΔG^*. Once a molecular cluster size exceeds g^*, even momentarily, it becomes stable and grows by condensation to larger particles. Therefore the nucleation rate is equal to the rate with which particles larger than g^* are created.

If A_i is a molecular cluster containing iA molecules, one can envisage the collisions between clusters and molecules of A as a series of reactions:

$$\begin{aligned} A_1 + A_1 &\leftrightarrows A_2 \\ A_2 + A_1 &\leftrightarrows A_3 \\ &\cdot \\ &\cdot \\ A_{g-1} + A_1 &\leftrightarrows A_g. \end{aligned} \tag{22}$$

Neglecting cluster to cluster collisions, the rate of change of the concentration of g-mers can be expressed as

$$\frac{dN_g}{dt} = J_{g-1} - J_g \tag{23}$$

where J_{g-1} is the net flow of nuclei from size $g-1$ to g, and J_g is the net flow of nuclei from size g to $g+1$ given by

$$J_g = \beta a_{g-1} N_{g-1} - \varepsilon_g a_g N_g \tag{24}$$

with a_g the surface area of a g-mer, β the rate of collision of vapor molecules with a unit surface area of the nucleus, and ε_g the rate of loss of single molecules per unit surface of a g-mer. The rate of collision β of monomers with a unit surface area of the nucleus can be calculated from kinetic theory, whereas ε_g is a generally unknown function of the embryo size. At equilibrium all net flows of nuclei are zero and therefore from Eq. (24):

$$\frac{N_g^e}{N_{g+1}^e} = \frac{\varepsilon_{g+1}\alpha_{g+1}}{\beta\alpha_g}. \tag{25}$$

This equation allows the calculation of the loss rate ε_g from the other known quantities, assuming that the distribution of nuclei number concentration follows a Boltzmann distribution,

$$N_g^e = N_1 \exp\left(-\frac{\Delta G}{kT}\right). \tag{26}$$

The major assumption of classical homogeneous nucleation theory is that during nucleation the nuclei number concentration attains pseudo-steady state in which their rate of appearance is just balanced by the rate of disappearance. Note that this assumption is not equivalent to the equilibrium assumption. At equilibrium $J_g = 0$, whereas at pseudo-steady state $J_g = J_{g+1} = \ldots = J$. Using this assumption and Eqs. (24) and (25) the nucleation rate J is equal to

$$J = J_{g-1} = J_g = \beta a_g N_g^e \left(\frac{N_g}{N_g^e} - \frac{N_{g+1}}{N_{g+1}^e}\right). \tag{27}$$

Summing Eq. (27) from $g=1$ to some sufficiently large $g=G-1$, successive terms cancel out and

$$\sum_{g=1}^{G-1} \frac{J}{\beta a_g N_g^e} = \frac{N_1}{N_1^e} - \frac{N_G}{N_G^e}. \tag{28}$$

We assume that the equilibrium and non-equilibrium monomer concentrations are approximately the same and therefore $N_1/N_1^e \approx 1$, while for sufficiently large G, $N_G/N_G^e \approx 0$. Rearranging Eq. (28) one obtains

$$J = \left(\sum_{g=1}^{G-1} \frac{1}{\beta a_g N_g^e} \right)^{-1}. \tag{29}$$

After approximating this sum by an integral, expanding N_g^e into a Taylor series around g^* [35] and using the relationships

$$\beta = \frac{p_A}{(2\pi m_A kT)^{1/2}} \quad a_g = 4\pi R_p^2 \quad N_g^e = N_1 \exp\left[- \frac{16\pi\sigma^3 v_1^2}{3\,k^3 T^3 (\ln S)^2} \right] \tag{30}$$

where m_A is the molecular mass and v_1 is the molecular volume of A, one obtains

$$J = \left[\frac{p_A}{(2\pi m_A kT)^{1/2}} \right] \left[\frac{2\sigma^{1/2} v_1}{(kT)^{1/2}} \right] N_1 \exp\left[- \frac{16\pi\sigma^3 v_1^2}{3k^3 T^3 (\ln S)^2} \right]. \tag{31}$$

The nucleation rate can be expressed in terms of measurable quantities as

$$J = \frac{1}{\rho_p} \left(\frac{2\sigma M_A}{\pi} \right)^{1/2} \left(\frac{p_A^0}{T} \right)^2 S^2 \frac{N_{av}^{3/2}}{R^2} \exp\left[- \frac{16\pi\sigma^3 M_A^2 N_{av}}{3R^3\, T^3 \rho_p^2 (\ln S)^2} \right] \tag{32}$$

where ρ_p is the particle density, M_A the molecular weight of A, N_{av} is Avogadro's number. The nucleation rate is a continuous function of the supersaturation S, but traditionally a critical supersaturation ratio S^* is defined as the one for which $J=1\ \text{cm}^{-3}\text{s}^{-1}$. For water under typical atmospheric conditions the critical supersaturation ratio is approximately ten. Homogeneous-homomolecular nucleation seldom occurs naturally because very high supersaturations are necessary.

Homogeneous-Heteromolecular Nucleation

When two gaseous species that can form a liquid solution are present as a vapor mixture, even if this mixture is subsaturated with respect to the pure substances, nucleation can occur if the vapor mixture is supersaturated with respect to their solution. The theory of heteromolecular nucleation is an extension of the monomolecular theory presented above. We shall restrict ourselves here in the discussion of binary nucleation.

The change in Gibbs free energy required for the formation of a droplet of radius R_p containing n_A molecules of A and n_B molecules of B from a vapor mixture of these two compounds is

$$\Delta G = n_A(\mu_A^l - \mu_A^g) + n_B(\mu_B^l - \mu_B^g) + 4\pi R_p^2 \sigma = -n_A kT \ln \frac{p_A}{p_{Asol}^o} - n_B kT ln \frac{p_B}{p_{Bsol}^o} + 4\pi R_p^2 \sigma \tag{33}$$

where μ^l and μ^g are the chemical potentials of the species in the liquid and gas phases, p_{Asol}^o and p_{Bsol}^o are the equilibrium vapor pressures of A and B over a flat solution with the same composition as the droplet. The three dimensional surface $\Delta G(n_A, n_B)$ has a saddle point representing the minimum height ΔG^* of the nucleation barrier [36]. ΔG^* is the free energy required to form a mixed critical nucleus from the vapor phase and has the following form:

$$\Delta G^* = \frac{16\pi\sigma^3 V^2}{3(RT \ln S^*)^2} \tag{34}$$

where σ and V represent the respective values of the surface tension and the mean molar volume of the droplet at the saddle point composition x_A^* (x_A^* is the mole fraction of species A). The critical supersaturation for the mixture S^* can be expressed as a function of S_A and S_B, the supersaturations of the components A and B [37]:

$$S^* = S_A^{X_A^*} S_B^{X_B^*} . \tag{35}$$

Evaluation of ΔG^* is complicated because it requires evaluation of the supersaturations S_A and S_B which in turn depend on the composition of the droplet. Calculation of this composition requires detailed data of the activity coefficients, surface tension and molar volumes of the solution as a function of composition for the temperature of interest [38] and of the vapor pressures of the pure species. The saddle point ΔG^* is usually found by discrete search that allows only integral numbers of A and B molecules in the critical cluster [39, 40].

One can show [35] that the binary nucleation rate for species A and B (where A is assumed to be in much higher concentration than B) can be expressed as

$$J = 4\pi R_p^{*2} \frac{N_B kT}{(2\pi m_B kT)^{1/2}} N_A \exp\left(-\frac{\Delta G^*}{kT}\right) \tag{36}$$

where R_p^* is the radius of the critical nucleus, ΔG^* is the Gibbs free energy change at the critical point, N_A and N_B are the vapor phase concentrations of A and B.

The most important atmospheric binary nucleating system is H_2SO_4/H_2O. This system has been the subject of several experimental [41–43, 40] and theoretical [44–49, 37, 38] studies. An additional complication arises in the H_2SO_4/H_2O system because the H_2SO_4 molecules in the gas phase form several hydrate species. The formation of these hydrates significantly changes the shape of the ΔG curve and has to be borne in mind in the calculation of ΔG^* [49]. Predicted nucleation

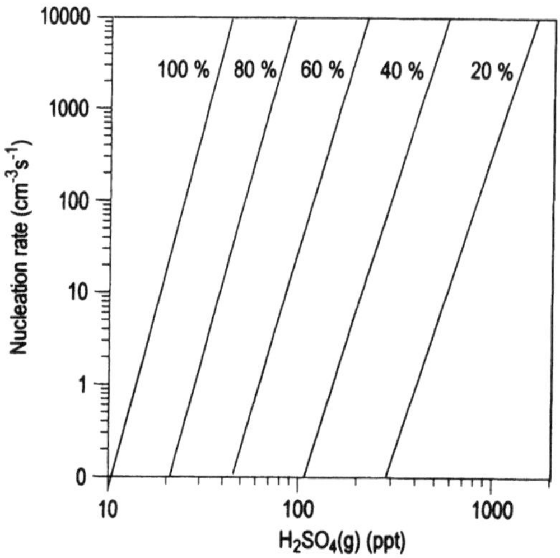

Fig. 2. Theoretical nucleation rates as a function of the H_2SO_4 vapor concentration for relative humidities 20, 40, 60, 80 and 100% at a temperature of 293 K [37]

rates for the H_2SO_4/H_2O system taking into account the influence of hydrate formation in the gas phase have been presented by Jaecker-Voirol and Mirabel [37] for relative humidities 20–100% and temperatures from −50 to 100 °C. The predicted nucleation rates for a temperature of 293 K and relative humidity ranging from 20 to 100% are presented in Fig. 2. Considerable disagreement still exists between the nucleation theory and the experimentally determined nucleation rates. Wyslouzil et al. [40] reported ratios of J_{exp}/J_{theor} varying between 10^{-9} and 10^{10} in the low relative humidity regime (r.h. $<30\%$). They also reported that this discrepancy is a strong function of the predicted critical nucleus composition with the smaller value of the ratio corresponding to 4 acid molecules and the larger to 30 acid molecules in the critical nucleus.

Condensation/Evaporation

The growth of aerosol particles by condensation is the most important pathway for material transfer between the gas and aerosol phase in the atmosphere. This process requires a supersaturated vapor and aerosol particles that serve as the condensation nuclei.

The Continuum Regime

The diffusion of a vapor A in air to a spherical particle of radius R_p is described by

$$\frac{\partial c_A}{\partial t} = \frac{D_A}{r^2}\frac{\partial}{\partial r}\left(r^2\frac{\partial c_A}{\partial r}\right) \tag{37}$$

where D_A is the diffusivity of A in air. The concentration of A is c_{As} at the particle surface and $c_{A\infty}$ far away from the particle. The time required for the concentration of A to reach steady-state around an aerosol particle ($R_p < 5$ μm) is $\tau = R_p^2/D_A < 10^{-7}$ s and therefore the above equation can be simplified, and after using the two boundary conditions solved as

$$c_{A\infty} - c_A(r) = \frac{R_p}{r}(c_{A\infty} - c_{As}) \tag{38}$$

and the mass flux of A towards the particle surface J_A is given by

$$J_A = 4\pi R_p D_A (c_{A\infty} - c_{As}) \tag{39}$$

where c_{As} can also be viewed as the concentration of A that is in equilibrium with the aerosol.

A major assumption implicit in the above derivation of the mass transfer rate of aerosol particles is that the mean free path of vapor molecules of A in air, λ_A, is small compared to the radius of the particle, that is when the Knudsen number, defined as $Kn = \lambda_A/R_p$, is much smaller than unity. When $Kn \ll 1$, the diffusion occurs in the so-called continuum regime. The mean free path of a gas can be calculated from kinetic theory of gases and for air at standard conditions is 65.1 nm. Thus particles with radii exceeding 0.2 μm or so can be considered, from the point of view of transport properties, to be in the continuum regime and Eq. (39) can be used for the description of the corresponding mass transfer processes.

Free Molecule Regime

Particles for which $Kn \gg 1$, that is particles having radii smaller than 10 nm, are considered to be in the free molecule regime and the corresponding mass transport rate is calculated applying the kinetic theory of gases. In this regime the particle appears to the vapor as just a large molecule. The number of A molecules hitting a unit area in unit time is equal to $N_A c/4$ where N_A is the number concentration of A molecules and c is the mean molecular speed of gas A. This mean molecular speed of A is equal to

$$c = \left(\frac{8kT}{\pi m_A}\right)^{1/2} \tag{40}$$

with m_A equal to the molecular mass of A. Therefore, the mass transport rate of A to the aerosol is

$$J_A = R_p^2\left(\frac{8\pi kT}{m_A}\right)^{1/2}(c_{A\infty} - c_{As}) \tag{41}$$

Transition Regime

In the region between the continuum and free molecule regimes (usually termed transition regime) neither continuum diffusion theory nor elementary kinetic theory of gases applies. Atmospheric aerosols in this regime have radii between approximately 10 and 200 nm. Expressions describing the mass transfer rates in this regime are derived using the flux matching technique, namely matching the flux predicted by continuum diffusion theory with that predicted by free molecule theory within a region beyond the particle surface [50]. Fuchs and Sutugin [51] proposed the following expression:

$$J = J_c \frac{1 + Kn}{1 + 1.71Kn + 1.33Kn^2} \tag{42}$$

where J_c is the continuum regime flux of A given by Eq. (39). Defining the free molecular path of A as $\lambda_A = 3D_A/c$, this expression reduces to $J = J_c$ for $Kn \to 0$ and Eq. (41) for $Kn \to \infty$.

The Accommodation Coefficient

In the discussion of mass transport to aerosols we have implicitly assumed that once a vapor molecule reaches the aerosol particle surface it always sticks. This assumption is equivalent to setting the value of sticking probability or accommodation coefficient to unity. Assuming an accommodation coefficient value of α one can repeat the calculations outlined in the previous paragraphs to derive the following expression:

$$J_A = 4\pi D_A R_p \frac{1 + Kn}{1 + 1.71Kn + 1.33Kn^2 + 1.33Kn(1 + Kn)(\frac{1}{\alpha} - 1)}(c_{A\infty} - c_{As}) \tag{43}$$

that for $\alpha = 1$ reduces to Eq. (42) and for $\alpha \to 0$ to $J_A \to 0$. Historically there has been much uncertainty in the magnitude of accommodation coefficients. Reported values for evaporation coefficients for experiments on several pure liquids have gradually approached unity as experimental techniques have improved [52]. Both experimental and theoretical evidence seem to indicate a universal value of unity for evaporation and condensation coefficients of a molecule on its own pure liquid. Some recent accommodation coefficient measurements on mainly clean water surfaces are summarized in Table 2.

All the above studies have concentrated on well defined water droplet surfaces and avoided the complexities of multicomponent aerosols. Further complications may arise from the existence of surfactant organic compounds such as long chain alcohols and fatty acids on aerosols which may form surface films limiting interfacial mass transport [57].

Table 2. Some accommodation coefficient measurements

Species	Substrate	Temperature	Accommodation Coefficient	Reference
SO_2	Water	273 K	0.11±0.02	[53]
H_2O_2	Water	273 K	0.18±0.02	[53]
HNO_3	Water	268 K	0.07±0.02	[54]
		293 K	0.19±0.02	
HCl	Water	274 K	0.18±0.02	[54]
		294 K	0.06±0.02	
N_2O_5	Water	271 K	0.06±0.01	[54]
		282 K	0.04±0.01	
HCl	40 wt% H_2SO_4/H_2O	283 K	0.15	[55]
	60 wt% H_2SO_4/H_2O	283 K	< 0.01	
O_3	Water		5.3×10^{-4}	[56]

Mass Transfer Timescales

There are several timescales governing the mass transport of material between the atmospheric aerosol and gas phases. They include the characteristic time for the gas phase concentration to relax to the steady state profile (which is less than 10^{-3} s for all particles of atmospheric interest), the characteristic time for mass transport, the time to achieve equilibrium at the air/aerosol interface [35] and the timescale to achieve equilibrium between the bulk aerosol and gas phases. The rate of change of the ambient gas phase concentration, C, of a single species due to its transport between the gas and aerosol phases can be written as

$$\frac{\mathrm{d}C}{\mathrm{d}t} = \int_0^\infty n(R_\mathrm{p})J(R_\mathrm{p})\,\mathrm{d}R_\mathrm{p} \tag{44}$$

where $n(R_\mathrm{p})$ is the number distribution of particles and $J(R_\mathrm{p})$ is the condensation/evaporation flux Eq. (43). The mass transport timescale, τ_m, is then given by

$$\tau_\mathrm{m}^{-1} = 4\pi D_\mathrm{A} \int_0^\infty \frac{R_\mathrm{p} n(R_\mathrm{p})(1+Kn)}{1+1.71Kn+1.33Kn^2+1.33Kn(1+Kn)(\frac{1}{\alpha}-1)}\,\mathrm{d}R_\mathrm{p}\,. \tag{45}$$

Wexler and Seinfeld [58] used ambient aerosol distributions measured in Los Angeles and, assuming an accommodation coefficient of 0.1, they calculated mass transfer time scales varying from less than one second to 20 min. Note that for smaller values of the accommodation coefficient or less polluted conditions the mass transfer time scale can be much larger. Wexler and Seinfeld [59] estimated that under cooler, less polluted conditions than Los Angeles this mass transport time scale could be of the order of hours.

Inorganic Aerosol Material Formation

NH_3, HNO_3, HCl, H_2SO_4 and H_2O are the major inorganic gas phase species involved in the formation and growth of aerosol species in the atmosphere. Most of the elements found in coarse aerosol particles over land or sea are involatile and chemically inert with the exception of a few elements such as chloride. Chloride is mainly emitted as NaCl, in primary sea-salt aerosol particles. However, these particles can react with sulfuric and nitric acids:

$$HNO_3(g) + NaCl(s) \rightleftarrows NaNO_3(s) + HCl(g)$$
$$H_2SO_4(g) + 2\,NaCl(s) \rightleftarrows Na_2SO_4(s) + 2\,HCl(g)$$
$$H_2SO_4(g) + NaCl(s) \rightleftarrows NaHSO_4(s) + HCl(g)$$

to produce gaseous HCl. These reactions should result in a chloride deficit compared to sodium in the aerosol phase, something that has often been observed in both polluted and remote environments [60, 61].

Gas phase ammonia can react with several acid vapors to form ammonium salts in the aerosol phase by the following reversible reactions:

$$NH_3(g) + HNO_3(g) \rightleftarrows NH_4NO_3(s)$$
$$2\,NH_3(g) + H_2SO_4(g) \rightleftarrows (NH_4)_2SO_4(s)$$
$$NH_3(g) + H_2SO_4(g) \rightleftarrows NH_4HSO_4(s)$$
$$3\,NH_3(g) + 2\,H_2SO_4(g) \rightleftarrows (NH_4)_3H(SO_4)_2(s)$$
$$NH_3(g) + HCl(g) \rightleftarrows NH_4Cl(s)\,.$$

The potential main reactive inorganic components of the solid aerosol phase are therefore: Na_2SO_4, $NaHSO_4$, NaCl, $NaNO_3$, NH_4Cl, NH_4NO_3, $(NH_4)_2SO_4$, NH_4HSO_4, and $(NH_4)_3H(SO_4)_2$ [5]. When the relative humidity is high and the aerosol particle forms an aqueous solution, the above salts dissociate into the corresponding ions, and the aerosol components are: H_2O, NH_4^+, SO_4^{2-}, NO_3^-, H^+, Na^+, Cl^-, HSO_4^- and H_2SO_4. The chemical reactions that occur in the NaCl–HCl–H_2SO_4–NH_3–HNO_3–H_2O system with the corresponding equilibrium constants are presented in Table 3.

Atmospheric Gas-Aerosol Equilibrium Models

Atmospheric gas-aerosol equilibrium models calculate the partitioning of the total sulfate, nitrate, chloride, ammonium and water in an air parcel assuming that the system is in thermodynamic equilibrium. Their output includes the gas phase and aerosol phase concentrations of all species of interest. Equation (5) is the base for these equilibrium calculations. Substituting Eq. (13) into Eq. (5) one obtains

Table 3. Chemical reactions occurring in the $NaCl$-HCl-H_2SO_4-NH_3-HNO_3-H_2O system

Reaction	Equilibrium constant[a]		
	A	B	C
$NaCl(s) + HNO_3(g) \rightleftarrows NaNO_3(s) + HCl(g)$	3.96	5.5	2.18
$NH_3(g) + HNO_3(g) \rightleftarrows NH_4^+ + NO_3^-$	3.99×10^{17}	64.7	11.51
$HCl(g) \rightleftarrows H^+ + Cl^-$	2.03×10^{6}	30.21	19.91
$NH_3(g) + HCl(g) \rightleftarrows NH_4^+ + Cl^-$	2.12×10^{17}	65.08	14.51
$Na_2SO_4(s) \rightleftarrows 2\ Na^+ + SO_4^{2-}$	0.481	0.98	39.57
$(NH_4)_2SO_4(s) \rightleftarrows 2NH_4^+ + SO_4^{2-}$	1.425	−2.65	38.55
$HSO_4^- \rightleftarrows H^+ + SO_4^{2-}$	1.031×10^{-2}	7.59	18.83
$HNO_3(g) \rightleftarrows H^+ + NO_3^-$	3.64×10^{6}	29.47	16.84
$NH_4Cl(s) \rightleftarrows NH_3(g) + HCl(g)$	1.04×10^{-16}	−71.04	2.40
$NH_3(g) + HNO_3(g) \rightleftarrows NH_4NO_3(s)$	3.35×10^{16}	75.11	−13.46
$NaNO_3(s) \rightleftarrows Na^+ + NO_3^-$	11.97	−8.22	16.01
$NaCl(s) \rightleftarrows Na^+ + Cl^-$	37.74	−1.57	16.89
$NaHSO_4(s) \rightleftarrows Na^+ + HSO_4^-$	2.44×10^{4}	0.79	4.59

[a]Source: Pilinis and Seinfeld [5]. The partial vapor pressures of the gases are in atm and the aqueous phase concentrations in mol kg^{-1}. The equilibrium constants as a function of temperature T (K) are given by the following expression:

$$K(T) = A \exp\left\{B\left(\frac{298.15}{T} - 1\right) - C\left(1 + \ln\left(\frac{298.15}{T}\right) - \frac{298.15}{T}\right)\right\} \tag{46}$$

$$\exp\left[-\frac{1}{RT}\sum_{i=1}^{k} v_{ij}\mu_i^o\right] = \prod_{i=1}^{k} \alpha_i^{v_{ij}} = \prod_{i=1}^{k} \gamma_i^{v_{ij}} m_i^{v_{ij}} \equiv K_j \tag{47}$$

where $K_j(T)$ is the equilibrium constant for the j-th reaction (values are given in Table 3). Consider for example the reaction

$$H_2SO_4(g) + 2NH_3(g) \leftrightarrows (NH_4)_2SO_4(s) \tag{48}$$

for which $v_{H_2SO_4} = 1, v_{NH_3} = 2$ and $v_{(NH_4)_2SO_4} = -1$, and the condition for equilibrium is

$$2\mu_{NH_3} + \mu_{H_2SO_4} - \mu_{(NH_4)_2SO_4} = 0 \tag{49}$$

or, replacing the suitable expressions for the chemical potentials,

$$\exp\left(\frac{\mu^*_{(NH_4)_2SO_4} - 2\mu^o_{NH_3} - \mu^o_{H_2SO_4}}{RT}\right) = K(T,P) = p^2_{NH_3}\, p_{H_2SO_4} \tag{50}$$

and therefore at equilibrium the product of the square of the ammonia partial pressure and of the sulfuric acid should equal a constant. This constant depends on temperature and pressure only.

With these given equilibrium constant values, the equilibrium problem is the calculation of the concentrations of the various species in the aerosol phase, m_i. The required activity coefficients are calculated from laboratory data and a suitable multicomponent activity coefficient method [62,63,6]. Comparison of the predicted activity coefficients by these methods and the available experimental data suggests that these methods perform well for the low concentration region (1–6 mol/l) [64]. Kim et al. [64] also concluded that for higher solution concentrations the Bromley method generally starts to deviate from the data for, notably, $(NH_4)_2SO_4$, NH_4NO_3, H_2SO_4, HNO_3, and HCl. The Pitzer method performs well for most cases but in general, the Kuzik–Meissner method performs as well or better than the Pitzer method.

The sulfate and sodium atmospheric aerosol compounds have extremely low vapor pressures and reside exclusively in the aerosol phase. Aqueous or solid Na_2SO_4 and $(NH_4)_2SO_4$ are the preferred aerosol forms of sulfate. When both ammonium and sodium exist in the aerosol phase, sulfate combines preferentially with sodium. If there is insufficient sodium and ammonium present to neutralize the available sulfate forming Na_2SO_4 and $(NH_4)_2SO_4$, other forms of sodium and ammonium sulfate, such as $NaHSO_4$, NH_4HSO_4, or $(NH_4)_3H(SO_4)_2$ are formed. Therefore the molar ratio of positive ion components (ammmonium and sodium) and sulfate

$$R_s = ([NH_4^+]_T + [Na^+])/[SO_4^{2-}]_T \tag{51}$$

is a key component determining the gas/aerosol equilibrium state [5]. If R_s exceeds two, the system is sulfate poor. The available sulfate forms Na_2SO_4 and $(NH_4)_2SO_4$ and the excess gas phase ammonia reacts with the available gas phase HCl and HNO_3 and drives them in the particulate phase in accordance with the corresponding equilibria. If R_s is less than two, the system is sulfate rich and sulfate exists as both HSO_4^- and SO_4^{2-}, most ammonium is in the aerosol phase and most nitrate and chloride remain in the gas phase. This understanding of the behavior of the atmospheric aerosol system is incorporated in the available thermodynamic models for the computationally efficient calculation of the thermodynamic equilibrium aerosol composition [65].

Figures 3 and 4 depict the behavior of the atmospheric aerosol system for a sulfate rich system ($R_s = 1.67$) and a sulfate poor system ($R_s = 5.77$) as a function of the relative humidity [64]. In the sulfate rich system the available sulfate is neutralized by the available sodium and all the available ammonia. For the low relative humidities (less than 80%) all the chloride resides in the gas phase and only traces of nitrate are transferred to the particulate phase. More than 90% of the nitrate exists as gas phase nitric acid. Further increases in relative humidity result in increasing aerosol water content (as much as 220 $\mu g\ m^{-3}$ for r.h. = 90%) and transfer of some of the available chloride and nitrate to the aerosol phase. In the sulfate poor case (Fig. 4) the available sulfate is neutralized by ammonia and for higher relative humidities more ammonium and nitrate are transferred to the aerosol phase.

The complicated interactions between the various inorganic aerosol components can also be investigated using a thermodynamic equilibrium model [64]. Figure 5 illustrates the phenomena occurring in an air parcel containing 5.9 $\mu g\,m^{-3}$ sodium, 10 $\mu g\,m^{-3}$ ammonia, 30 $\mu g\,m^{-3}$ nitric acid and 9.36 $\mu g\,m^{-3}$ hydrochloric acid during the oxidation of SO_2 to H_2SO_4 at 70% r.h and 298 K.

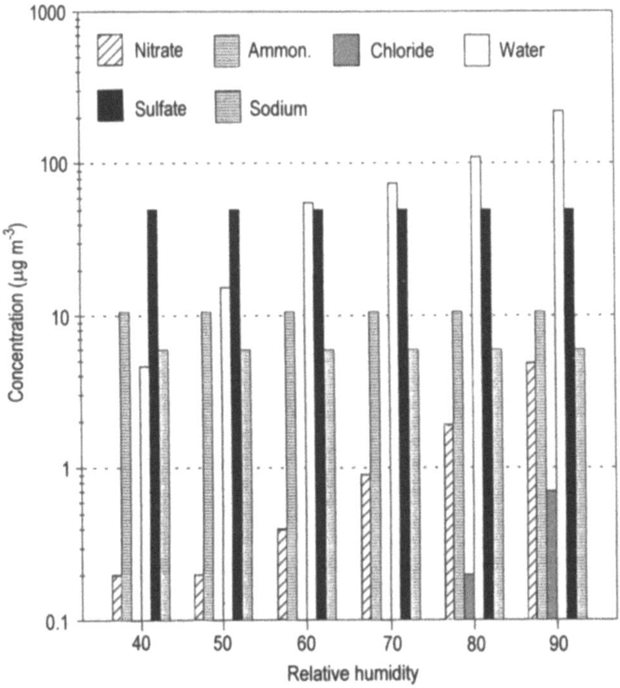

Fig. 3. Predicted aerosol composition as a function of the relative humidity for an air parcel containing 50 $\mu g\,m^{-3}$ total sulfate as H_2SO_4, 30 $\mu g\,m^{-3}$ total nitrate as HNO_3, 9.36 $\mu g\,m^{-3}$ total chloride as HCl, 10 $\mu g\,m^{-3}$ total ammonium as NH_3, and 5.9 $\mu g\,m^{-3}$ sodium, using the method of Kim et al. [64–65]

In the absence of sulfuric acid the total aerosol mass is 80.2 $\mu g\,m^{-3}$ with half of it consisting of water. Half of the available ammonia remains in the aerosol phase together with more than 75% of the nitric acid and 50% of the chloride. As sulfuric acid is produced it is exclusively transported to the aerosol phase due to its low vapor pressure. The ammonia equilibrium is shifted towards the aerosol phase in an effort to neutralize the sulfuric acid and nitrate and chloride evaporate. After the formation of 20 $\mu g\,m^{-3}$ of sulfate, the aerosol mass increases to 105.6 $\mu g\,m^{-3}$, with the addition of 10.5 $\mu g\,m^{-3}$ of water, 4.3 $\mu g\,m^{-3}$ of ammonia and the evaporation of 6.7 $\mu g\,m^{-3}$ of nitrate and 2.2 $\mu g\,m^{-3}$ of chloride. Production of additional quantities of sulfuric acid leads to the transfer of all the available ammonia to the aerosol phase and the evaporation of all nitrate and chloride.

The aerosol transformations during the production of nitric acid in air parcel containing 5.9 $\mu g\,m^{-3}$ sodium, 10 $\mu g\,m^{-3}$ ammonia, 10 $\mu g\,m^{-3}$ sulfuric acid and 9.36 $\mu g\,m^{-3}$ hydrochloric acid at 70% r.h. and 298 K are depicted in Fig. 6. Without any nitric acid present the aerosol mass concentration is 66.6 $\mu g\,m^{-3}$ consisting of 37.7 $\mu g\,m^{-3}$ of water, less than 30% of the available total ammonia

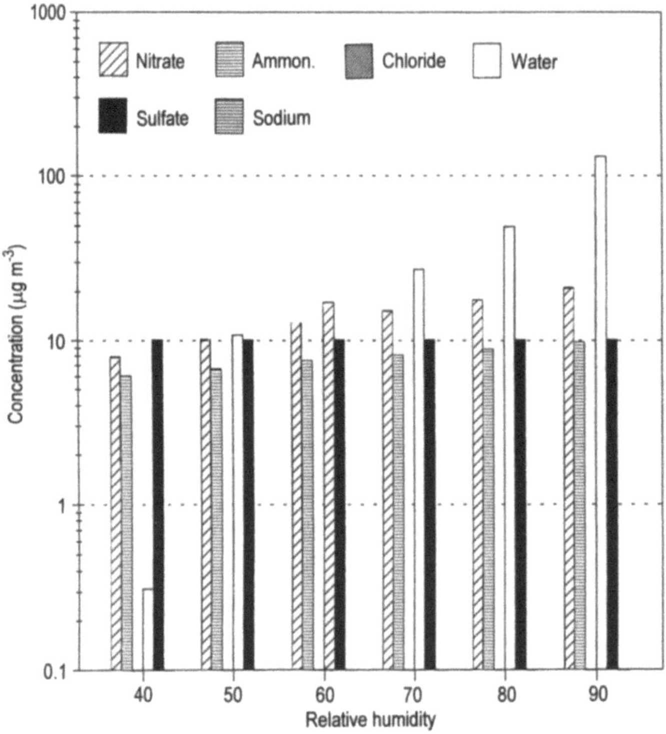

Fig. 4. Predicted aerosol composition as a function of the relative humidity for an air parcel containing 10 μg m^{-3} total sulfate as H_2SO_4, 30 μg m^{-3} total nitrate as HNO_3, and 10 μg m^{-3} total ammonium as NH_3, using the method of Kim et al. [64–65]

and half the available total chloride. Production of nitric acid causes an increase in the water and ammonia concentrations in the aerosol phase and after a while transformation of some SO_4^{2-} to HSO_4^-. The amount of nitrate in the aerosol phase increases rapidly initially, but after reaching a limit value determined by the ammonia availability it increases very slowly. Chloride is displaced by the nitrate to the aerosol phase, but this depletion is slower compared to the corresponding sulfate addition in the previous example.

Fig. 5. Predicted aerosol composition as a function of the sulfuric acid concentration for an air parcel containing 5.9 μg m^{-3} sodium, 10 μg m^{-3} total (gas plus aerosol) NH_3, 30 μg m^{-3} total HNO_3, and 9.36 μg m^{-3} total HCl assuming thermodynamic equilibrium and using the method of Kim et al. [64–65]. The R.H. is 70% and the temperature 298 K

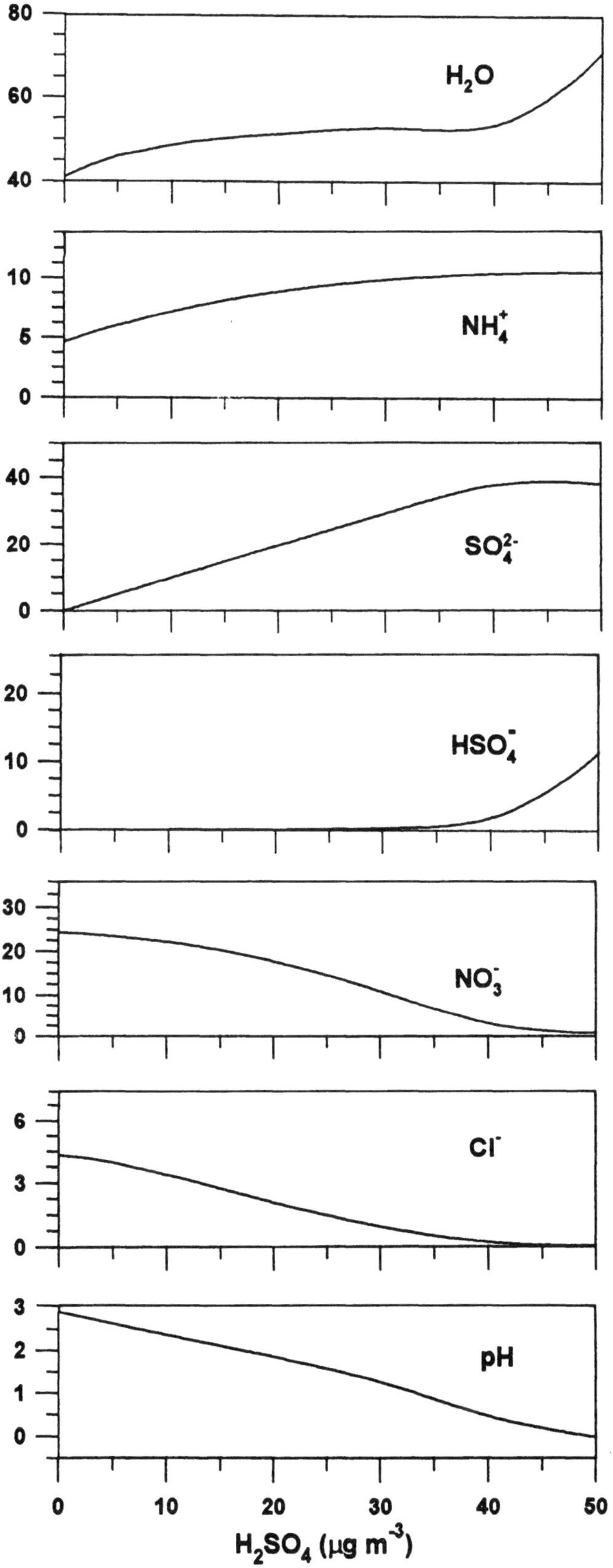

80
60
40
H2O
10
5
0
NH4+
40
20
0
SO42-
20
10
0
HSO4-
30
20
10
0
NO3-
6
3
0
Cl-
3
2
1
0
pH
0
10
20
30
40
50
H2SO4 (µg m-3)

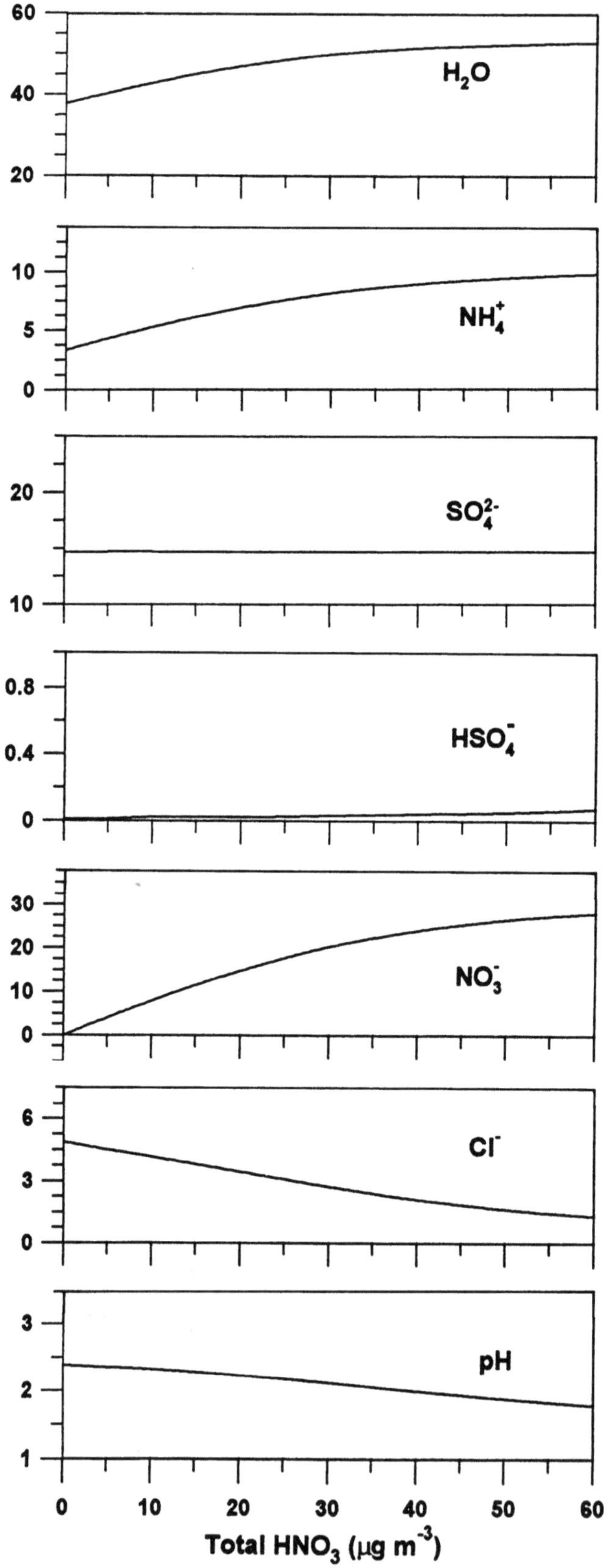

H2O
NH4+
SO42-
HSO4-
NO3-
Cl-
pH
0
10
20
30
40
50
60
Total HNO3 (µg m-3)

Organic Aerosol Material Formation

Organic aerosols formed by gas phase photochemical reactions of hydrocarbons, ozone and nitrogen oxides have been identified in both urban and rural atmospheres. Aliphatic organic nitrates [66], dicarboxylic acids (adipic and glutaric acids) [67], carboxylic acids derived from aromatic hydrocarbons (benzoic and phenylacetic acids), etc. have been identified.

Biogenic Aerosol

Natural hydrocarbons like the monoterpenes ($C_{10}H_{16}$) and isoprene (C_5H_8) are emitted by various types of trees and plants. In the U.S. the biogenic hydrocarbon sources are estimated to produce 30–60 Mt of carbon per year (isoprene and monoterpenes combined) whereas anthropogenic hydrocarbon sources have been estimated to account for 27 Mt of carbon per year [68–70]. Laboratory investigations have indicated that biogenic hydrocarbons are very reactive under typical atmospheric conditions [71]. The aerosol forming potential of biogenic hydrocarbons has been investigated in a series of smog chamber studies [32] and it has been suggested that isoprene photooxidation does not contribute to the production of secondary aerosol under ambient conditions. On the contrary, the pinenes form secondary aerosol in their reactions with O_3 and OH and have the potential to contribute significantly to aerosol in areas with high vegetation coverage. Monoterpenes were estimated to contribute around 15% to the secondary organic aerosol in urban areas with low vegetation like Los Angeles, while they dominate the secondary organic aerosol in areas with high aerosol coverage like Atlanta [32].

Anthropogenic Aerosol

The calculated contribution of the main individual secondary organic aerosol precursors to the secondary organic aerosol concentration in Los Angeles on August 28, 1987 is presented in Table 4 [31]. Toluene, the nonmethane hydrocarbon with the highest emission rate in the Los Angeles area (165 t day^{-1}), was predicted to contribute 28% of the secondary organic aerosols. Toluene is due to exhaust

Fig. 6. Predicted aerosol composition as a function of the nitric acid concentration for an air parcel containing 5.9 $\mu g\, m^{-3}$ sodium, 10 $\mu g\, m^{-3}$ total (gas plus aerosol) NH_3, 10 $\mu g\, m^{-3}$ sulfuric acid and 9.36 $\mu g\, m^{-3}$ total HCl assuming thermodynamic equilibrium and using the method of Kim et al. [64–65]. The R.H. is 70% and the temperature 298 K

Table 4. Predicted per cent contribution to secondary organic aerosol concentrations at Los Angeles

Species	Contribution	
	Grosjean and Seinfeld [27]	Pandis et al. [31]
Aromatics	58	65
Biogenic Hydrocarbons	10	16
Alkanes	21	15
Olefins	11	4

Table 5. Amount of secondary aerosol produced in a typical Los Angeles smog episode according to functional groups [72]

	Aerosol produced (kg day^{-1})			
Precursor	Carbonyls	Aliphatic Acids	Nitrophenols	Aliphatic Nitrate
Alkenes	–	608	–	–
Cyclic olefins	62	131	–	9
Terpenes	295	623	–	41
Alkanes	243	–	–	121
Cycloalkanes	72	–	–	72
Aromatics	–	–	3118	–
Total	672	1362	3118	243

and hot soak evaporative emissions of gasoline powered vehicles and to surface coatings.

Grosjean [72] calculated the daily production rates of various chemical functionalities of the secondary organic aerosol formed in situ during a smog episode in Los Angeles using the precursor hydrocarbon emission inventory and the results of smog chamber studies. His estimates are presented in Table 5.

The Equilibrium Assumption

One of the most intriguing aspects of atmospheric aerosols concerns the behavior of volatile inorganic species like ammonium nitrate. Stelson et al. [73] first suggested that aerosol ammonium nitrate levels could be determined from the equilibrium among ammonia, nitric acid, and aerosol ammonium nitrate. The assumption of thermodynamic equilibrium has been employed to partition the volatile compounds between the gas and aerosol phases [74–82] and to predict the aerosol size distribution [77, 5, 83, 84]. Although the equilibrium assumption has been supported by ambient data [85–87], other data indicate that it may not always hold [88–90].

Timescales for Equilibrium

Wexler and Seinfeld [59] have predicted that, under certain atmospheric conditions, the volatile inorganic components of atmospheric aerosol may not be in equilibrium with their gas phase counterparts due to transport limitations. These investigators suggested that the timescale for equilibration between a monodisperse aerosol population of radius R_p and the gas phase is

$$\tau_p = \frac{\rho_p R_p^2}{3 D_i m_p}(1 + \beta_i) \tag{52}$$

where ρ_p is the aerosol density, D_i is the gas phase diffusivity of species i, m_p is the particle mass concentration, and β_i is the surface accommodation factor. This expression suggests that, for small aerosol mass concentrations and large aerosol sizes (e.g. conditions that could be found near the coast of a polluted area), this equilibrium timescale can be of the order of several hours and the aerosol is not necessarily in equilibrium with the surrounding gas phase. In this case, the aerosol mass concentration is governed by transport and not thermodynamics. On the other hand, when the aerosol mass concentration is high and the aerosol size small, the equilibrium timescale is of the order of a few seconds and thermodynamic equilibrium is expected to be a valid assumption.

Equilibrium Assumption and Aerosol Distribution

Even when the aerosol is in equilibrium with the gas phase, the size distribution of the volatile components of atmospheric aerosols often cannot be determined from thermodynamic considerations alone, and thus mass transport has to be included to determine the size distribution of the volatile components [59]. The inability of thermodynamics to determine always uniquely the size distribution of an aerosol population can be illustrated by a simplified example. Let us assume that our aerosol population consists of NH_4NO_3 particles of different sizes. Thermodynamic equilibrium requires that the product of the gas phase concentrations of NH_3 and HNO_3 should be equal to a constant. Assume now that a certain amount of HNO_3 is produced in the system. Return to an equilibrium state requires the condensation of some HNO_3 and an equimolar amount of NH_3 to the aerosol particles so that the product of their gas phase concentrations is again equal to the equilibrium constant value. But this thermodynamic driving force is the same for all the solid ammonium nitrate particles and thermodynamic considerations alone actually result in infinite ways of distributing this ammonium nitrate among the various particles. The aerosol distribution is determined by transport considerations and, as Wexler and Seinfeld [59] point out, often "thermodynamics is not enough". These authors describe several other cases where thermodynamics alone cannot assist us in determining the aerosol size distribution and suggest that in general both transport and thermodynamics should be used in our effort to predict the ambient aerosol size composition distribution.

References

1. Rogge WF, Mazurek MA, Hildemann LM, Cass GR, Simoneit BRT (1993) Atmos Environ 27:1309
2. Denbigh K (1981) The principles of chemical equilibrium, 4th edn. Cambridge University Press
3. Harrison LP (1965) Humidity and moisture, Vol. 3 A. Wexler (ed), p 105, Prentice-Hall, Englewood Cliffs, NJ
4. Pruppacher HR, Klett JD (1980) Microphysics of clouds and precipitation. D Reidel, Hingham, Mass
5. Pilinis C, Seinfeld JH (1987) Atmos Environ 21:2453
6. Kusik CL, Meissner HP (1978) Am Inst Chem Eng Symp Ser 173:14
7. Cullis SF, Hirschler MM (1980) Atmos Environ 14:1263
8. Möller D (1984) Atmos Environ 18:19
9. Finlayson-Pitts BJ, Pitts JN Jr (1986) Atmospheric Chemistry, Wiley, New York
10. Atkinson R, Lloyd AC (1984) J Phys Chem Ref Data 11:179
11. Stockwell WR, Calvert JG (1983) Atmos Environ 17:2231
12. Möller D (1984) Atmos Environ 18:29
13. Charlson RJ, Lovelock JE, Andreae MO, Warren SG (1987) Nature 326:655
14. Atkinson R, Pitts JN Jr, Aschmann SM (1984) J Phys Chem 88:1584
15. Hatakeyama S, Akimoto H (1983) J Phys Chem 87:2387
16. Saltzman ES, Savoie DL, Zika RG, Prospero JM (1983) J Geophys Res 88:10897
17. Berresheim H, Andreae MO, Ayers GP, Gillett RW, Merrill JT, Davis VJ, Chameides WL (1990) J Atmos Chem 10:341
18. Berresheim H (1987) J Geophys Res 92:13245
19. Berresheim H, Eisele FL, Tanner DJ, McInnes LM, Ramseybell DC (1993) J Geophys Res 98:12701
20. Bandy AR, Scott DL, Blomquist BW, Chen SM, Thornton DC (1992) Geophys Res Lett 19:1125
21. Lawrence MG (1993) J Geophys Res 98:20663
22. Jones CL, Seinfeld JH (1983) Atmos Environ 17:2370
23. Russell AG, McRae GJ, Cass GR (1985) Atmos Environ 19:893
24. Atkinson R (1991) J Phys Chem Ref Data 20:459
25. Tuazon EC, Atkinson R, Plum CN, Winer AM, Pitts Jr JN (1983) Geophys Res Lett 10:953
26. Grosjean D (1977) in Ozone and Other Photochemical Oxidants, p 45 NRC
27. Grosjean D, Seinfeld JH (1989) Atmos Environ 23:1733
28. Izumi K, Fukuyama T (1989) Abstracts of PACIFICHEM, Honolulu, HI
29. Grosjean D (1984) Sci Total Environ 37:195
30. Wang SC, Paulson SE, Grosjean D, Flagan RC, Seinfeld JH (1992) Atmos Environ 26A:403
31. Pandis SN, Harley RA, Cass GR, Seinfeld JH (1992) Atmos Environ 26A:2269
32. Pandis SN, Paulson SE, Seinfeld JH, Flagan RC (1991) Atmos Environ 25A:997
33. Shaw MA, Pandis SN, Flagan RC, Seinfeld JH (1991) Abstracts of the 10th Annual Meeting of the American Association for Aerosol Research, Traverse City, MI
34. Izumi K, Fukuyama T (1990) Atmos Environ 24A:1433
35. Seinfeld JH (1986) Atmospheric Chemistry and Physics of Air Pollution, John Wiley, New York
36. Reiss H (1950) J Chem Phys 18:840
37. Jaecker-Voirol A, Mirabel P (1989) Atmos Environ 23:2053
38. Jaecker-Voirol A, Mirabel P (1988) J Phys Chem 92:3518
39. Wyslouzil BE, Seinfeld JH, Flagan RC, Okuyama K (1991a) J Phys Chem 94:6827
40. Wyslouzil BE, Seinfeld JH, Flagan RC, Okuyama K (1991b) J Phys Chem 94:6842
41. Reiss H, Margolese DI, Schelling FJ (1976) J Coll Int Sci 56:511
42. Boulaud D, Madelaine G, Vilga D, Bricard J (1977) J Chem Phys 66:4854
43. Mirabel P, Clavelin JL (1978) J Chem Phys 68:5020
44. Doyle GJ (1961) J Chem Phys 35:795

45. Kiang CS, Stauffer D (1973) Faraday Symp Chem Soc 7:26
46. Mirabel P, Katz JL (1974) J Chem Phys 60:1138
47. Ithoh M, Takahasi K, Kasahara M (1977) J Aerosol Sci 8:183
48. Shugard WH, Reiss H (1976) J Chem Phys 65:2827
49. Jaecker-Voirol A, Mirabel P, Reiss H (1987) J Chem Phys 87:4849
50. Fuchs NA (1964) The Mechanics of Aerosols, Pergamon, New York
51. Fuchs NA, Sutugin AG (1971) in International Reviews of Aerosol Physics and Chemistry vol. 2, Hidy M, Brock JR, eds. p 1–60, Pergamon, New York
52. Lednouich SL, Fenn JB (1977) AICHE J 23:454
53. Worsnop DR, Zahniser MS, Kolb CE, Gardner JA, Watson LR, Van Doren JM, Jayne JT, Davidovids P (1989) J Phys Chem 93:1159
54. Van Doren JM, Watson LR, Davidovits P, Worsnop DR, Zahniser MS, Kolb CE (1990) J Phys Chem 94:3265
55. Watson LR, Van Doren JM, Davidovits P, Worsnop DR, Zahniser MS, Kolb CE (1990) J Geophys Res 95:5631
56. Tang IN, Lee JH (1987) Chemistry of Acid Rain: Sources and Atmospheric Processes (edited by Johnson RW, Gordon GE), p 109–117, ACS Symposium Series, Washington
57. Gill PS, Graedel TE, Weschler CJ (1983) Rev Geophys 21:903
58. Wexler AS, Seinfeld JH (1992) Atmos Environ 26A:579
59. Wexler AS, Seinfeld JH (1990) Atmos Environ 24A:1231
60. Wall SM, John W, Ondo JL (1988) Atmòs Environ 22:1649
61. Eldering A, Solomon PA, Salmon LG, Fall T, Cass GR (1991) Atmos Environ 25A:2091
62. Bromley LA (1973) AICHE J 19:313
63. Pitzer KS (1986) Pure Appl Chem 58:1599
64. Kim YP, Seinfeld JH, Saxena P (1993) Aeros Sci Tech 19:157
65. Kim YP, Seinfeld JH, Saxena P (1993) Aeros Sci Tech 19:182
66. Grosjean D, Friedlander SK (1975) J Air Pollut Control Assoc 25:1038
67. O'brien RJ, Crabtree JH, Holmes JR, Hoggan MC, Bockian AH (1975) Env Sci Technol 9:577
68. Lamb B, Guenther A, Gay D, Westberg H (1987) Atmos Environ 21:1695
69. Zimmerman PR (1979) EPA Report 450/4–4–79–004
70. Altshuller AP (1983) Atmos Environ 17:2131
71. Arnts RR, Gay BG (1979) EPA Report 600/3–79–081
72. Grosjean D (1992) Atmos Environ 26A:953
73. Stelson AW, Friedlander SK, Seinfeld JH (1979) Atmos Environ 13:369
74. Stelson AW, Seinfeld JH (1982a) Atmos Environ 16:983
75. Stelson AW, Seinfeld JH (1982b) Atmos Environ 16:993
76. Bassett M, Seinfeld JH (1983) Atmos Environ 17:2237
77. Bassett M, Seinfeld JH (1984) Atmos Environ 18:1163
78. Russell AG, McRae GJ, Cass GR (1983) Atmos Environ 17:949
79. Russell AG, McCue KF, Cass GR (1988) Environ Sci Technol 22:263
80. Saxena P, Seigneur C, Peterson TW (1983) Atmos Environ 17:1315
81. Saxena P, Hudischewskyj AB, Seigneur C, Seinfeld JH (1986) Atmos Environ 20:1471
82. Russell AG, Cass GR (1986) Atmos Environ 20:2011
83. Pilinis C, Seinfeld JH, Seigneur C (1987) Atmos Environ 21:943
84. Pilinis C, Seinfeld JH (1988) Atmos Environ 22:1985
85. Doyle GJ, Tuazon EC, Graham RA, Mischke TM, Winer AM, Pitts JN (1979) Environ Sci Technol 13:1416
86. Grosjean D (1982) Sci Total Environ 25:263
87. Hildemann LM, Russell AG, Cass GR (1984) Atmos Environ 18:1737
88. Cadle SH, Countess RJ, Kelly NA (1982) Atmos Environ 16:2501
89. Tanner RL (1982) Atmos Environ 16:2935
90. Allen AG, Harrison RM, Erisman J (1989) Atmos Environ 23:1591

Sources, Particle Size Distribution and Transport of Aerosols

Jozef M. Pacyna

Norwegian Institute for Air Research (NILU), P.O. Box 100, 2007 Kjeller, Norway

Introduction 70
Natural Sources of Aerosols 72
Anthropogenic Sources of Aerosols 76
Sources of Primary Aerosols 76
Sources of Secondary Aerosols 82
Transport of Aerosols within Air Masses 83
Transport on Global Scale 84
Transport on Regional Scale 84
Removal of Particles 89
Global Distribution of Aerosols and Their Trends 92
Concluding Remarks 94
References 95

Summary

Various modes of aerosols with respect to their size are described. Natural and anthropogenic sources of aerosols are presented and their fluxes are evaluated. Gas-to-particle conversion processes, including transformations of gaseous hydrocarbons from vegetation, seem to generate the largest amounts of natural aerosols followed by sea spray and desert dust. The annual anthropogenic flux of aerosols contributes about 60% of the total aerosol burden in the atmosphere. Major anthropogenic sources of aerosols include direct emissions from stationary and mobile sources, and formation of secondary aerosols from their gaseous precursors released from power stations, industrial plants, and mobile sources. Atmospheric transport of aerosols on global and regional scale is then discussed together with their removal processes from the atmosphere. Major focus is placed on particle size as this parameter governs the atmospheric transport and dry and wet deposition of aerosols. Major dispersion models applied in Europe and North America to study the aerosol transport are presented. Results of several receptor models used to assess the origin of aerosols at various remote locations are also discussed. Information from a number of measurement programs is used, together with the results of dispersion and receptor models, to discuss global distribution of aerosols and their trends. While fluxes of aerosols from natural sources are quite unchanged over the years, anthropogenic aerosols, after dramatic increase during the 1970s and 1980s, seem to show a slight decrease in the last few years. General description of research needs is presented in the final part.

Introduction

The atmospheric aerosol, defined as a multiphase system containing gases with suspended particles, participates in various physical and chemical processes in the atmosphere. As a result, aerosol is responsible for changing the chemical composition of the atmosphere and causes numerous environmental and human health effects. As condensed phase material, aerosols function as absorbers, emitters, and scatterers of light, and as convenient surfaces for the deposition of gas phase molecules. The aerosols not only serve as good collectors of condensable reaction products but are also active partners for surface chemistry and for aqueous phase chemistry in humid atmospheres.

Condensation and chemical processes occur during the formation of aerosols as well as during their transport within air masses. It has been found that aerosols in the size range from 0.01 μm to 1.0 μm diameter grow principally by gas-to-particle conversion, the process by which molecules diffuse to the surface of a particle and subsequently are incorporated into the particle [1]. The other portion of the aerosols in the above mentioned size range enters the atmosphere passing through control equipment installed in various industrial sources of particle emissions.

Size of the particles is one of the most important parameters affecting the transport scale of aerosols, their atmospheric deposition and migration through the environment, and their environmental and human effects. Thus, size is the main determinant of the behavior of an aerosol and is usually presented in terms of the aerodynamic diameter related to unit density spherical particles with the same aerodynamic properties. In practice, the shape and density of particles differ and the size of aerosols is expressed as equivalent aerodynamic diameter (EAD).

There are various modes of particles with respect to their sizes and origin including:

— nucleation mode, consisting of < 0.2 μm diameter particles generated by processes involving condensation of hot vapors or formed through gas-to-particle conversion,
— accumulation mode, consisting of 0.2 to 2.0 μm diameter particles formed from nucleation mode particles through coagulation or condensation of vapors, and
— coarse mode, consisting of > 2.0 μm diameter particles generated by various mechanical attrition processes.

Nucleation and accumulation mode particles are also defined as fine particles in contrast to coarse particles. Various modes of particles with indication of some formation pathways are presented in Fig. 1 [2].

There are also other terms to classify particulate matter in the air including suspended particulate matter (SPM), total suspended particles (TSP), PM10 describing particles which pass through a size selective inlet with a 50% efficiency cut-off at 10 μm aerodynamic diameter, smoke, black smoke, inhalable and respirable particles, and thoracic particle mass describing particles which penetrate

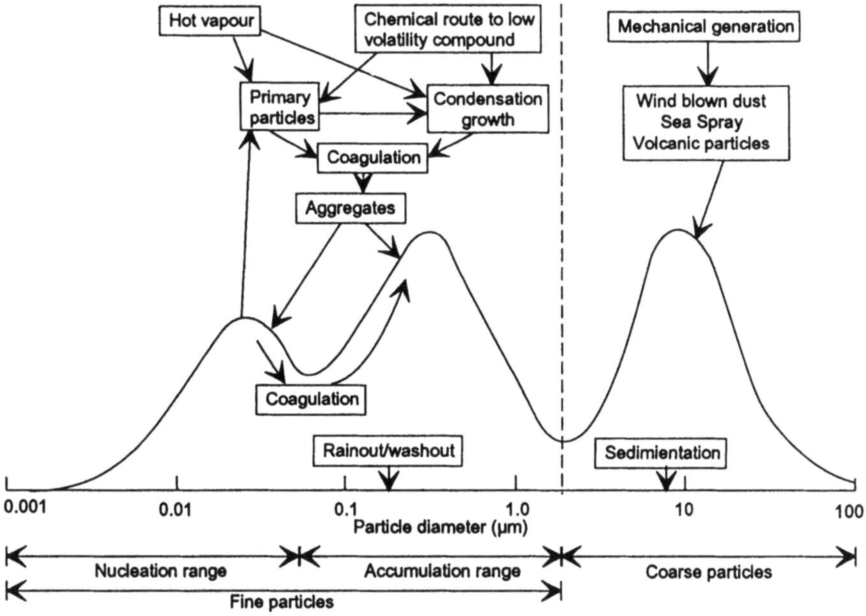

Fig. 1. Schematic representation of a typical size distribution for atmospheric particles, indicating some formation pathways [2]

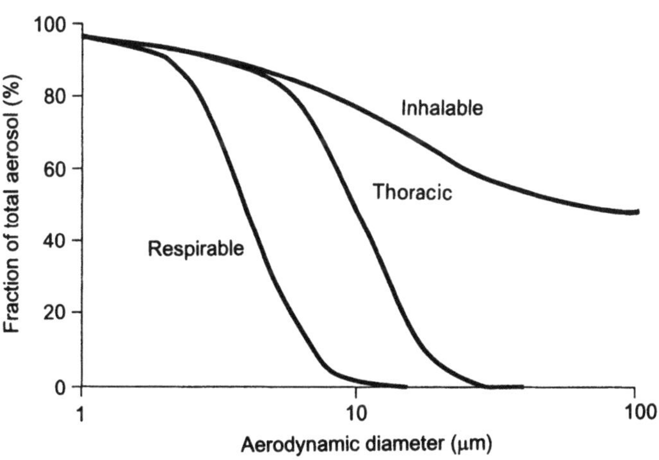

Fig. 2. Typical curves defining respirable, thoracic and inhalable particle fractions [2]

beyond the nasopharynx and larynx. Typical curves defining respirable, thoracic and inhalable particle fractions are shown in Fig. 2 [2].

The size of aerosols transported within air masses depends on the size and chemical composition of particles generated by various anthropogenic activities and natural emission sources, modified in the ambient air by meteorological conditions and the presence of various chemical compounds.

The aim of this account is to provide a brief overview on emission sources of particles from anthropogenic and natural sources with focus on particle size distribution of emitted material, and regional and global transport of atmospheric aerosols. Major dispersion and receptor models to study long-range transport are presented with emphasis on estimation of dry and wet deposition processes. Size-differentiated chemical composition of aerosols measured at various remote locations is presented as a way to verify modeled concentrations of several chemical compounds on particles.

Natural Sources of Aerosols

The major sources of natural aerosols include:

— crustal soils (e.g. dust) and the ocean surface (e.g. sea salt),
— products of gas phase chemical reactions in the atmosphere, with the precursor gases originating from combustion or from biological activity (various gas-to-particle conversions),
— volcanic eruptions, which inject sulfur gases into the atmosphere, with the subsequent formation of sulfuric acid aerosols,
— soot from natural fires, and
— atmospheric water clouds.

Aerosols generated from the above sources are often defined as continental and marine aerosols. The continental aerosol may consist of crustal aerosol, secondary aerosol formed by gas-to-particle conversions of combustion products, volcanic aerosol, and carbonaceous aerosol, including organic hydrocarbons and elemental carbon. The marine aerosol is an aerosol with oceanic origin, including sea salt aerosol and secondary aerosol formed by gas-to-particle conversions of biological activity products.

There have been several approaches to estimate emission fluxes of aerosols to the troposphere from the above mentioned sources [3,4]. A summary of these estimates is presented in Table 1. Gas-to-particle conversion processes, including transformations of gaseous hydrocarbons from vegetative emissions seem to generate the largest amounts of natural aerosols. Cyclic olefins are considered to be the most important class of organic aerosol precursors although aromatics, alkenes and terpenes from forest may also be important. It has been pointed out [5] that aerosol growth during conversions of olefins was consistent with the formation of supersaturated compounds in the gas phase followed by condensation on pre-existing particles. These processes are very poorly understood and the estimate in Table 1 should be used with caution.

More studies have been carried out to explain the formation and behavior of secondary marine aerosols. The quantity of these aerosols seems to be larger than the amount of sea spray aerosol produced by bursting bubbles at the water surface. Dimethyl sulfide (DMS), produced from the biological activity of phytoplankton,

Table 1. Global estimates of the major natural aerosol sources in the troposphere, as summarized in [3, 4]

Source	Strength (10^6 tonnes × $year^{-1}$)
Crustal aerosol	100–500
Sea spray	1000
Secondary aerosol formed through gas-to-particle conversion of combustion and biological activity products	300–1100
Other sources	3–150

has been identified as the main source of gaseous sulfur over the ocean. Emitted to the atmosphere by sea-air exchange DMS is photo-oxidized to methanesulfonic acid (MSA) and sulfur dioxide, and further to particulate sulfate. This aerosol is the major component of non-sea-salt sulfate and an important contributor to the acidity of natural precipitation. The reactions that create sulfate aerosol can be divided into clear-sky and in-cloud processes. In the former process, gaseous sulfuric acid is produced from sulfur dioxide and DMS reacting in the presence of water vapor. This product forms fine particles. Then ammonium sulfate is formed through reaction of sulfuric acid with ammonia. In clouds, the sulfate aerosol is produced by oxidation of sulfuric oxide, dissolved in cloud droplets, in the presence of hydrogen peroxide to form sulfuric acid and its ammonium salts in solution. Submicron aerosol droplets are formed through evaporation of water from this solution.

Most recent studies on the air-sea exchange of various components have been carried out within one of the European Experiment on Transport and Transformation of Environmentally Relevant Trace Constituents in the Troposphere over Europe (EUROTRAC) projects on Air-Sea Exchange (ASE) [6]. In addition to the DMS transformations in various regions, photochemical production of carbonyl sulfide in sea water has been studied, as well as formation and variability of hydrogen peroxide, assumed to be the result of disproportionation of radical species photochemically formed. A numerical simulation of marine aerosol dynamics has been proposed [7].

The production mechanisms for sea spray aerosols are quite well known [8–10]. At the end of the 19th century it was already recognized by researchers studying the formation of clouds that oceans should be a major source of cloud condensation nuclei, a fraction of atmospheric aerosols which play an important role in the hydrologic cycle by forming clouds at relative humidities slightly above the saturation level with respect to water. In general, the sea spray aerosol is produced in the following way. A bubble rises to the surface and penetrates the surface; a bobble cap and central water jet are formed. The jet breaks up into droplets; the bobble cap also breaks up into many smaller droplets, having salt concentration similar to that of sea water. Then the particles are transported upward by turbulent transfer and the relative humidity decreases from the saturated values immediately above the surface to values more characteristic of the boundary layer. The salt concentration in the droplet also changes and chemical reactions start involving gas phase compounds and the dissolved salt material. The above mentioned droplet mechanisms form a bimodal distribution of sea spray aerosols with supermicron mode from the jet drops and submicron mode from the bobble cap drops. Wind speed strongly affects the production of sea spray aerosols.

An important contribution to the global aerosol comes from crustal aerosol formed by erosive processes. The description of generation mechanisms is quite simple. In fairly strong winds a fraction of soil can become airborne. How much of the soil can become airborne depends on the structure, moisture content and other soil parameters, which determine the threshold wind speed at which the dust is entrained in the air stream. After this threshold is exceeded, the vertical flux of dust increases rapidly with wind speed [11], and fine particles can be produced by the sandblasting effect of saltation, which disaggregates fine particles from the surface of larger particles. A considerable portion of the soil material entrained in the air remains near the surface and is soon re-deposited again. Measurements made at 1 m to 2 m above the surface during erosion show a biomodal size distribution in which the large particle mode ranging from 10 to 100 µm consists of quartz and other sand-size grains, and the small particle mode below 10 µm contains primarily the clay particles and smaller quartz grains [4]. The fractionation process from soil to clay aerosol occurs continuously in the air with increasing distance from the source. One method used to quantify this process is to study the ratio of Al to Si in aerosols during their transport within air masses [12]. Crustal aerosols generally have values of this ratio which are intermediate between those for average crustal material and those for the clay minerals. Measurements in the Arctic show that the Si/Al ratio in crustal aerosols can be two or more times lower than in the average crust and clay minerals [13].

The dust component in atmospheric aerosols is highly variable as shown in Fig. 3 on the basis of measurements by various researchers, summarized in [3].

There have been a few approaches to estimate the global tropospheric dust cycle. Estimates presented in [3] are included in Table2. The Sahara desert seems to play an important role in the northern hemispheric dust cycle, providing about half of it [3]. It is interesting to compare the particle size distribution of various fractions of sand from physical weathering processes. This comparison is shown in Fig. 4. and indicates that eolian dust and loess, an eolian deposit, can be a subject of long-range transport within air masses. The data presented in this figure demonstrate how the major fractions are derived from the original source.

Volcanoes are another source of natural aerosol, releasing into the atmosphere large amounts of gases and particles of ash and cinders. Both primary and secondary aerosols are formed during and after volcanic eruptions. Particles quickly settle and are washed out of the troposphere by precipitation. The efficiency of these processes depends primarily on the particle size. The longer-term effects of volcanic eruptions are due to injections of sulfur dioxide into the stratosphere,

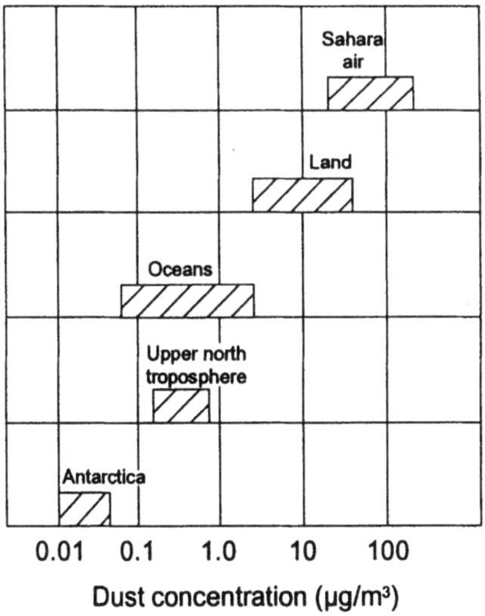

Fig. 3. Survey of the concentration ranges of mineral dust in the troposphere based on numerous studies by – among others – the following authors: Blifford, Chesselet, Duce, Ferguson, Hoffman, Gillette, Goldberg, Griffin, Jaenicke, Prospero, Rahn, Schutz, Winchester, and Zoller [3]

Table 2. Estimates of the global tropospheric dust cycle, as presented in [3]

Part of troposphere	Dust burden 10^6 tons	Source strength 10^6 tons yr^{-1}
Northern hemisphere	3.0	150
Southern hemisphere	1.0	50
Whole troposphere	4.0	200
Sahara plume	1.2– 4.0	60–200
Total troposphere (plus Sahara)	5.2– 8.0	260–400
Total troposphere (plus Sahara)*1	3.2–12.0	130–800

*1 After application of the uncertainty factors.

where the gas oxidizes and reacts with water to form sulfuric acid aerosols [14]. Because the stratosphere is above the convective weather currents of the troposphere, it is very stable and aerosol clouds can persist for more than a year. The fluxes of particles from volcanoes are very difficult to quantify but are believed to be below 100 million tonnes per year. The total sulfur flux can be as high as 10 million tonnes per year although these data are very uncertain and vary from one year to another.

In most ambient atmospheric situations, carbon contributes 10–20% of total aerosol mass [2]. Carbonaceous aerosols, originating from natural and anthropogenic continental sources, contribute significantly to the atmospheric burden over various regions of the world [15]. They are composed of two major components, organic matter (60–80% of total carbon mass) and elemental carbon, the latter often being referred to as graphitic or soot carbon [16]. Natural emissions of carbonaceous aerosols from biomass burning processes, particularly in the tropics are considered to be a major source of soot carbon over the remote tropical

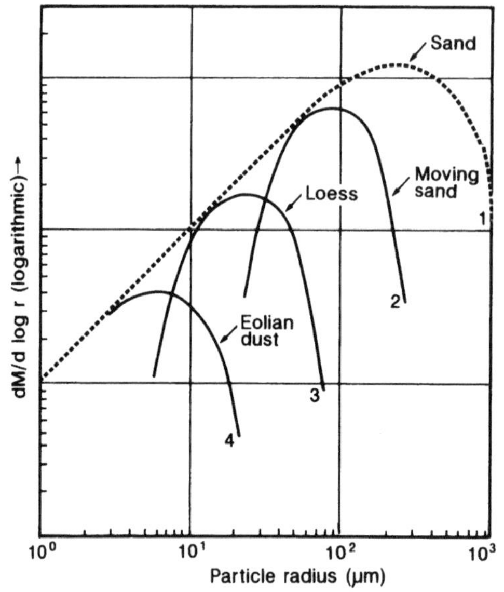

Fig. 4. Sand fractionation processes by wind, schematic. The original sand distribution is fractionated into major fraction 2, 3, and 4 as a function of distance from the source [3]

Atlantic and Pacific [17, 18] but are still poorly documented [19]. It should be mentioned that the bulk of biomass burning is human initiated and appears to be increasing with time. In general, the annual flux of 20 million tonnes of fine particulate carbon (elemental carbon) from the tropical regions into the global atmosphere was calculated [15], which can be compared to the flux of fine particulate carbon emitted from industrial sources estimated to be 25 million tonnes per year [20]. Between 50 and 85% of the measured concentrations of elemental carbon are attached to aerosols smaller than 1 μm diameter.

The total particle organic carbon flux (including elemental carbon) was estimated 180 million tonnes per year with almost 40% from biomass burning [21].

Anthropogenic Sources of Aerosols

Anthropogenic sources of emissions of primary and secondary aerosols are numerous and difficult to quantify. In general, the annual anthropogenic flux of aerosols contributes about 60% to the total aerosol burden in the atmosphere.

Sources of Primary Aerosols

There are two major groups of anthropogenic sources of primary aerosols: stationary and mobile sources. Stationary sources include emissions from chimney stacks and fugitive emissions.

A number of mechanical processes, such as grinding or spraying, generate coarse particles with diameters larger than a few micrometers. Various high temperature processes, such as coal and oil combustion to generate electricity and heat, industrial operations to produce industrial goods, and waste incineration emit large amounts of fine particles into the atmosphere.

The quantity of atmospheric emissions of particles from various sources depends upon:

— the contamination of fossil fuels and other raw materials, e.g. a content of non-combustible matter in coals and wastes,
— the technology of the industrial processes, and
— the type and efficiency of control equipment.

Detailed discussion as to what extent the above parameters affect particle emissions and emissions of chemical compounds on the particles, e.g. trace metals, is presented in [22]. The main conclusion was that contamination of raw materials and the type and efficiency of control equipment are the most important parameters.

Electrostatic precipitators (ESPs), fabric filters, and wet scrubbers are the most frequently used emission control installations in major stationary sources. Cyclones are also used. The collection efficiency for various size particles is presented in Fig. 5. The ESPs are particularly efficient in removing all types of particles $< 0.01\ \mu m$ diameter, and can tolerate temperatures as high as 400°C [22]. The applicability of conventional precipitators is, however, limited by the electric resistivity of the particles. Differential mass particle size distribution at the outlet of the ESP is presented in Fig. 6 [23] showing bimodal distribution.

Fabric filters are used mostly in power plants and waste incinerators. The collection efficiency is always very high, and even for particles of $0.01\ \mu m$ diameter it exceeds 99%. However, the lifetime of fabric filters is very dependent upon the working temperature and their resistance to chemical attack by corrosive elements in the flue gas. The bimodal particle size distribution is again observed at the outlet of the fabric filters as seen in Fig. 7 [24].

The removal of particles by wet scrubbers is practically independent of temperature, although high operating temperatures increase the water consumption and result in the formation of steam plume. Generally, wet scrubbers are more efficient in removing trace metals on fine particles than the ESPs [25, 26]. This comparison is presented in Table 3.

Fugitive emissions are difficult to measure and even more difficult to control. They include the following groups of sources:

— wind blown dusts from open storage areas, uncontrolled landfills, and spoil heaps,
— emissions from buildings through natural and fan assisted ventilation, and
— emissions from roadways as a result of local air turbulence.

Obviously, a wide range of particles with respect to their size distribution is emitted from fugitive sources depending on the emission generation processes.

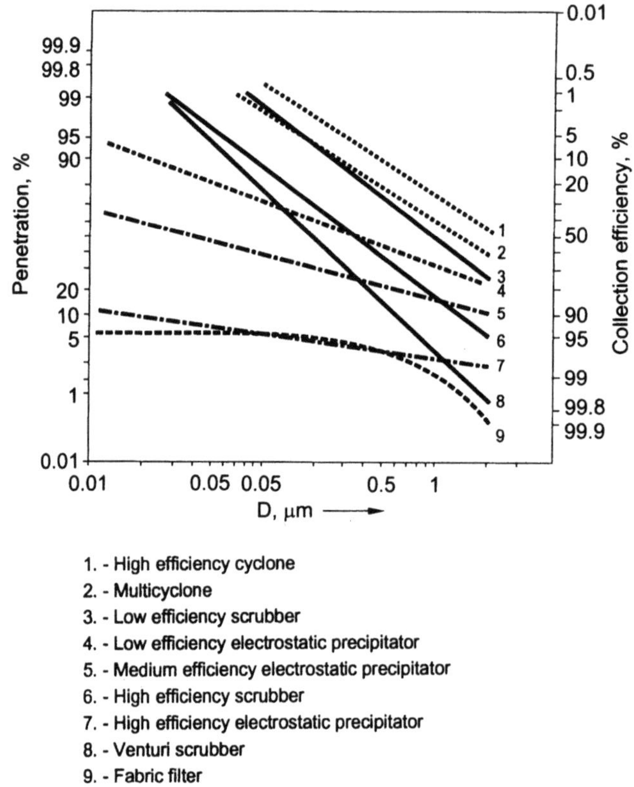

Fig. 5. Average collection efficiencies and penetrations of control equipment for various particle sizes

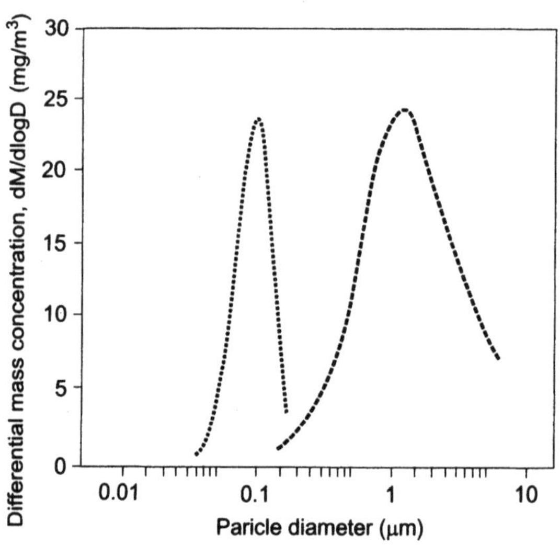

Fig. 6. Differential mass particle size distribution at the outlet of an electrostatic precipitator (ESP) [23]

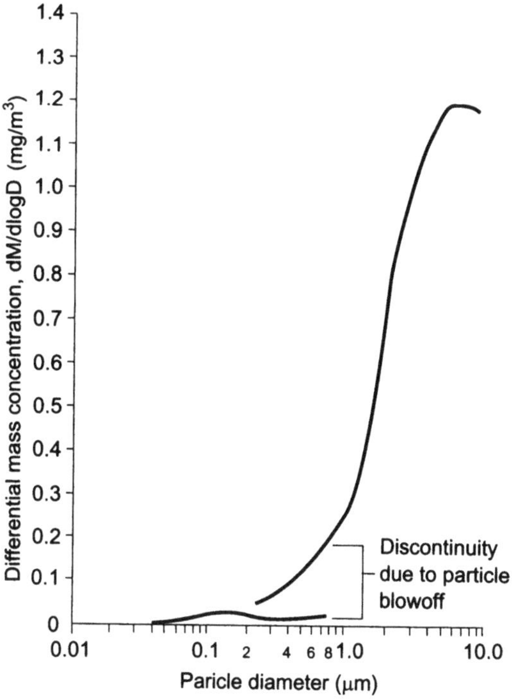

Fig. 7. Differential mass particle size distribution at the outlet of fabric filters [24]

Table 3. Penetration (in %) of elements contained in particles

Element	Penetration Range ESP	Venturi	Element	Penetration Range ESP	Venturi
As	4.3–11.5	2.5– 7.5	Pb	2.2–5.5	
Be	0.5– 0.9		Sb	3.1–7.7	3.0–6.6
Cd	3.3– 8.8		Se	3.8–8.1	10–21
Co	1.2– 3.2	0.06– 2.1	V	1.6–3.7	0.5–1.1
Cr	1.2–12.1	0.6–36	W	3.1–7.2	1.7–3.5
Mo	1.8– 6.8	0.9– 2.2	Zn	2.3–6.3	0.3–8.6
Mn	0.3– 1.6	0.07– 4.6	Zr	0.5–1.6	0.05–0.14

Particles are also emitted during combustion of gasoline and diesel oil, the latter being responsible for the majority of black smoke. Several parameters, such as fuel composition, engine design and maintenance, operating conditions, and emission control devices may influence the physical characteristics of particles emitted from diesel engines. These carbonaceous particles are composed in part of high-molecular-weight polycyclic aromatic hydrocarbons and their size ranges from 0.2 to 0.3 μm diameter.

With the exhaust from gasoline engines, particle size depends on fuel composition. Unleaded fuel results in the emission of larger particles. It was estimated that about 90% of particles emitted during combustion of leaded fuels are less than 0.35 μm in diameter [27]. The particle size distribution with leaded and unleaded fuels suggests that the weight of the small particles (<0.35 μm diameter) is much higher with leaded than unleaded fuels. The diameter of emitted particles depends not only on fuel composition but also on engine operating conditions. With leaded fuel, cyclic operations yield larger particles than steady-state operation [27].

There have been a few approaches to estimate fluxes of various components which enter the atmosphere on particles. The first quantitative worldwide estimate of the annual industrial input of 16 trace metals into the air has been published recently [28]. The summary of this estimate for arsenic, cadmium, lead, and zinc is presented in Table 4. The above study used information collected during the estimation of atmospheric emissions of trace metals on particles in Europe and North America. Recently, global emission inventory of lead was updated for 1989. The estimates show that these emissions range from 150000 to 209000 tonnes with 62% of lead from gasoline combustion, followed by non-ferrous metal production with 26% [29]. The upper range of these estimates is presented in Table 5.

Pyrometallurgical processes in the primary non-ferrous metal industries are the major sources of atmospheric arsenic, cadmium and zinc. Combustion of leaded gasoline is still the major source of lead. Little information is available on emission of trace metals from various diffuse sources.

The size of trace metal-containing particles emitted from various sources has also been a subject of studies, particularly in connection with the long-range transport of trace metals within the air masses and the inhalation of toxic compounds. An example of chromium distribution on particles emitted from coal-fired power plants, electric-arc furnaces, plants producing chromate chemicals and chromate refractories is shown in Fig. 8 [30]. A pronounced relationship is observed between the chromium concentration and particle size for the coal-fired power plant emissions, with chromium concentrated predominantly in the <1.0 μm diameter particles. Chromium from electric-arc furnace steel plants is found mainly in coarse particles.

Table 4. Global emissions of trace metals to the atmosphere in 1983 (in 10^3 tonnes)

	Source category	As	Cd	Pb	Zn
1.	Fossil fuel combustion	0.4– 3.7	0.2– 1.1	2.7– 18.4	3.1– 23.4
2.	Gasoline combustion	–	–	248.0	–
3.	Non-ferrous metal industry	9.6–15.1	2.6– 8.2	30.0– 69.6	51.0– 93.8
4.	Other anthropogenic sources	2.0– 6.8	0.3– 2.7	8.0– 40.0	15.9– 76.3
A.	Total anthropogenic emissions	12.0–25.6	3.1–12.0	288.7–376.0	70.0–193.5
	(Mean values)	18.8	7.6	332.0	132.0
B.	Natural emissions	7.8	1.0	19.0	4.0

Table 5. Global emission inventory for lead in 1989 – Maximum scenario (in tonnes)

	Gasoline combustion	Fossil fuel combustion	Non-ferrous metal production	Waste incineration	Cement production	Steel and iron	Total
Africa	12 316	628	4421		85	86	17 536
Asia	44 006	4209	21 323	207	867	3 713	74 325
Australia	2 000	411	2775	62	12	118	5 378
Europe	47 579	3477	13 041	537	641	4 278	69 553
North America	14 192	993	7613	2498	177	1 316	26 789
South America	8 796	195	5422		98	555	15 066
Total	128 889	9913	54 595	3 304	1 880	10 066	208 647

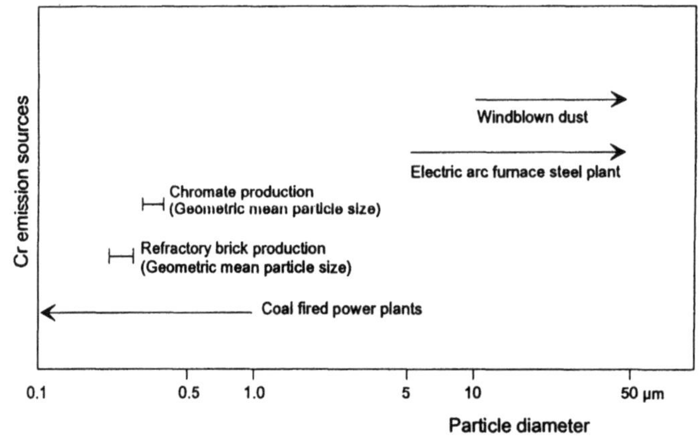

Fig. 8. Chromium distribution on particles emitted from various source categories [30]

Sources of Secondary Aerosols

So far much attention has been directed to study the formation of secondary aerosols from their gaseous precursors emitted from power stations, major industrial plants, and mobile sources. About 500 smog chamber studies have been carried out in both indoor and outdoor chambers over the past 15 years to investigate chemical reaction mechanisms for urban air pollution [31].

Major field experiments in the plumes were carried out in North America in the 1970s, including Midwest Interstate Sulfur Transport and Transformation Program-MISTT, Regional Air Pollution Study-RAPS, Sulfur Regional Experiment-SURE and Multistate Atmospheric Power Production Pollution Study-MAP3S. In Europe aircraft measurements were carried out within the OECD programme on Long-range Transport of Air Pollutants. Major emphasis in these studies was placed on the oxidation of sulfur dioxide to atmospheric aerosol sulfate under various meteorological and topographical conditions.

More recent studies, such as the 1987 Southern California Air Quality Study (SCAQS) have been carried out to improve our knowledge on the chemistry of other species, including hydrocarbons and nitrogen compounds [31].

The following conclusions have been drawn from the above mentioned studies on the formation of aerosols from gas-to-particle formation. Aerosol sulfate is formed through the oxidation of sulfur dioxide at the rate ranging from 1 to 3% per h. The oxidation rate depends on the interaction of chemical kinetics, e.g. the molecular rates, and physical kinetics, e.g. the processes of dispersion and mixing of the plume with ambient air. The sulfur dioxide oxidation rates are highly dependent on the presence of oxidants, sunlight, catalysts, and co-pollutants.

The oxidation of NO and further conversion to nitrates and nitric acid have also been investigated in various industrial plumes of pollutants. Aerosol nitrate can be identified mostly as ammonium nitrate but also as sodium nitrate. Unlike sulfuric acid, however, nitric acid has a higher vapor pressure and does not readily condense on aerosols. The formation of aerosol nitrate requires the presence of ammonia, or other alkaline species to form salts such as ammonium nitrate. The equilibrium between the nitric acid and ammonium nitrate strongly depends on temperature, humidity and acidity [31].

Less information is available on the gaseous HCl conversions in the air. Generally HCl is relatively stable towards oxidative attack. Homogeneous oxidation of HCl occurs through reaction with OH at a rate of 1–3% per h to form atomic chlorine. This product is known to react either with methane (or other hydrogen containing trace gases) to regain HCl or with ozone to produce ClO radicals, subsequently converted to stable molecule nitrate ClONO2.

Ammonia is the only common alkaline gas in the atmosphere and as such it plays a major role in the neutralization of acids generated by the oxidation of sulfur and nitrogen oxides. As a result, ammonium is a major aerosol component. The conversion rate of ammonia to ammonium is a function of height, as the major sources of ammonia are at ground level, while the aerosol acids are formed through the oxidation of their precursors throughout the whole atmosphere. This conversion rate also depends on the time of day, season, concentrations of acidic components, and temperature and relative humidity of the atmosphere.

When the gas-to-particle conversion occurs on existing particles, it is important to understand to what extent this process is affected by the chemical composition of the particles. This applies mostly to the role of the elemental and organic carbon of the carbonaceous fraction of the aerosol. Although hydrophobic at a source, carbonaceous aerosols may become hydrophilic when transported over several hundred kilometers [32]. Knowledge of the extent to which carbon particles can act as catalysts in gaseous reactions or as nuclei show that the potential for this can be large [33, 34].

Transport of Aerosols Within Air Masses

Particles entering the atmosphere from primary sources of emission or formed in the atmosphere through gas-to-particle conversions are a subject of transport within air masses. The scale of this transport depends on the effective height of emission source, meteorological conditions, and the chemical and physical characteristics of aerosols. Obviously, the above parameters affect the growth of particles during transport and thus the efficiency of their removal from the air by dry and wet deposition. The process of growth and ageing of aerosols is different in wet and dry atmospheres having a direct impact on the atmospheric residence time for aerosols. Both, source oriented and receptor oriented approaches have been applied to study the transport of aerosols on global and regional scale.

Transport on Global Scale

Various definitions are given in the literature to describe the global scale of pollution transport. In general, the transport of pollutants over a few thousand kilometers can be regarded as global, e.g. hemispheric transport.

Several studies have been carried out to explain long-range transport of dust from major deserts to various receptors in the northern hemisphere. African dust events reaching northwestern Europe have been documented and examined using a variety of techniques, such as satellite imagery, upper-air soundings, surface observations, X-ray analyses of the dust composition, low-level dust concentration measurements, and objectively calculated air trajectories [35]. It was concluded that minor falls of African dust in northwestern Europe must occur several times per year. The African dust fall over other parts of Europe had been studied in the 1970s. Westward-moving plumes are even more common and often reach the Caribbean [36].

Dust from deserts in Asia and Africa was also measured at high altitudes (about 5 km) in the Norwegian Arctic [13]. Concentration ratios for several crustal elements indicated that fractions of eolian dust from these deserts can be identified in the Arctic as fine fraction particles.

A broad evidence of very long-range transport of aerosols generated during biomass burning has been collected [19]. Smoke plumes can rise up to several kilometers, although the height in the tropics, where much of the biomass burning occurs, is limited by the trade-wind inversion present at an altitude of some 3 to 5 km [37]. These plumes, trapped in an inversion layer, can be transported on a few thousand kilometers in various directions. For example, air masses from the biomass burning regions in South America are usually moving toward the south and southeast due to the effect of the Andes barrier on large-scale circulation.

Sporadic releases of particles into the atmosphere and their subsequent transport on a global scale occur during volcano eruptions and industrial accidents. Two prominent examples of such processes include El Chichon volcanic eruption and the Chernobyl accident. The Mexican eruption injected massive amounts of sulfur dioxide into the stratosphere [38]. Slow decrease of this gas over several months was measured as the chemical conversion to sulfate aerosol occurred. The Chernobyl accident provided a good illustration of the transport of radioactive aerosols which can survive for several days.

Transport on Regional Scale

Regional scale of pollution transport is often defined as a travel distance of up to a couple of thousand of kilometers and is therefore also referred to as continental. It was postulated in the 1960s that some pollutants such as acidic compounds can be transported within air masses on a regional scale. Transport of pollutants originating from sources in the United Kingdom as well as in Eastern Europe was studied in various receptor areas in Scandinavia. Later on the transport phenomenon was studied in other parts of the globe. Various dispersion models and specially designed measurements have been performed to assess this transport quantitatively. Concerning aerosols, major emphasis was placed on assessing transport of various chemical compounds on to particles, particularly trace elements and sulfur and nitrogen species into various receptors.

The results of long-term measurements of pollution of the North Sea and Baltic Sea waters lead to the conclusion that as much as 50% of lead and mercury and 30–50% of arsenic, cadmium, chromium, copper, nickel, and zinc enter these waters through atmospheric deposition. Atmospheric mercury is primarily in the vapor phase with more than 80% in an elemental form [39].

Using the best "available" data, quantitative estimates have recently been made on the fluxes of many toxic contaminants into the Great Lakes [40]. The general conclusion that can be drawn on the basis of this and other studies [41] is that 35–50% of the annual load of the toxic compounds enter the Great Lakes waters through atmospheric deposition of particles. Concerning lead, it was concluded that the atmospheric input was responsible for 35–47% of all inputs to the Great Lakes during the period from 1977 to 1981. Long-range transport of trace metals from source regions as far as 800 km contributes about 70% of the total annual atmospheric deposition in the area of Lake Superior [42]. Both, the significance of long-range transport and total annual atmospheric deposition to the Lake increase for the smaller 0.5 μm particle size. Except for arsenic, the total atmospheric deposition decreases 10–25% when all the particles are 5.0 μm in diameter.

Recent studies of the origin of the Arctic air pollution have concluded that 3–14% of the total emissions of arsenic, cadmium, lead, zinc, vanadium, and antimony in all of Eurasia is deposited in the Arctic [43].

Major progress has been made in the development of dispersion models during the last decade. A summary of information on models used to study the transport of trace elements on particles is presented in Table 6 for Europe [44] and in Table 7 for North America [45]. In general, two versions of these models are used. The Eulerian models compute chemical species concentrations at a fixed geographical location. The alternative approach is termed Lagrangian model. In this approach an air parcel is moving with air motions, thus allowing one to follow the chemistry of an individual parcel. Source and sink terms vary as the parcel moves over different source regions.

Temporal and spatial resolution requirements for regional scale models have been defined and discussed [46, 47]. Initial experiments to elaborate the basic meteorological factors affecting long-range aerosol transport have been conducted, mostly in one (vertical) and two dimensions (vertical and one horizontal scale).

Various receptor models, made as a hybrid between multivariate techniques and dispersion models, have also been applied to study long-range transport of particles containing trace elements and other pollutants. Through the use of multielement chemical analyses and multivariate statistical procedures it is possible

Table 6. A comparative summary of models applied for heavy metals in Europe.

Model			Application			
			Input		Output	
Name	Type	Institution	Meteorological input	Calculated substances	Receptor area	Time scale
EMEP (European Monitoring and Evaluation Programme)	Lagrangian one-layer trajectory model, sulphur model modified for inert particle associated substances	NILU (Norsk Institutt for Luftforskning) Lillestrøm, Norway	6-hourly from the Norwegian Numerical Weather Prediction Model	V, Mn, Cd, Pb, As, Sb, Se	Birkenes, Southern Norway	Daily averages
EMEP /MSC-East	Eulerian-Lagrangian combined model, calculations are provided for a number of emitters and receivers at EMEP standard form	Meteorological Synthesizing Centre - East of EMEP, Moscow, Russia	6-hourly from Numerical Weather Prediction Model with corrections by observed data (Russian Hydro-meteorological Centre)	Pb, Cd, Zn, As	Hemispherical area; EMEP area. Spatial resolution of 150 km × 150 km	Daily, monthly and annual averages
TREND	Statistical approach of a Gaussian plume model and a trajectory model	RIVM (Rijksinstituut voor Volksgesondheit en Milieuhygiene) Bilthoven, The Netherlands	Long-term averages from the Dutch Meteorological Office	As, Sb, Cd, Cr, Cu, Ni, Pb, Zn	North Sea	Annual averages
OPS (Operational Priority Substances)	More operational version of TREND	TNO (Netherlands Organization for Applied Scientific Research) Delft, The Netherlands		23 contaminants Cd, Cr, Cu, Pb, Hg, Ni, Zn, N chlorinated hydrocarbons, PAH benzene PCB, pesticides	North Sea Dutch Wadden Sea	
				As above plus 6 additional pesticides	River Rhine drainage basin	
		RIVM		PCDD, PCDF	North Sea and surrounding area	

HHLRT (Hamburg Long Range Transport Model)	Three-dimensional stochastic model	Meteorological Institute, University of Hamburg, Germany	6-hourly analyses of the European Centre for Medium Weather Forecast (ECMWF)	Pd, Cd	North Sea	Monthly averages
EMEP	Lagrangian one-layer trajectory model, sulphur version modified for inert particle associated substances	GKSS Research Centre Geesthacht, Germany	6-hourly from the Norwegian Numerical Weather Prediction Model and from the European Monitoring Network	Pb	North Sea Baltic Sea	Monthly and annual averages
				Pb, Cd, Zn, As		
	Sulphur version modified for reactive mercury species in gaseous and particulate form			Hg		
TRACE (TRace toxic Air Concentrations in Europe)	Improved climatological type model	IIASA (International Institute for Applied System Analysis) Laxenburg, Austria	Improved climatological input	Pb, Cd, Zn, As	Europe	Annual averages
GESIMA (GEesthacht SImulation Model of the Atmosphere)	High-resolution non-hydrostatic mesoscale model	GKSS Research Centre Geesthacht, Germany	Numerical Weather prediction models, e.g. "Deutschland" Model of the German Weather Service	Pb	North Sea coastal area in Germany	Days

Table 7. A comparative summary of models applied for heavy metals in North America

			Model			
			Application			
			Input		Output	
Name	Type	Institution	Meteorological input	Calculated substances	Receptor area	Time scale
ASTRAP (Advanced Statistical TRajectory Air Pollution)	Statistical Lagrangian, trajectory 9 vertical layers	Argonne National Laboratory, USA; Environment Canada	6-hourly analysis from Canadian Meteorological Center	Pb	North America	Seasonal
				Pb, Cd	Great Lakes	
				Toxaphene		
RELMAP	Lagrangian one-layer trajectory	NOAA/US EPA	12-hourly wind	As, Cd, Cr, Pb, Ni and 14 organic pollutants	Lake Michigan	Daily
				As, Cd, Cr, Pb, Ni	Lake Superior	
AES-LRT	Lagrangian one-layer trajectory	Environment Canada	6-hourly analysis from Canadian Meteorological Center		Arctic	Daily
ADOM	Eulerian, 12 vertical layers	Ontario Ministry of the Environment, Environment Canada	6-hourly analysis from Canadian Meteorological Center processed through PBL model	Hg	Eastern North America	Hourly

to discriminate the components and sources of fine and coarse mode aerosols measured at remote locations. The applicability of multivariate techniques for resolving sources and source regions for aerosols measured at several remote locations far from major emission regions has been tested with the use of the absolute principal component analysis (APCA) and the chemical mass balance (CMB) methods. The APCA method determines the composition of the major source components, such as coal combustion related, crustal, or sea-salt related which contributed to the measured concentrations at the receptors. In the past, the APCA method was applied to total suspended particulate concentrations (particles measured in both the fine and coarse fractions). Further improvement of this receptor modeling method was obtained by applying APCA to aerosol elemental concentration measurements in separate particle size fractions. One important limitation of the APCA methods is, however, that their results do not allow one to obtain a fine resolution of the contributions from various distant source regions to the chemical composition of the remote aerosol. To attempt this resolution, CMB source apportionment must be performed using either a set of emission source profiles or a set of elemental signatures. A few examples of the APCA and CBM applications to study the origin of aerosols at various locations are given below.

The aerosol in the Antarctic Peninsula is dominated in the fine and coarse modes by the sea-salt component, but there is a small soil dust contribution [48]. In the fine mode, a sulfate component was also observed.

Results from measurements carried out in the Norwegian Arctic concluded that, on average, 70% of the concentrations of anthropogenic trace metals in the fine mode of winter aerosols was attributed to Asian source regions and the rest to European sources [49]. The contribution of European sources is higher than the contribution from Asian sources in summer.

Studies carried out in Ivory Coast concluded that the burning of biomass in the area remobilizes dust particles which were deposited on the plants by both dry and wet deposition [50]. However, the largest fraction of the aerosol mass originated from the burning of biomass itself. An interesting conclusion was also that rapid gas-to-particle conversion reactions take place in the cooling of burning plumes.

Aerosol chemistry in Amazonia during the wet season was found to be strongly influenced by long-range transport of soil dust, marine aerosol, and possibly biomass combustion products advected into the area by large scale tropospheric circulation [51].

Removal of Particles

Deposition processes determine the persistence of aerosol particles in the atmosphere. They also transfer pollutants to the ecosystems at the planetary surface. Extended research has been carried out to assess the mechanisms of particle removal from the atmosphere by dry and wet deposition [52]. It is not intended here to describe in detail sink processes of particles from the air, but a short summary is given based on the review presented in [52]. This summary should assist in the understanding of global distribution of aerosols.

In general, large particles with diameters above 10 μm are removed from the air mainly by dry deposition. Their residence time is of the order of hours. The finest particles with diameters below 0.03 μm are often the subject of coagulation to larger particles and thus they are removed together. Some of these fine particles may also be attached to cloud droplets. The atmospheric residence time for these particles is up to 10 h.

Particles in the accumulation mode are too small to be very efficiently removed by dry deposition and too large to accumulate on large particles. It is concluded that they are removed from the air primarily through various precipitation scavenging mechanisms. A fraction of these particles become cloud condensation nuclei and then cloud droplets. They are removed by nucleation scavenging. If precipitation forms, these particles are the subject of removal through the collision-coalescence process. This is the primary removal process for particles in the 0.1 to 1.0 μm diameter range. The atmospheric residence time of these particles is of the order of ten days, depending on many parameters, particularly occurrence of precipitation in the time from particle formation.

There are a number of parameters affecting the efficiency of particle removal processes from the atmosphere. The parameters having impact on dry deposition have been reviewed in [53, 54]. Various models of dry deposition have been developed where dry deposition velocity is a function of particle diameter. An example of these models is presented in Fig. 9 [53]. Ranges of dry deposition

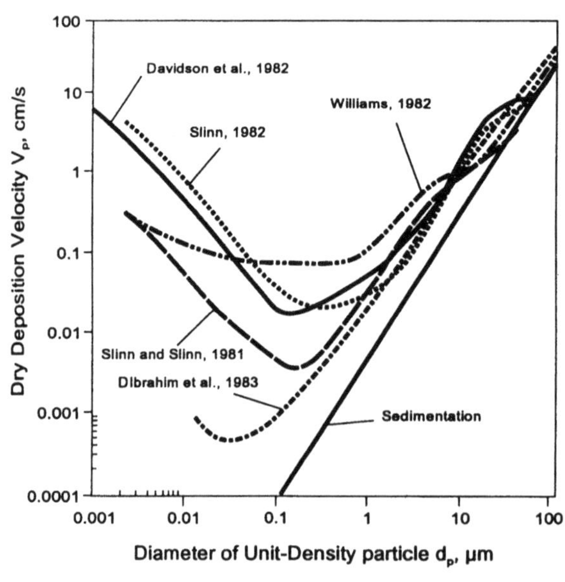

Fig. 9. Dry deposition velocity as a function of particle size, obtained from models developed by various authors [53]

velocities for a few trace elements emitted on particles are given in Fig. 10. These values are used in dispersion models studying regional transport of pollutants.

Parameters affecting wet deposition of particles and various trace elements on them were reviewed in [55]. An example of variations of scavenging ratios, defined as the ratios of concentrations of a given substance in precipitation to those in the air, is presented in Fig. 11 [56]. The data are based on measurements at Enewatok Atoll in the Marshall Islands. Comparison of pollution scavenging in the marine regions with that in the continental areas was made. It was concluded that the particle-size dependent rain scavenging in continental regions does not apply in tropical marine regions, where precipitation is often dominated by trade wind showers.

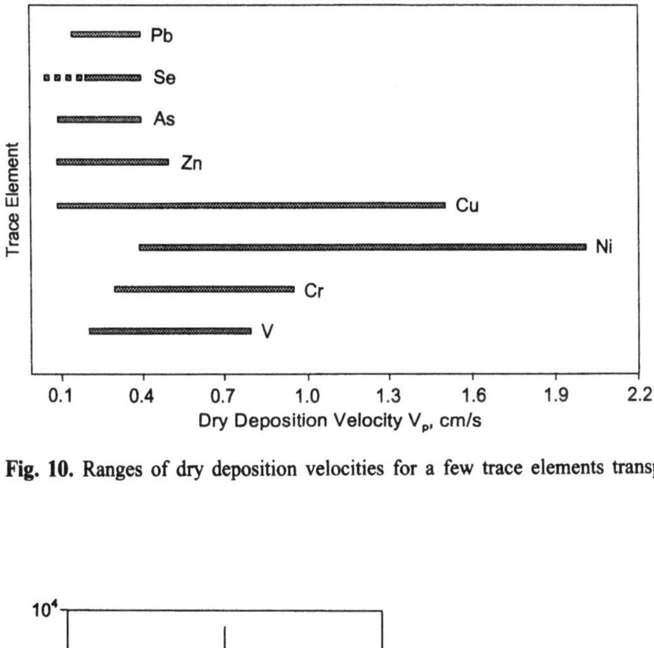

Fig. 10. Ranges of dry deposition velocities for a few trace elements transported on particles

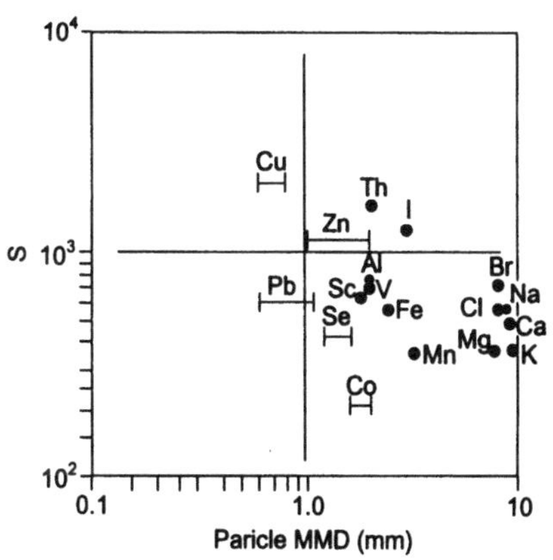

Fig. 11. Variation of scavenging ratio (S) vs. particle mass median diameter at Enewatok Atoll based on rain event and 2- to 5-day average aerosol data [56]

Global Distribution of Aerosols and their Trends

Several studies have been carried out to assess the size distribution of particles measured at continental and marine locations. The overall conclusion can be that the tropospheric aerosol consists of two main components: a submicrometer aerosol with mass mode below 1 µm diameter, and coarse mode between 5 and 20 µm [57]. An example of particle size distribution measured at the marine station in Norway in 1988 is presented in Fig. 12. The two components of the aerosol are clearly seen. It should be added that particle size distribution data are often presented assuming the spherical shape of the particles. Even if the particles are not spherical they may become so during their ageing, e.g. through attachment of water. However, this may not occur if particles consist completely of nonhygroscopic substances.

Aerosols from the marine boundary layer were also collected during a cruise from the Peru/Ecuador shelf through the equatorial and tropical Pacific to the Hawaiian Islands [58]. Further evidence was obtained on the bimodal distribution of particle sizes as well as on the chemical composition of these particles.

The desert aerosol is present over the deserts, but extends over the adjacent oceans due to its potential for even intercontinental transport [59]. Three dimensional general circulation models (3D-GCM) were applied to study this transport [60]. The size distribution of the desert aerosols shows bimodal distribution with major contribution from coarse particles.

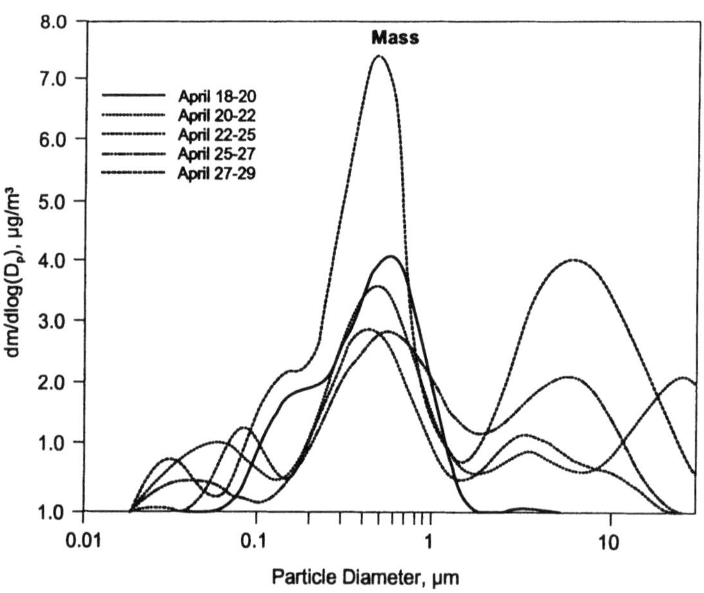

Fig. 12. Particle size distribution measured at the marine station Birkenes in Norway in 1988

The remote continental aerosol is often defined as the aerosol formed deep in continents where human impact is limited. Two examples of regions with such aerosols are rain forest and Siberia. High contribution of condensation nuclei to the total spectrum of particles is expected for this type of aerosol.

The polar aerosol belongs to the group of aerosols that seems to be the final stage of a production occuring at other locations. The concentrations and enrichment factors of several elements measured in the Arctic aerosol have confirmed that the chemical nature of particles in various size fractions can be a very useful tool for assessing the origin of aerosols [61]. During episodes of long-range transport, the concentrations and enrichment factors of some elements increased substantially in the particle fractions below 0.5 μm diameter, as seen in Fig. 13. At the same time, in fractions above 0.5 μm diameter they remain unchanged or change only insignificantly. The increase of concentrations in the finest fractions is probably caused by long-range transport of pollutants from anthropogenic sources, while the changes in fractions above 2 μm diameter are probably the result of variations in earth's crust erosion.

A few studies have been carried out to obtain size distribution of the biogenic aerosol [62] which contains pollen, spores, fungi, bacteria, broken plant parts, insects and abrasions. The cascade impactor data from the Amazon basin showed that there are large differences in the size distributions of the soil and biogenic related elements, with sulfur and potassium exhibiting a clearly bimodal character [63].

Size distributions of particles formed during biomass burning are similar to those of other urban/industrial pollution aerosol, with three distinguishable parts:

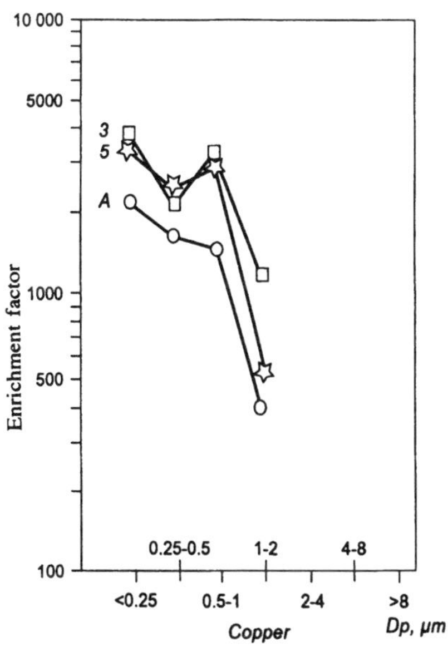

Fig. 13. Enrichment factors for copper, calculated during an episode of pollution transport to Ny Alesund (Spitsbergen) (*samples 3 and 5*) and during non-episodic conditions (*sample A*)

a nucleation mode, an accumulation mode, and a coarse mode, with the accumulation mode dominating [64]. The fraction of accumulation mode particles contains primarily tarry, condensed hydrocarbons that are typically spherical in shape.

Concerning trends, natural aerosols have probably remained roughly constant in their concentration, distribution, and properties for at least a century [65], with some exception for volcanic aerosols.

In contrast, anthropogenic aerosols have increased dramatically during the last several decades following the growth of the industrial capacity of the western free market economies and those of the former Soviet Union and other socialist nations. Subsequently, the "third world" transition into modern society and industrialization during the 1970s and 1980s added to the problem of air contamination including the increase of particle concentrations in the air. In the late 1970s, these concentrations began to decrease in the western countries, particularly in the U.S., following the installation of more efficient control devices in major industrial sectors.

A small decrease of fine particle concentrations in the Arctic became evident in the late 1980s [66]. Very preliminary explanations of this decrease can probably be related to the implementation of the UN ECE emission reduction protocol for sulfur in Western Europe and partial replacement of coal by natural gas to produce heat and electricity, particularly in the former Soviet Union.

Alternative explanation of the lower concentrations of fine particles in the Arctic can be related to lower emissions in the Eastern European countries due to their transition from centrally planned to market-oriented economies at the beginning of the 1990s. This process resulted in closing down several poorly managed factories and at the same time eliminating a number of polluting technologies.

Concluding Remarks

Aerosols play an important role in the global change of our environment. Their impact on human and environmental health has been recognized through multidisciplinary research programs carried out around the globe. However, more monitoring programs are needed in order to provide further information on the sources, transformations, sinks, and lifetimes for different aerosol components. Particle size distribution and size-differentiated chemical composition are the parameters which should be focused on in these programs.

Sources and behavior of natural aerosols are less understood than sources and behavior of anthropogenic aerosols. Of particular interest is formation of aerosols through gas-to-particle conversion processes. Information on these processes, leading to the production of acidic compounds on aerosols, is incomplete. Major scientific questions one may ask are to what extent do the physical and chemical transformations of selected atmospheric constituents contribute to the formation of acidic aerosol and what are the parameters affecting these transformations? To what extent do aerosols, formed by various processes, affect climate forcing [67]?

The areas of further research on aerosol formation and transformations include the kinetics of homogeneous, heteromolecular nucleation and heteromolecular condensation on preexisting surfaces, and aqueous phase atmospheric chemistry and reaction in liquid aerosols [52].

The adaptation of existing, and the development of new, nested hemispherical and/or global scale models are needed to study the impact of aerosols emitted in major source regions on the quality of air in remote locations and to assess the effects of these source-receptor relationships.

Removal of aerosols from the air during their long-range transport needs further investigation. More information should be collected on the particle size distribution of ambient aerosols and on parameters influencing the size distribution, including both meteorological and topographical factors. Direct observations of dry deposition, coupled with measurements of the particle size distribution are needed in order to obtain parameterization of dry deposition models.

Acknowledgement. I dedicate this work to the memory of the late Terry L. Clark.

References

1. Seinfeld JH (1986) Atmospheric chemistry and physics of air pollution. Wiley, New York
2. The U.K. Department of the Environment (1993) Urban air quality in the United Kingdom. First Report of the Quality of Urban Air Review Group, London
3. Junge C (1979). In: Morales C (ed) Saharan dust. Wiley, Chichester
4. Patterson EM (1982). In: Technical proceedings of a workshop on atmospheric aerosols, Baltimore, MD, 6–8 Nov 1979
5. Grosjean D (1977). In: Ozone and other photochemical oxidants, National Academy of Sciences, Washington, DC, pp 45–125
6. EUROTRAC (1993) 1992 Annual Report, Part 3, Air-Sea Exchange, Garmisch-Partenkirchen, Germany
7. Mestayer PG (1993). In: 1992 Annual Report of EUROTRAC, Part 3, Air-Sea Exchange, Garmisch-Partenkirchen, Germany
8. Mason BJ (1957) Geof Pura Appl (Milano), 36, 148–155
9. Blanchard DC (1963) Progress Oceanog, 1: 71–202
10. Winchester JW, Duce RA (1975). In: Proc the Symp on Fates of pollutants in the air and water environments, Philadelphia, PA, 6–11 April 1975
11. Gillette DA (1979). In: Morales C (ed) Saharan dust. Wiley, Chichester
12. Rahn KA (1976) Atmos Environ, 10, 597–601
13. Pacyna JM, Ottar B (1989) Atmos Environ, 23, 809–815
14. Zurer S (1984) C&EN Special Report, September 24, 26–40
15. Cachier H, Buat-Menard P, Fontugne M (1985) J Atmos Chem, 3, 469–489
16. Wolff GT, Groblicki PJ, Cadle SH, Countess RJ (1982). In: Wolff and Klimisch (eds), Particulate carbon atmospheric life cycle, Plenum, New York
17. Marty JC, Saliot A (1982) Nature, 298, 144–147
18. Gagosian RB, Zafiriou OC, Peltzer ET, Alford JB (1982) J Geophys Res, 87, 11133–11144
19. Levine JS (ed.) (1991) Global biomass burning: atmospheric, climatic, and biospheric implications, The MIT Press, Cambridge, Massachusetts
20. Penner JE, Eddleman H, Novakov T (1992) Atmos Environ
21. Duce RA (1978) Pure Appl Geophys, 116, 244–273

22. Pacyna JM (1989). In: Pacyna JM, Ottar B (eds) Control and fate of atmospheric trace metals, Kluwer Academic Publishers, Dordrecht, the Netherlands
23. Ensor DS, Markowski G, Woffinden G, Legg R, Cowen S, Murphy M, Shendrikar AD (1983) Evaluation of electrostatic precipitator performance at San Juan unit 1. EPRI CS-3252 Report, Electric Power Research Institute, Palo Alto, CA
24. Ensor DS, Cowen S, Shendrikar A, Markowski G, Waffinden G (1981) Kramer station fabric filter evaluation. EPRI CS-1669 Report, Electric Power Research Institute, Palo Alto, CA
25. Ondov JM, Ragaini RC, Bierman AH (1979) Environ Sci Technol, 13, 588–601
26. Pacyna JM (1980) Coal-fired power plants as a source of environmental contamination by trace metals and radionuclides. Habilitation thesis, Technical University of Wroclaw, Poland
27. Brown RD (1981) Health and environmental effects of oil and gas technologies: Research needs. A report to the federal interagency committee on the health and environmental effects of energy technologies. MITRE Corp., McLean, VA
28. Nriagu JO, Pacyna JM (1988) Nature, 333, 134–139
29. Pacyna JM, Shin BD, Pacyna P (1993) Global emissions of lead. A report for the Atmospheric Environment Service, Environment Canada
30. Pacyna JM, Nriagu JO (1988). In: Nriagu JO and Nieboer E (eds), Chromium in the natural and human environments, Wiley, New York
31. Seinfeld JH (1989) Science, 243, 745–752
32. Heintzenberg J, Covert DS (1990) J Atm Chemistry, 10, 383–398
33. Gundel LA, Guyot-Sionnest NS, Novakov T (1989) Aerosol Sci and Technol, 10, 343–351
34. Dlugi R (1989) Aerosol Sci and Technol, 10, 93–105
35. Reiff J, Forbes GS, Spieksma FTM, Reynders JJ (1986) J Climate and Appl Meteor, 25, 1543–1567
36. Carlson TN, Prospero JM (1972) J Appl Meteor, 11, 283–297
37. Andreae MO (1991). In: Levine JS (ed.), Global biomass burning: atmospheric, climatic, and biospheric implications, The MIT Press, Cambridge, Massachusetts
38. ApSimon HM, Wilson JJN, Simms KL (1989) Analysis of the dispersion and deposition of radionuclides from Chernobyl across Europe. Proc R Soc Lond A425, 365–405
39. Pacyna JM (1992) The baltic sea environmental programme. The topical area study for atmospheric deposition of pollutants. Final Report NILU OR 46/92. Norwegian Institute for Air Research, Lillestrøm, Norway
40. Eisenreich SJ, Strachan WMJ (1992) Estimating atmospheric deposition of toxic substances to the Great Lakes-an update. Report from Workshop in Burlington, Ontario, Jan. 31–Feb. 2, 1992, sponsored by the Great Lakes Protection Fund and Environment Canada
41. ICF (1992) Atmospheric deposition of toxic chemicals to surface waters: identification and summary of the recent literature, ICF Incorporated, Fairfax, VA
42. Clark TL (1993). In: Pacyna JM, Voldner E, Keeler GJ, Evans G (eds), Proc the First Work. on emissions and modelling of atmospheric transport of persistent organic pollutants and heavy metals, The U.S. EPA, Durham, NC, 6–7 May, 1993
43. Akeredolu F, Barrie LA, Olson MP, Oikawa KK, Pacyna JM, Keeler GJ (1993) The flux of anthropogenic trace metals into the Arctic from mid-latitudes in 1979/80. Atmospheric Environment.
44. Petersen G (1993). In: Pacyna JM, Voldner E, Keeler GJ, Evans G (eds), Proc the First Work. on emissions and modelling of atmospheric transport of persistent organic pollutants and heavy metals, The U.S. EPA, Durham, NC, 6–7 May, 1993
45. Voldner E (1993). In: Pacyna JM, Voldner E, Keeler GJ, Evans G (eds), Proc the First Work. on emissions and modelling of atmospheric transport of persistent organic pollutants and heavy metals, The U.S. EPA, Durham, NC, 6–7 May, 1993
46. Pielke RA, Lyons WA, McNider RT, Moran MD, Moon DA, Stocker RA, Walko RL, Uliasz M (1990). In: Proc 18th NATO/CCMS International Technical Meeting on Air Pollution Dispersion Modelling and Its Application, Vancouver, BC, 13–17 May, 1990
47. Moran MD, Pielke RA, McNider RT (1990). In: Proc. 18th NATO/CCMS International Technical Meeting on Air Pollution Dispersion Modelling and Its Application, Vancouver, BC, 13–17 May, 1990
48. Artaxo P, Andrade F, Maenhaut W (1990) Nucl Instrum and Meth Phys Res
49. Maenhaut W, Cornille P, Pacyna JM, Vitols V (1989) Atmos Environ, 23, 2551–2569
50. Maenhaut W (1993). In: Proc. Belgian Global Change Symp, Brussels, 17–18 May, 1993
51. Talbot RW, Andreae MO, Berresheim H, Artaxo P, Garstang M, Harriss RC, Beecher KM, Li S-M, (1990) J Geophys Res, 95, 16955–16969

52. Hegg DA (ed.) (1991) Report of the meeting of experts on aerosol physics and chemistry. WMO Global Atmosphere Watch, No. 74, Hampton, VA, 30–31 May, 1991
53. Davidson CI, Wu Y-L (1989). In: Pacyna JM, Ottar B (eds), Control and fate of atmospheric trace metals. Kluwer Academic Publishers, Dordrecht, the Netherlands
54. Nicholson KW (1988) Atmos Environ, 22, 2653–2666
55. Buat-Menard P, Duce RA (1986) Nature, 321, 508–510
56. Barrie LA, Schemenauer RS (1989). In: Pacyna JM, Ottar B (eds), Control and fate of atmospheric trace metals. Kluwer Academic Publishers, Dordrecht, the Netherlands
57. Deepak A, Vali G (1991) The international global aerosol programme (IGAP) plan: overview. WMO Global Atmosphere Watch, No. 76, Geneva
58. Raemdonck H, Maenhaut W, Andreae MO (1986) J Geophys Res, 91, 8623–8636
59. Prospero JM (1990). In: Knap AH (ed.), The long-range atmospheric transport of natural and contaminant substances, Kluwer Academic Publishers, New York, NY
60. Wefers M, Jaenicke R (1990). In: Masuda S, Takahashi K (eds), Aerosols Proc 3rd Int Aerosol Conf. Kyoto. Japan. 24–27 September. 1990
61. Pacyna JM, Vitols V, Hanssen JE (1984) Atmos Environ, 11, 2447–2459
62. Jaenicke R, Matthias S (1988). In: Hobbs PV, McCormick MP (eds), Aerosols and climate. A. Deepak Publishing, Hampton, VA
63. Artaxo P, Hansson H-C, (1990). In: Masuda S, Takahashi K (eds), Aerosols. Proc 3rd Int Aerosol Conf, Kyoto, Japan, 24–27 September, 1990
64. Radke LF, Hegg DA, Hobbs PV, Nance JD, Lyons JH, Laursen KK, Weiss RE, Riggan PJ, Ward DE (1991). In: Levine JS (ed.), Global biomass burning: atmospheric, climatic, and biospheric implications, The MIT Press, Cambridge, Massachusetts
65. Charlson RJ, Wingley TML (1994) Scientific American, 2, 28–35
66. Bodhaine B (1993) private communication.
67. Charlson RJ, Schwartz SE, Hales JM, Cess RD, Coakley JA Jr, Hansen JE, Hofmann DJ (1992) Science, 255, 423–430

Physical, Chemical and Optical Properties of Atmospheric Aerosols

Christodoulos Pilinis[1] *and Spyros N. Pandis*[2]

[1] Rosenstiel School of Marine and Atmospheric Science, University of Miami, Miami, FL 33149-1098
[2] Department of Chemical Engineering and Engineering & Public Policy, Carnegie Mellon University, Pittsburgh, PA 15213

Summary

Atmospheric aerosols are complicated suspensions of multicomponent particles of different sizes in air. This section provides an overview of the fundamental physical, chemical and optical properties of these particles. We begin with the discussion of the aerosol shape and density and we then focus on the ability of these particles to deliquesce, that is to absorb moisture from the atmosphere and grow from solid particles to aqueous droplets. The effect of the aerosol chemical composition and temperature on this hygroscopic behavior is outlined based on first thermodynamic principles. The ability of a lot of these particles to act as cloud condensation nuclei and form cloud droplets is also discussed. Aerosol particles can be viewed as atmospheric micro-reactors, that is as the medium where heterogeneous chemical reactions can take place. Our present understanding of these heterogeneous reactions is reviewed. Finally the fundamental principles governing atmospheric aerosol scattering and absorption of radiation are reviewed, including a brief description of the instrumentation used to measure these optical properties.

List of Symbols and Abbreviations 100
Introduction 101
Physical Properties of Atmospheric Aerosols 101
  Atmospheric Aerosol Shape 101
  Density of Atmospheric Aerosol 102
  Deliquescence 104
    Relative Humidity of Deliquescence of Multicomponent Aerosols 105
  Temperature Dependence of the DRH and Solubility 107
  Aerosols as Cloud Condensation Nuclei (CCN) 108
Chemical Properties of Atmospheric Aerosols 112
Optical Properties of Atmospheric Aerosols 113
  Scattering and Extinction Coefficients of Single-Component Particles ... 114
  Scattering and Extinction Coefficients of Multicomponent Particles 117
  Visibility Degradation – Climate Effects of Atmospheric Aerosols 120
  Visibility Measurements 122
References 123

List of Symbols and Abbreviations

α_i	Activity of Species *i* in an Aqueous Solution
α_{ws}	Water Activity of Saturated Solution
b_{sca}	Scattering Coefficient
b_{ext}	Extinction Coefficient
$C(x)$	Contrast
γ	Activity Coefficient
D_p	Particle Diameter
DRH	Deliquescence Relative Humidity
h_s	Enthalpy of an Electrolyte as a Solid
λ	Wavelength of Light
L_s	Latent Heat of Precipitation of an Electrolyte from a Saturated Solution
μ_i^0	Standard Chemical Potential of Species *i*
μ_2	Chemical Potential of Electrolyte 2
$\mu_\bullet$	Chemical Potential of Water
m	Complex Refractive Index
m_2	Molality of Electrolyte 2
m_w	Molality of Water
M_w	Molecular Weight of Water
m_s	Molality of an Electrolyte in a Saturated Solution
n_2	Number of Moles of Electrolyte 2
n_w	Number of Moles of Water
n_s	Number of Moles of Solute
p_w	Vapour Pressure of Water
p_w^f	Vapour Pressure of Water over Flat Surface
p_w^0	Vapour Pressure of Pure Water over Flat Surface
Q_{sca}	Scattering Efficiency
Q_{ext}	Extinction Efficiency
R	Gas Constant
ρ	Density of Particle
σ	Surface Tension of Solution
S_c	Critical Supersaturation
T	Temperature
v_w	Molar Volume of Water
v_s	Molar Volume of Solute
x_w	Mole Fraction of Water in a Solution

Introduction

Atmospheric aerosols are multiphase solid or liquid particles with diameters smaller than 10 μm. These particles usually consist of an insoluble organic core containing both elemental and organic carbon, surrounded by either an aqueous solution of inorganic salts or solid inorganic material. The relative contributions of the organic and inorganic material to the total aerosol mass depend on several factors including the origin of the particulate matter, the atmospheric conditions, and the size of the particle. Fine particulate matter ($D_p < 2$ μm) in polluted urban areas, like Los Angeles, may contain up to 40% particulate carbon [1], while coarse particles are predominantly inorganic. Marine aerosol, on the other hand, is an aqueous solution of mainly NaCl and $(NH_4)_2SO_4$.

Here we discuss the physical, chemical and optical properties of atmospheric aerosols. The atmospheric aerosol shape and density as a function of chemical composition and their deliquescence, that is their ability to absorb water when the relative humidity exceeds a threshold value, are examined first. The theoretical framework for the activation of aerosols to cloud droplets is presented. Our understanding of the chemical processes taking place on the surface or inside atmospheric aerosols remains incomplete. The current state-of-science knowledge regarding possible reactions in the particulate phase is summarized. Finally we describe the optical properties of atmospheric aerosols. The methodology for the derivation of the scattering and extinction efficiencies is presented. A brief description of the instrumentation used to measure the optical properties is also included.

Physical Properties of Atmospheric Aerosols

Atmospheric Aerosol Shape

A common assumption in the study of atmospheric aerosols is that these particles are spherical with a diameter D_p. While liquid atmospheric particles nearly always satisfy this assumption, solid atmospheric aerosol can have quite irregular shapes for which the geometric diameter D_p is not meaningful. Most of the atmospheric aerosol properties (e.g. volume, surface area, settling velocity, Brownian diffusion, etc.) depend on particle size and therefore some means of defining an equivalent particle diameter is needed. An equivalent diameter is the diameter of the sphere that has the same value of a particular property as that of the irregular particle. For non-spherical particles one can define several types of particle diameters with the most common being as below.

1. The mass equivalent diameter, D_m, that is the diameter of a solid sphere which has the same mass and density as the irregular particle. This diameter

is equal to: $D_m = (6\,m_p/\pi\rho_p)^{1/3}$ where m_p and ρ_p are the mass and density of the particle respectively.

2. The classical aerodynamic diameter, D_a, defined as the diameter of a sphere of unit density (1 g cm^{-3}) which has the same terminal falling speed in air as the particle under consideration. For a spherical particle the aerodynamic diameter is related to its geometric diameter D_p by: $D_a = D_p(\rho_p/\rho_o)^{1/2}$, where ρ_o is the reference density of 1 g cm^{-3}. Because of this density effect the aerodynamic diameter of a spherical particle of high density is higher than its geometric diameter. The aerodynamic diameter is extremely useful because it can be measured directly and is associated with the particle residence time in the atmosphere, its deposition in the human respiratory system, etc. [2].
3. The Stokes aerodynamic diameter, D_s, is the diameter of sphere that has the same density and settling velocity as the particle. The Stokes diameter standardizes particles of various shapes to spheres having the same aerodynamic property, settling velocity.
4. The electrical mobility equivalent diameter, D_b, is the diameter of a sphere that has the same electrical mobility as the aerosol particle. The electrical mobility, Z, expresses the ability of a particle to move in an electrical field and for a spherical particle it is equal to: $Z=(\mathrm{neC_c}/4\pi\eta D_g)$, where n is the number of charges of the particle, e is the charge of an electron, $C_c = 1 + 2.52\,\lambda/\mathrm{d}$ is the Cunningham correction factor, λ is the air mean-free path, η is the air viscosity, and D_g the geometric diameter of the particle. A popular aerosol instrument, the differential mobility analyzer (DMA), classifies aerosol by electrical mobility diameter. The electrical mobility of a particle depends not only on the particle size, but also on its charge. If a particle is not charged its electrical mobility equivalent diameter is zero.

Density of Atmospheric Aerosol

Density plays an important role in determining several aerosol transport properties. For example, dry deposition and cloud scavenging depend on settling velocities and inertial characteristics, which are functions of the density of the aerosol particles. When the aerosol is an aqueous solution of various salts, it is practically spherical and its density can be theoretically calculated as the density of a non-ideal solution [3]. If insoluble organic material is also present in the aerosol phase, a weighted average of the densities of the aqueous solution and the organic material is used.

A major problem arises when the aerosol is dry, e.g. low relative humidity, the particle consists of elemental carbon, etc. Coagulation of dry particles results in the formation of aerosol of irregular shape and non-uniform density. Satisfactory methods for the theoretical estimation of the resulting densities are not currently available and consequently measurements are required for the accurate calculation of the density of particles of complex morphology.

Laboratory measurements on single aerosol particles can be performed by using the improved Millikan cell [4], shown in Fig. 1. Particles are suspended between the centers of the capacitor plates, submerged in a temperature controlled paraffin oil bath. The particle under consideration is illuminated by a He-Ne laser beam aligned with the gravitational and electrical fields and can be viewed with a telemicroscope. Aerosol particles are alternately raised by the electric field against the force of gravity and then forced downward by an electric field of equal strength but opposing polarity. The size of the particles can be found from their rates of fall at various pressures of the gas in the cell using the following equation [5]:

$$D = 2\frac{V_2\, p_2 - V_1\, p_1}{V_1 - V_2}\frac{A\, p_0\, l_0}{p_1\, p_2} \tag{1}$$

where V_1 and V_2 are the falling speeds at pressures p_1 and p_2, respectively. A is a numerical value depending on the reflection of gas molecules from the

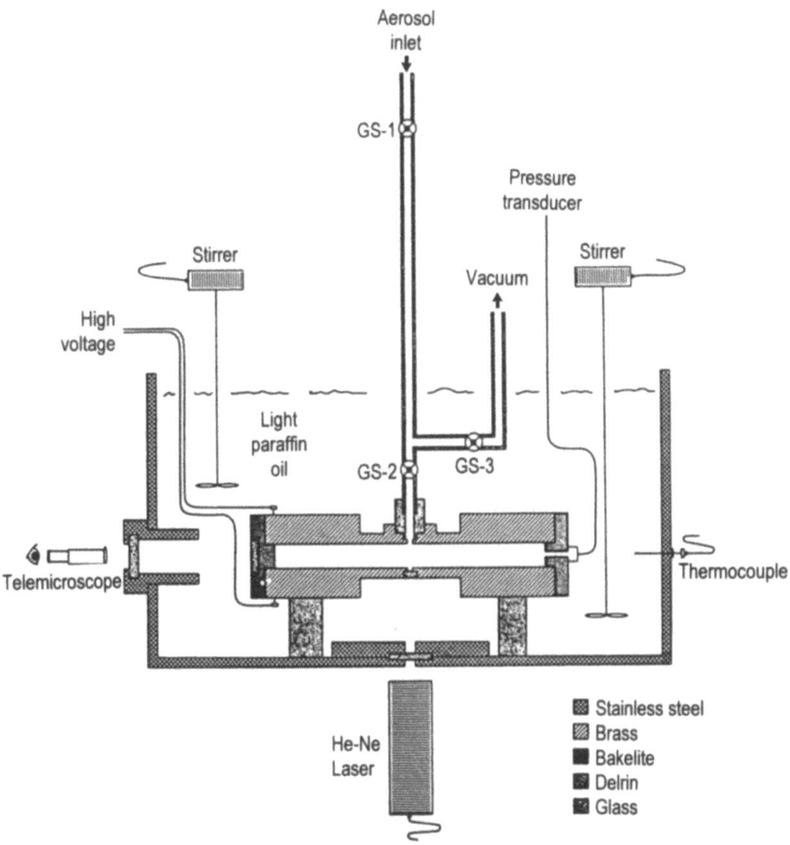

Fig. 1. Improved Millikan apparatus for the laboratory measurement of atmospheric aerosol density [4] "Reprinted from J. Aerosol Sci., 16:57, 1985, with kind permission from Elsevier Science Ltd."

particle surface with values between 0.70 and unity: l_0 is the mean free path of air at pressure p_0. After determining the diameter of the particle using the above equation, and the particle mass from the value of the balancing electrical field, the density of the particle can be calculated. This approach is accurate for spherical particles. For non-spherical particles the dynamic shape factor, k, must be introduced in the equation and a similar approach can be followed to determine the particle characteristics, including k [4].

Millikan cells, while very accurate, are not practical for the measurement of the density of large numbers of atmospheric aerosols. An alternative method for measurement of particle density is that of the Differential Mobility Analyzer-Impactor Density Measurement System [6]. The Differential Mobility Analyzer (DMA) is used to select particles of predefined size from a polydisperse aerosol. At the exit of the DMA the aerosol flows through a single stage MOUDI impactor having four orifice plates. This technique is capable of measuring particle densities with errors less than 8% and can be used to provide direct measurements of ambient particle densities in the submicron size range.

Deliquescence

Let us consider initially the growth an aerosol particle consisting of a single component, e.g. NaCl. At very low relative humidities atmospheric aerosol particles are solid. As the ambient relative humidity increases the particles remain solid until the relative humidity reaches a threshold value characteristic of each salt, at which the substance absorbs atmospheric moisture and produces a saturated aqueous solution. The relative humidity at which this phase transition occurs is called relative humidity of deliquescence (DRH). Further increase of the ambient relative humidity leads to additional water condensation onto the salt solution to maintain thermodynamic equilibrium. On the other hand, as the relative humidity over a salt solution decreases, evaporation of water occurs. However, the solution generally does not crystallize at the deliquescence relative humidity, as thermodynamic equilibrium requires, but may remain supersaturated until a much lower humidity, at which crystallization finally occurs [7–9]. This hysteresis phenomenon with different deliquescence and crystallization points is illustrated schematically in Fig. 2. The relative humidities of deliquescence for a few inorganic salts, which are common constituents of ambient aerosols, are given in Table 1.

The deliquescence behavior of atmospheric aerosols can be described using thermodynamic principles. Water equilibrium between the gas and aerosol phases at the point of deliquescence requires that the relative humidity of deliquescence of a salt, DRH, satisfies

$$DRH = \alpha_{ws} \tag{2}$$

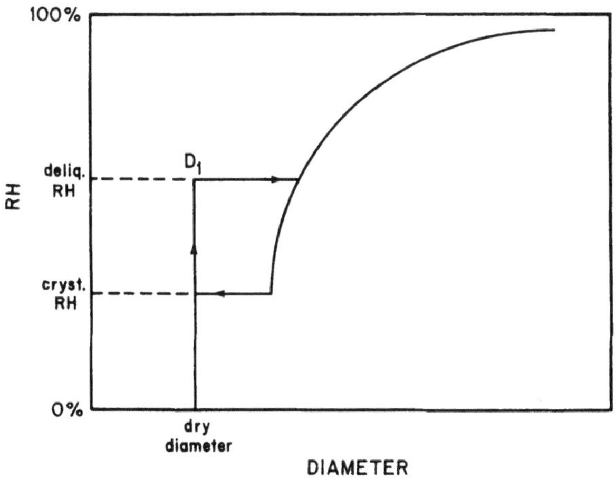

Fig. 2. Variation in particle size with relative humidity "Reprinted from J. Atmos. Environ., 23:1601, 1989, with kind permission from Elsevier Science Ltd."

Table 1. Relative humidity of deliquescence of electrolyte solutions at T = 298 K

Salt	DRH
KCl	0.842
Na_2SO_4	0.842
NH_4Cl	0.800
$(NH_4)_2SO_4$	0.799
NaCl	0.753
$NaNO_3$	0.743
$(NH_4)_3H(SO_4)_2$	0.690
NH_4NO_3	0.618
$NaHSO_4$	0.520
NH_4HSO_4	0.400

where *DRH* is in the 0–1 scale, and a_{ws} is the water activity of the saturated solution of the salt at that temperature. The water activity values can be calculated from thermodynamic arguments using aqueous salt solubility data [10–12].

Relative Humidity of Deliquescence of Multicomponent Aerosols

Multicomponent aerosol particles exhibit similar behavior to single component salts. As the ambient relative humidity increases the salt mixture is solid, until the relative humidity reaches the deliquescence point of the mixture, at which the aerosol absorbs atmospheric moisture and produces a saturated solution. For educational purposes, let us consider a solution of two electrolytes in equilibrium in the atmosphere. The change of the DRH of a single solute aqueous solution when another electrolyte is added can be calculated using the Gibbs-Duhem equation [13]:

$$n_1 d\mu_1 + n_2 d\mu_2 + n_w d\mu_w = 0 \quad (3)$$

where n_1, n_2, and n_w are the number of moles of electrolytes 1, 2 and water respectively, while μ_1, μ_2, and μ_w are the corresponding chemical potentials. Let us assume that initially electrolyte 1 is in equilibrium with solid salt 1 in a solution that does not yet contain electrolyte 2. As electrolyte 2 is added to the solution, the chemical potential of electrolyte 1 does not change, because it remains in equilibrium with its solid phase. Thus $d\mu_1 = 0$ in the above equation. The chemical potentials of electrolyte 2 and water can be expressed in terms of their activities by the equation

$$\mu_i = \mu_i^0 + RT \ln\alpha_i \tag{4}$$

where μ_i^0 is the standard chemical potential of species i. Combining the last two equations and taking into consideration that $n_2/n_w = M_w m_2/1000$ where m_2 is the molality of electrolyte 2 and M_w is the molecular weight of water, we have the following equation:

$$m_2\, d\ln\alpha_2 + \frac{1000}{M_w} d\ln\alpha_w = 0\,. \tag{5}$$

Integration of the last equation from $m_2' = 0$ to $m_2' = m_2$ gives

$$\ln\frac{\alpha_w(m_2)}{\alpha_w(0)} = -\frac{M_w}{1000}\int_0^{m_2} \frac{m_2'}{\alpha_2}\frac{d\alpha_2}{dm_2'} dm_2'\,. \tag{6}$$

Wexler and Seinfeld [13] have proven that the integral in the right hand side of the last equation is greater than zero and therefore $a_w(m_2) < a_w(0)$. Hence the activity of water decreases as electrolyte 2 is added into the system, until the solution becomes saturated in that electrolyte too. The aerosol is exposed to the atmosphere and therefore its DRH also decreases (Eq. 2).

The solution of Eq. 6 for the case of the system NH_4NO_3–NH_4Cl is depicted graphically in Fig. 3. Four relative humidity regimes can be defined as below.
(a) r.h. > DRH(NH_4Cl) > DRH(NH_4NO_3). When the relative humidity exceeds the relative humidity of deliquescence of both compounds (regime a), the aerosol is an aqueous solution of NH_4^+, NO_3^- and Cl^-.
(b) DRH(NH_4Cl) > r.h. > DRH(NH_4NO_3). If the aerosol contains enough NH_4NO_3 so that the aerosol composition is to the right of line 1 the aerosol is an aqueous solution of the corresponding ions. In this regime the addition of NH_4NO_3 to the aerosol results in the complete dissolution of NH_4Cl. If there is not enough NH_4NO_3 and the aerosol composition is to the left of line 1 the aerosol consists of solid NH_4Cl in equilibrium with an aqueous solution of NH_4^+, NO_3^- and Cl^-.
(c) DRH(NH_4Cl) > DRH(NH_4NO_3) > r.h. > DRH*. In this regime the ambient r.h. is below the DRH of both salts and one would expect that the aerosol would be solid. However, the interaction between the two electrolytes results in three different subregimes of behavior. When the aerosol composition is to the right of line 2 (NH_4NO_3 is more than 80%) the aerosol consists of solid NH_4NO_3

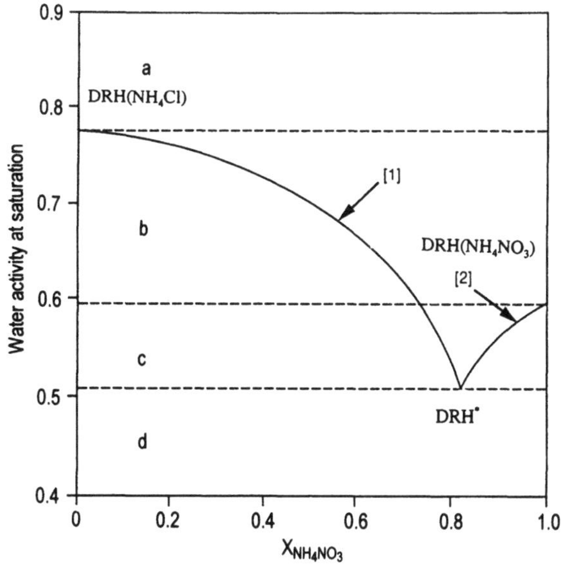

Fig. 3. Water activity at saturation of an aqueous solution of NH_4NO_3 and NH_4Cl "Reprinted from J. Atmos. Environ., 25A:2731, 1991, with kind permission from Elsevier Science Ltd."

in equilibrium with an aqueous solution of NH_4^+, NO_3^- and Cl^-. If the aerosol composition is to the left of line 1, the aerosol consists of solid NH_4Cl in equilibrium with an aqueous solution of NH_4^+, NO_3^- and Cl^-. Finally there is a small third regime between lines 1 and 2 where there is no solid phase and the aerosol consists exclusively of an aqueous solution of NH_4^+, NO_3^- and Cl^-.
(d) DRH* >r.h. If the relative humidity is below DRH* = 0.51 the aerosol consists of a mixture of the two solid salts.

The above example shows that in general one cannot define a single relative humidity of deliquescence for a NH_4NO_3–NH_4Cl (or any other multicomponent) aerosol. The DRH will depend strongly on the aerosol composition and will vary between the highest single component DRH to a DRH* that is lower than the lowest single component DRH.

Temperature Dependence of the DRH and Solubility

Both the relative humidity of deliquescence and the solubility of electrolytes are functions of the ambient temperature. The solubility dependence on temperature is described [14] by

$$\left[\frac{\partial \ln(m_s^2)}{\partial T}\right]_p = \frac{H_o - h_s}{RT^2} \tag{7}$$

where m_s is the molality of the electrolyte in a saturated solution, H_o is the molal enthalphy at infinite dilution and h_s is the enthalpy of the electrolyte as a solid.

Table 2. Temperature dependence of the relative humidity of deliquescence

	Temperature (K)		
Electrolyte	273	298	323
$(NH_4)_2SO_4$	0.816	0.799	0.782
$NaNO_3$	0.800	0.743	0.673
KCl	0.877	0.842	0.802

The difference $(H_o - h_s)$ is the enthalpy of crystallization. Integrating the above equation one obtains the temperature dependence of the aqueous solubility of an electrolyte

$$\ln \frac{m_s(T)}{m_s(T_o)} = -\frac{H_o - h_s}{2R}\left(\frac{1}{T} - \frac{1}{T_o}\right) \tag{8}$$

where the reference temperature T_o is usually 298.15 K.

The temperature dependence of the relative humidity of deliquescence is given [13] by

$$\left[\frac{\partial \ln(\alpha_w)}{\partial T}\right]_p \simeq \frac{M_w m_s L_s}{1000RT^2} \tag{9}$$

where L_s is the latent heat of crystallization of the electrolyte from a saturated solution of molality m_s. An assumption associated with this equation is that the latent heat of vaporization of water is not significantly affected by the presence of the electrolyte. The integrand of Eq. (9) is difficult to evaluate, because the molality of the saturated solution is also a function of temperature (see Eq. 8). For small changes in the ambient temperature, m_s may be assumed to be constant, and the temperature dependence of the water activity can be approximated by [13, 14].

$$\ln \frac{\alpha_w(T)}{\alpha(T_o)} \simeq -\frac{M_w\, m_s\, L_s}{1000R}\left(\frac{1}{T} - \frac{1}{T_o}\right) . \tag{10}$$

Thermodynamic data for most atmospheric aerosol electrolytes can also be found in [13–16]. Table 2 presents the DRH dependence of some salts on temperature.

Aerosols as Cloud Condensation Nuclei (CCN)

Our discussion up to this point has focused on the behavior of atmospheric aerosol for relative humidities less than 100%. When the atmosphere becomes supersaturated (the relative humidity exceeds 100%) some aerosol particles become activated and grow to become droplets, thus forming a cloud or a fog. Let us consider a particle of diameter D_p containing n_w moles of water and n_s moles of solute. The mole fraction of the solvent (water) is equal to

$$x_w = \frac{n_w}{n_w + n_s} \tag{11}$$

and the total volume of the particle is:

$$\frac{\pi}{6} D_p^3 = n_w v_w + n_s v_s \tag{12}$$

where v_w and v_s are the molar volumes of water and solute, respectively. Assuming that the ambient r.h. exceeds the DRH of the solute and that the solute mass remains constant, the diameter of the particle is only a function of its water content.

The ratio of the vapor pressure of water over a particle of diameter D_p, p_w, to the vapor pressure of water over the flat surface of a solution with the same composition, p_w^f, is given by the Kelvin equation [17, 18]

$$\ln \frac{p_w}{p_w^f} = \frac{4 v_w \sigma}{RTD_p} \tag{13}$$

where σ is the surface tension of the solution. The vapor pressure of the solvent over a flat surface is given by

$$p_w^f = p_w^o \gamma x_w \tag{14}$$

where p_w^o is the vapor pressure of water over a flat pure water surface and γ is the water activity coefficient. Combination of the last two equations leads to

$$\ln \frac{p_w}{p_w^o} = \frac{4 \sigma v_w}{RTD_p} + \ln\gamma + \ln x_w \ . \tag{15}$$

Assuming that the solution is ideal ($\gamma \approx 1$) and replacing x_w in the previous equation one obtains

$$\ln \frac{p_w}{p_w^o} = \frac{4 \sigma v_w}{RTD_p} - \ln\left(1 + \frac{n_s v_s}{\pi D_p^3/6 - n_s v_s}\right) . \tag{16}$$

The left hand side of this equation is the natural logarithm of the ratio of the vapor pressure of water over a particle containing n_s moles of solute, to that over a flat water surface. Expanding the second term of the right hand side in a Taylor series, keeping only the first term of the expansion, and assuming that $n_s V_s \ll \pi D_p^3/6$, Eq. (16) can be rewritten as

$$\ln \frac{p_w}{p_w^o} = \frac{4 \sigma v_w}{RTD_p} - \frac{6 n_s v_s}{\pi D_p^3} . \tag{17}$$

This equation is called the Köhler equation [19–21]. The first term on the right hand side corresponds to the curvature (or Kelvin) effect on the pure solvent vapor pressure, while the second term is the solute contribution to the lowering of the solvent vapor pressure. The combination of the two terms for a NaCl particle with dry diameter of 0.4 μm is presented schematically in Fig. 4. The diameter corresponding to the maximum is called *critical diameter*, D_{pc}, and the maximum supersaturation *critical supersaturation*, S_c. The critical diameter and supersaturations for any aerosol particle are functions of its dry diameter and chemical composition.

Particles with diameter larger than their critical diameter are in unstable equilibrium with their environment. Let us assume that a particle of size larger than its critical diameter is initially in equilibrium with the atmosphere and a small perturbation (e.g. condensation of some water molecules) causes an increase of its diameter. The new water vapor pressure at the particle surface is lower than the original one (Fig. 4); hence additional water will condense on the particle which will continue growing. On the other hand, if some water molecules evaporate, the particle's diameter decreases, and the water vapor pressure at the particle surface will be higher than the atmospheric pressure. As a result the particle will continue to evaporate until it reaches the corresponding equilibrium particle size at the left of the critical diameter (ascending branch). Note that particles with diameters smaller than the corresponding criticial diameter are in stable equilibrium with the surrounding atmosphere. Perturbations in their diameter either due to water condensation or evaporation cause vapor equilibrium pressure changes that oppose the perturbation and the particles return to their original size.

If the aerosol particle consists of an aqueous solution of a salt and also contains a solid insoluble substance, as is usually the case in the atmosphere, a

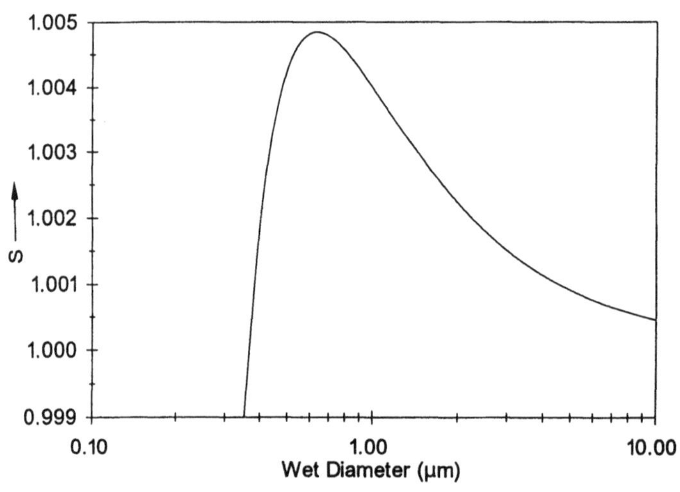

Fig. 4. Supersaturation as a function of the particle diameter for aqueous NaCl droplets at 293 K assuming complete dissociation of the salt in solution for dry diameter 0.04 μm

similar approach can be followed to derive the corresponding Köhler equation, [18]

$$\ln\frac{p_w}{p_w^o} = \frac{(4\sigma v_w / RT)}{D_p} - \frac{(6n_s v_s/\pi)}{D_p^3 - D_{pl}^3} \quad (18)$$

where D_{pl} is the diameter of the insoluble core of the particle. This equation implies that the smaller the water soluble fraction, the higher the supersaturation needed to activate the particle. Figure 5 shows the critical supersaturation necessary for the activation of aerosol particles, as a function of their dry particle diameter. The behavior is similar for both NaCl and $(NH_4)_2SO_4$ particles. The larger the dry particle diameter the lower the supersaturation needed to activate the particle. For a supersaturation of 0.1%, typical of marine stratus clouds, all NaCl and $(NH_4)_2SO_4$ particles larger than approximately 0.1 μm (dry diameter) will be activated and produce cloud droplets. Smaller particles will remain as interstitial aerosol. Aerosol particles which are capable of initiating drop formation are called Cloud Condensation Nuclei (CCN). The initial droplet chemical composition will therefore be dominated by the CCN chemical composition and is expected to vary with droplet size and from droplet to droplet of the same size [22, 23]. The chemical composition of the aerosol also influences the activation process. Smaller aerosol particles can be CCN, while others with larger diameter are not, depending on the hygroscopicity of the particles and their chemical composition.

In summary, at a specific relative humidity, DRH, which depends on the composition of the dry particle, the soluble components of the aerosol deliquesce into an aqueous solution. As the relative humidity increases the aerosol particle grows in size by absorbing water, maintaining thermodynamic equilibrium with the atmosphere. If the relative humidity exceeds the critical supersaturation for that particle, it gets activated and grows rapidly, by absorbing water, into a cloud

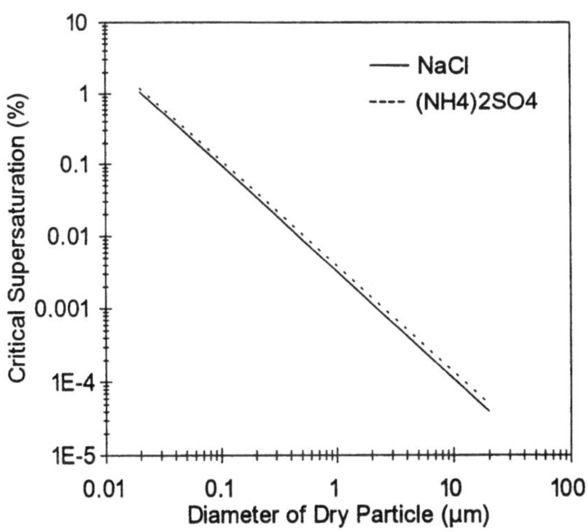

Fig. 5. Critical supersaturation (%) for the activation of NaCl and $(NH_4)_2SO_4$ aerosol particles as a function of their dry diameter

drop. The higher the ambient supersaturation the lower the cutoff diameter of the particles that can be activated, and thus the higher the number of CCN.

A wide range of CCN concentrations has been observed ranging from several thousand per cm^3 [24] to less than 10 cm^{-3} [25]. The importance of this subset of the aerosol population for precipitation formation has been a subject of research for the last 40 years [18, 26–27]. Recently the direct influence of the CCN number concentration on the cloud albedo and the corresponding impact on the planet's radiative balance has been receiving increased attention [27]. Twomey et al. [28–30] suggested that increased emission of anthropogenic particles that can act as CCN could result in an increase of the cloud droplet concentrations and an increase of the cloud albedo causing global cooling. The global cooling effect of anthropogenic CCN, now referred to as the Twomey effect, has recently surfaced as a competing cooling mechanism in global climate models [31–32]. An overview of the various aspects of the CCN/clouds/climate system has been presented by Fouquart and Isaka [33] and Hudson [27].

Chemical Properties of Atmospheric Aerosols

Aerosol particles can also be viewed as atmospheric micro-reactors, i.e., as the medium where heterogeneous chemical reactions such as oxidation of dissolved SO_2 to sulfate, can take place. The importance of chemical processes in atmospheric aerosols has been the subject of a series of investigations. Modeling studies based on laboratory kinetic data have suggested that substantial amounts of sulfate can be produced in the aqueous phase via a variety of chemical reactions. The main oxidation paths, in the case of fog and cloud droplets, are the reactions of aqueous H_2O_2 and ozone with dissolved SO_2:

$$H_2O_2(aq) + SO_2(aq) \rightarrow H_2SO_4$$

$$O_3(aq) + SO_2(aq) + H_2O \rightarrow H_2SO_4 + O_2 \ .$$

Despite the well established importance of these reactions for clouds or fogs, they are not expected to play as important a role in the aerosol phase for the following reasons.

1. The liquid water content of atmospheric aerosols is at least three orders of magnitude smaller than the liquid water content of a typical cloud or fog. Therefore, the effective size of the "aerosol reactor" is much smaller than the size of a typical "cloud/fog aqueous-phase reactor".
2. SO_2 is a weak acid and its aqueous-phase solubility decreases dramatically with the aqueous-phase pH. Most of the aerosol particles are acidic and the resulting aerosol-phase concentrations of SO_2 are expected to be extremely low.

3. Recent experimental work [34] has shown that the presence of organic species in the particulate phase may inhibit reactions like the catalytic oxidation of SO_2 by Fe and Mn. This inhibition can be attributed to the presence of species that act as complex formers for iron.

Even if kinetic calculations indicate that the gas and aqueous phase oxidation pathways are the predominant means for sulfate production in the atmosphere [35–36], there have been indications that some production of sulfate does take place in the aerosol phase. Significant amounts of sulfate can be produced in alkaline aerosol particles, e.g. the fresh sea-salt aerosol in the marine atmosphere [37–38]. Analysis of experimental data collected during the Coordinated Air Sea Experiment / Western Atlantic Ocean Experiment (CASE / WATOX) has shown that homogeneous oxidation of SO_2 could not explain the observed sulfate concentrations, especially in the large sea-salt particle mode. Luria and Sievering [37] explained quantitatively the observed sulfate formation on sea-salt aerosol by the heterogeneous reaction of dissolved SO_2 with dissolved O_3. Despite this finding, the chemical behavior of inorganic and organic species in the aerosol phase remains practically unknown. The oxidation rates that have been used in the cases of dilute fog and cloud droplets may not be applicable for the concentrated aerosol solutions, while other catalytic reactions could play an important role in the particulate phase. Because of the present lack of understanding of the chemical processes in aerosol particles, most researchers assume that aerosol particles consist of species that are either emitted directly into the atmosphere or are produced in the gas phase by photochemical reactions and are then transported to the particulate phase by either nucleation or condensation.

Optical Properties of Atmospheric Aerosols

Visibility degradation is the most readily perceived impact of air pollution. Atmospheric particulate matter is the major contributor in the visibility problems encountered today in several regions of the world (Los Angeles, Tokyo, Mexico City, Athens, etc.). When aerosol particles interact with light, two different types of processes can occur. The energy received by the particle can be reemitted in all directions, but with a direction depending intensity (light scattering). Alternatively, the radiation energy can be transformed into other forms of energy, such as heat or chemical energy (light absorption).

The quantity that describes the scattering and absorption properties of a medium is the complex refractive index, m. The complex refractive index consists of a real part, n, and an imaginary part, κn, $m = n - n\kappa i$. The real refractive index n of a medium for a certain wavelength is equal to the ratio of the wavelength in free space to the wavelength inside the medium. The imaginary part of m, κn, is the damping factor of the wave, and κ is called the index of absorption or the index of attenuation. The imaginary part of the refractive index is zero for

Table 3. Refractive index of some common materials at λ=0.589 μm

Material	Index of Refraction
Vacuum	1.0
Water Vapor	1.00025
Air	1.0003
Water	1.33
Ice (water)	1.31
NaCl	1.55
SiO_2(quartz)	1.55
Iron	1.51−1.63i
$(NH_4)_2SO_4$	1.53
Soot	1.96−0.66i
Model Aerosol	1.54−0.015i

nonabsorbing particles. The index of absorption of a medium is a measure of the absorbed intensity, as described by the Beer-Lambert Law [39]:

$$I = I_0 \exp\left(-\frac{4\pi\kappa}{\lambda} z\right) \tag{19}$$

Where the intensity of the radiation decreases from I_0 to I over the path length z inside the medium. Refractive indices for various materials of interest are given in Table 3.

Scattering and Extinction Coefficients of Single-Component Particles

Scattering by particulate matter of sizes comparable to the wavelength of visible light is mostly responsible for the observed atmospheric visibility degradation. Scattering and absorption of light are complex functions of both the composition and size distribution of atmospheric aerosol [40–44]. The fraction of the incident light intensity I scattered and absorbed per unit length s is defined as the extinction coefficient:

$$b_{\text{ext}} = -\frac{1}{I}\frac{dI}{ds} \,. \tag{20}$$

The extinction coefficient consists of two parts, corresponding to scattering and absorption [45]:

$$b_{\text{ext}} = b_{\text{sca}} + b_{\text{abs}} \,. \tag{21}$$

The scattering and extinction coefficients of homogeneous, single species, spherical particles are functions of the mass distribution of the particles $f(\mathrm{x})$, the complex refractive index of the particles m, and the wavelength of the incident light λ [40]. Specifically

$$b_{sca}(\lambda) = \int_0^\infty \frac{3}{2\rho D_p} Q_{sca}(m, \alpha) f(x)\, dx \tag{22}$$

$$b_{ext}(\lambda) = \int_0^\infty \frac{3}{2\rho D_p} Q_{ext}(m, \alpha) f(x)\, dx \tag{23}$$

Where D_p is the particle diameter, ρ is the particle density, $x = \ln(D_p/D_{p0})$, D_{p0} being the diameter of the smallest particles, and $a = \pi D_p/\lambda$, a dimensionless size parameter. The quantities Q_{sca} and Q_{ext} are the scattering and extinction efficiencies of a single particle, respectively [39, 46–47]. These efficiencies have been derived theoretically by Mie's solution to Maxwell's equation. The Mie formulae for Q_{sca} and Q_{ext} are

$$Q_{sca} = \frac{2}{\alpha^2} \sum_{k=1}^{\infty} (2k+1)[|a_k|^2 + |b_k|^2] \tag{24}$$

$$Q_{ext} = \frac{2}{\alpha^2} \sum_{k=1}^{\infty} (2k+1) Re[a_k + b_k] \tag{25}$$

where

$$a_k = \frac{\alpha \psi_k'(y)\psi_k(\alpha) - y\psi_k'(\alpha)\psi_k(y)}{\alpha \psi_k'(y)\zeta_k(\alpha) - y\zeta_k'(\alpha)\psi_k(y)} \tag{26a}$$

$$b_k = \frac{y \psi_k'(y)\psi_k(\alpha) - \alpha\psi_k'(\alpha)\psi_k(y)}{y \psi_k'(y)\zeta_k(\alpha) - \alpha\zeta_k'(\alpha)\psi_k(y)} \tag{26b}$$

with

$$y = \alpha m\,. \tag{27}$$

The functions $\psi_k(z)$ and $\zeta_k(z)$ are the Ricatti-Bessel functions defined in terms of the Bessel functions J_k as

$$\psi_k(z) = \left(\frac{\pi z}{2}\right)^{1/2} J_{k+1/2}(z) \tag{28}$$

$$\zeta_k(z) = \left(\frac{\pi z}{2}\right)^{1/2} [J_{k+1/2}(z) + i(-1)^k J_{-k-1/2}(z)]\,. \tag{29}$$

The absorption efficiency can be calculated as the difference of the extinction minus the scattering efficiencies:

$$Q_{abs} = Q_{ext} - Q_{sca}\,. \tag{30}$$

Calculation of the scattering and extinction efficiencies using Mie theory requires estimation of the coefficients a_k and b_k and the evaluation of the infinite sums in Eqs. (24) and (25). The method of Wickramasinghe [48], presented below, converges rapidly and, for most practical cases, requires the summation of less than twenty terms. If one defines

$$A_k(y) = \psi'_k(y)/\psi_k(y) \tag{31}$$

Eqs. (26) become

$$a_k = \frac{\left(\frac{A_k(y)}{m} + \frac{k}{\alpha}\right) Re(\zeta_k(\alpha)) - Re(\zeta_{k-1}(\alpha))}{\left(\frac{A_k(y)}{m} + \frac{k}{\alpha}\right) \zeta_k(\alpha) - \zeta_{k-1}(\alpha)} \tag{32}$$

$$b_k = \frac{\left(A_k(y)m + \frac{k}{\alpha}\right) Re(\zeta_k(\alpha)) - Re(\zeta_{k-1}(\alpha))}{\left(A_k(y)m + \frac{k}{\alpha}\right) \zeta_k(\alpha) - \zeta_{k-1}(\alpha)} . \tag{33}$$

To generate $A_k(y)$ one can use the recurrence relation

$$A_k(y) = -\frac{k}{y} + \frac{1}{\frac{k}{y} - A_{k-1}(y)} \tag{34}$$

with

$$A_0(y) = \cos y / \sin y . \tag{35}$$

For $\zeta_k(\alpha)$, one can use the relation

$$\zeta_k(\alpha) = \frac{2k-1}{\alpha} \zeta_{k-1}(\alpha) - \zeta_{k-2}(\alpha) \tag{36}$$

with

$$\zeta_{-1}(\alpha) = \cos\alpha - i \sin\alpha \tag{37}$$

and

$$\zeta_0(\alpha) = \sin\alpha + i \cos\alpha . \tag{38}$$

The extinction and scattering efficiencies as functions of the dimensionless parameter a for different values of the complex refractive index are presented in Fig. 6. If the imaginary part of the refractive index is zero, the absorption is zero and the extinction efficiency is equal to the scattering efficiency. Both the scattering and the absorption efficiencies each approach limiting values, as a increases. The extinction efficiency approaches 2 for large spheres (e.g. cloud droplets). As the absorption increases, the resonances producing the primary maximum are highly damped and the curve approaches its asymptotic limit sooner than in the zero absorption case.

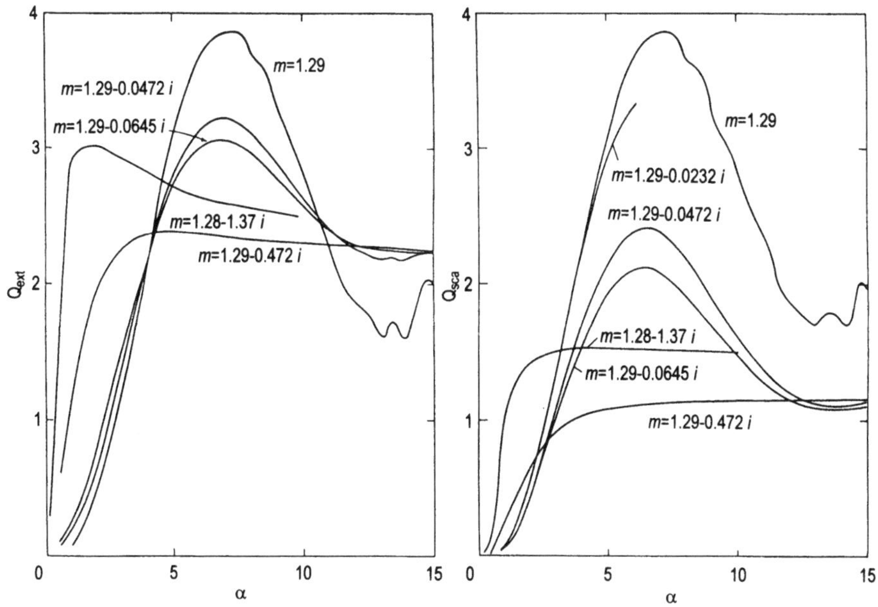

Fig. 6 a, b. Extinction, Q_{ext}, and scattering, Q_{sca}, efficiencies for various refraction indices plotted against the size parameter α "Reprinted From Kerker, M., The scattering of light and other electromagnetic radiation, with kind permission of Academic Press"

Scattering and Extinction Coefficients of Multicomponent Particles

Multicomponent atmospheric aerosol particles are assumed to be spherical with an insoluble core surrounded by an aqueous solution [41–43]. The formulae for the calculation of the efficiencies for such composite spheres are very similar to the Mie formulae for homogeneous spheres described in the previous paragraph. The coefficients a_k and b_k are even more complex functions of the Riccatti-Bessel functions. The reader is referred to Wickramasinghe [48] for the derivation of the coefficients a_k and b_k in this case. An alternative approach is to use Eqs. (25) and (26) with an index of refraction equal to the volume average of the indices of refractions of all the species in the particulate phase.

$$\overline{m}(x) = \left(\sum_{i=1}^{N} m_i \frac{f_i(x)}{\rho_i}\right)\left(\sum_{i=1}^{N} \frac{f_i(x)}{\rho_i}\right)^{-1} \tag{39}$$

where N is the number of species in the particulate phase, m_i, $f_i(x)$ and ρ_i are the index of refraction, the mass distribution function and the density of species i, respectively. A comparison of the theoretical predictions of the scattering coefficients using this approach with the observed values at Mead View of Nevada in the United States [44] is presented in Fig. 7.

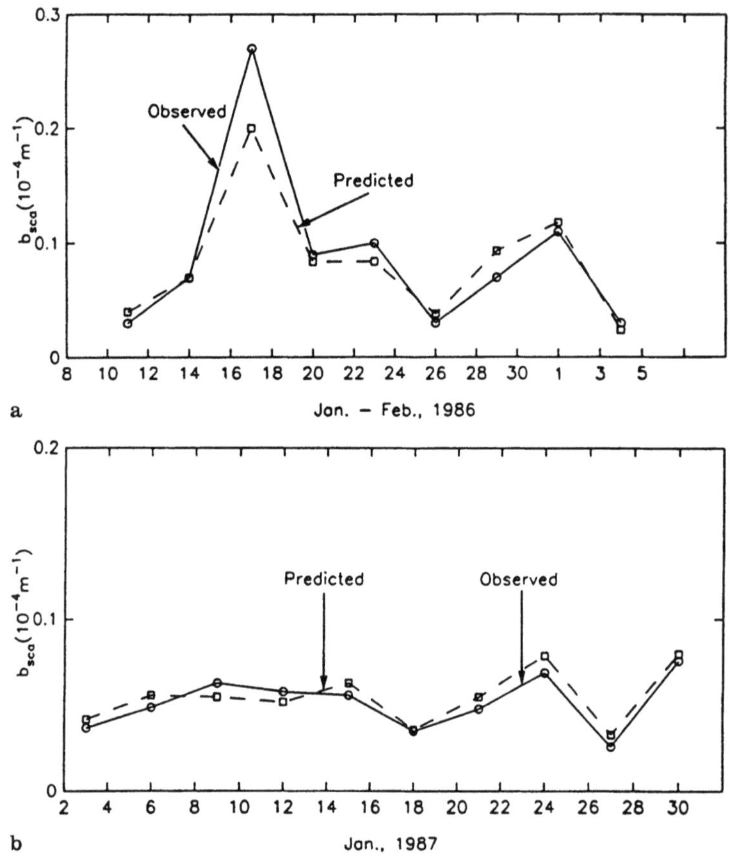

Fig. 7 a, b. Predicted and observed scattering coefficients at Mead View, Nevada, USA: **a** 11 January-4 February, 1986; **b** 3–30 January 1987 "Reprinted from J. Geophys. Res., 94:9937, 1989, with kind permission from the American Geophysical Union "

The magnitudes of the extinction and scattering coefficients of the atmosphere are complex functions of the aerosol composition and atmospheric conditions. A change in ambient relative humidity may cause a substantial increase in the aerosol mass due to water condensation on the aerosol. This increase of the aerosol mass will shift the aerosol distribution towards the coarse particle regime. Coarse particles, as we have seen, are not efficient in scattering light. The overall result is a highly nonlinear dependence of atmospheric scattering and extinction on relative humidity [44]. Figure 8 shows the predicted scattering coefficient as a function of the relative humidity for two cases. In the first case the atmosphere is sulfate rich, while in the second case it is sulfate deficient. In both cases, the scattering coefficient does not depend on the relative humidity, as long as the latter remains below the deliquescence point of the aerosol. In that regime the aerosol is dry and its mass and size distribution does not depend on the

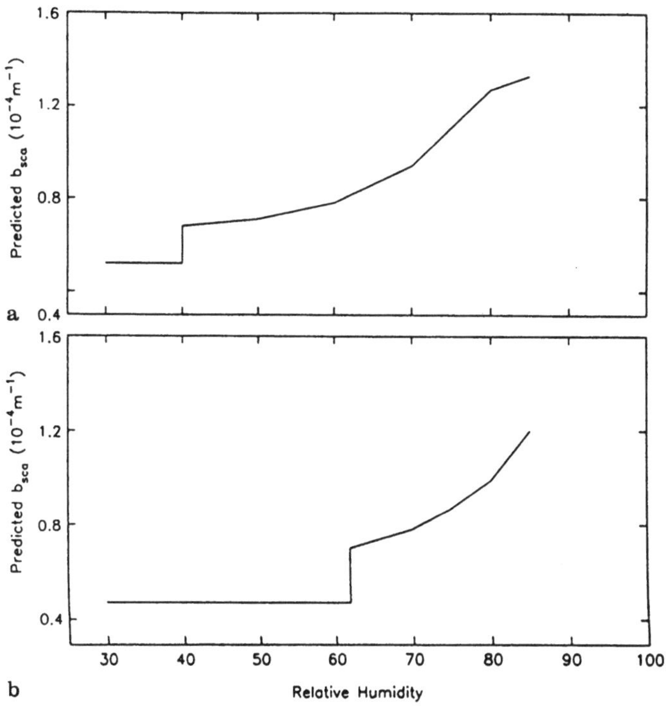

Fig. 8 a, b. Predicted scattering coefficients, as a function of the ambient relative humidity: **a** sulfate rich, ie $H_2SO_4 = 14.0$, $NH_3 = 4.5$, $HCl = 7.0$, $HNO_3 = 5.5$, $OC = 5.0$ and $EC = 3.0$ μg m^{-3}; **b** sulfate deficient, ie $H_2SO_4 = 4.0$, $NH_3 = 4.5$, $HCl = 7.0$, $HNO_3 = 5.5$, $OC = 5.0$ and $EC = 3.0$ μg m^{-3} "Reprinted from J. Geophys. Res., 94:9937, 1989, with kind permission from the American Geophysical Union "

relative humidity. When the relative humidity reaches the deliquescence point of the aerosol the particles absorb water and their mass increases substantially. In both cases in Fig. 8 this results in an increase of the scattering coefficient. Further increase of the relative humidity leads to further water condensation on the aerosol and further increase of the scattering coefficient.

The change of the concentrations of aerosol precursors may or may not affect proportionally the scattering and extinction coefficients. In environments, for example, where there is abundance of both HNO_3 and H_2SO_4, reducing the H_2SO_4 does not necessarily result in significant visibility improvement. This is demonstrated schematically in Fig. 9. For both 80% and 90% relative humidities the total aerosol mass remains relatively unchanged even for a 500% increase of the sulfate concentration. This behavior may be explained with the help of thermodynamic equilibrium. For low sulfate concentrations considerable nitrate exists in the particulate phase neutralizing the available ammonia. As more and more H_2SO_4 is added into the system, sulfates replace nitrate ions in the aerosol phase, and water associated with nitrates is replaced by water associated with sulfates,

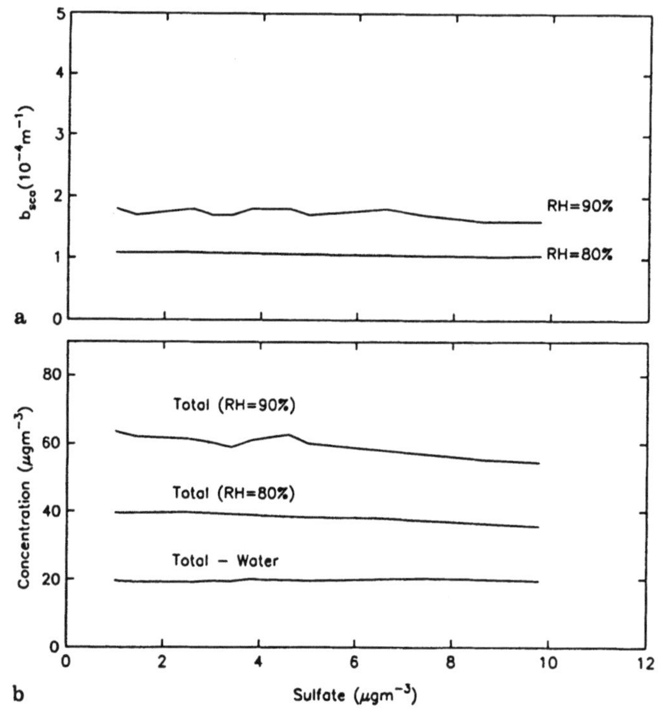

Fig. 9 a, b. Predicted scattering coefficients and total fine and dry aerosol concentrations as a function of the ambient sulfates for 60% and 90% relative humidity and 273 K temperature. The rest of the conditions are: $NH_3 = 4.5$, $HCl = 7.0$, $HNO_3 = 5.5$, $OC = 5.0$ and $EC = 3.0$ μg m^{-3} "Reprinted from J. Geophys. Res., 94:9937, 1989, with kind permission from the American Geophysical Union"

so that the total fine particulate mass remains relatively unchanged. As a result the optical properties of the aerosol do not change substantially either. If, on the other hand, HNO_3 is not available in the atmosphere, sulfates control the particulate phase concentration and reduction of the sulfate concentrations will result in substantial improvement in visibility.

Visibility Degradation – Climate Effects of Atmospheric Aerosols

Prevailing visibility is defined as the greatest distance in a given direction at which it is just possible to see and identify a prominent dark object in the daytime and an unfocused light source at night; and which is attained or surpassed for at least half the horizon circle but not necessarily in continuous sectors [17].

The prevailing visibility is reduced in polluted atmospheres due to scattering and absorption of light by gases and particulate matter.

Another useful quantity is the contrast at a distance x from an object, $C(\mathrm{x})$, which is defined as the relative difference between the light intensity of the target, $I_o(\mathrm{x})$, and the background, $I_b(\mathrm{x})$,

$$C(x) = \frac{I_b(x) - I_o(x)}{I_b(x)} . \tag{40}$$

Assuming that a perfectly black object (absorbs all incident light) is perceived against an ideal white background, $I_o(0)=0$. If we assume that the atmosphere is homogeneous so that scattering and absorption are the same everywhere, that the sky is cloudless and the viewing distance is horizontal, thus ignoring the earth's curvature [17, 49] then the contrast, C, decreases exponentially with distance from the object:

$$C(x) = e^{-b_{ext}x} \tag{41}$$

where the extinction coefficient b_{ext} is the total extinction coefficient, i.e., it accounts for the extinction of light by both gases and particles.

The lowest value of the contrast that allows recognition of an object is called the threshold contrast, and is usually assumed to be equal to 0.02 in visibility calculations. The distance corresponding to the threshold contrast is called visual range, x, and is given by the Koschmeider equation:

$$x = \frac{3.912}{b_{ext}} . \tag{42}$$

In addition to visibility degradation, atmospheric aerosols affect the radiation balance of the earth. Radiative influences of aerosols on climate are distinguished as direct and indirect. Direct radiative influence refers to the change of the solar radiation that reaches the surface of the earth, due to the scattering and absorption of atmospheric aerosol particles [50]. The net effect of aerosols on the radiation balance thus depends on the relative magnitude of the amount of light scattered back into space, vs the amount absorbed by the aerosol. The available experimental and theoretical evidence suggests that scattering of light dominates [50, 51]. Hence, atmospheric particles tend to cool down the atmosphere, thus reducing the effects of accumulation of greenhouse gases. Aerosols, as CCN, can also modify the cloud characteristics (droplet number and size distribution, cloud lifetime, etc.) and therefore the amount of solar radiation scattered by clouds. A substantial fraction of the incident solar radiation. around 27%. is reflected by clouds, which cover on average about 40% of the earth's surface. Sulfate aerosols appear to dominate anthropogenic influences on CCN number concentrations [51].

Visibility Measurements

Visibility instruments usually measure either the energy scattered out of the direct path of the beam or the energy that remains in the beam after it passes through the atmosphere. Two classes of instruments have been developed, the nephelometers which measure the scattered light, and the transmissometers which measure transmitted light.

The integrating nephelometer consists of a light-tight container that is fitted with a light source and a photodetector. The interior of the instrument is painted black and contains baffles, so that the detector is not directly illuminated by the source. As a result the detector only sees the light that is scattered from the light path. Ambient air is drawn through the instrument. The increase in the signal from the detector, with respect to the signal of the clean air, is proportional to the scattering coefficient due to particulate matter. Because of its potential for high accuracy, portability, and low cost, the nephelometer has been widely used for measurements of light scattering coefficients. There are, however, several potential sources of measurement uncertainties.

1. The contribution of coarse particles, $D_p > 5$ μm in diameter, is underestimated, because they tend to deposit at the inlet.
2. The optics of the instrument do not permit measurements of light scattered between 0 and 7 degrees, which is the area in which coarse particles scatter the light.
3. Heating of the instrument may reduce the relative humidity inside the instrument, causing evaporation of water and other volatile species like nitrates and ammonium. This change may distort both the composition and the size distribution of atmospheric aerosols.

Despite these limitations, nephelometers play a major role in measurements of scattering coefficients, especially in cases in which routine measurements are needed.

Transmissometers measure the transmittance of light from an artificial light source over a measured path to a detector. Path lengths of about 15 km are required to obtain accurate data. The average extinction coefficient along the path can be determined directly from the transmittance measurements. Some of the potential problems associated with these instruments are:

- the absence of any definitive field calibration;
- data are not reliable when clouds, rain and fog obscure the sight of path;
- transmissometers are very expensive compared to nephelometers.

The absorption coefficient may be measured by many methods, including photoacoustic spectroscopy, the calorimeter method, the integrating sphere method and filter techniques.

Photoacoustic spectroscopy measures the absorption coefficients of suspended particles in real time [52]. The air stream, from which NO_2 has been removed because it absorbs in the visible, is drawn into an acoustic cell, where it is illuminated by light that is modulated at the resonant frequency of the cell. Light energy absorbed by the particles heats the carrier gas, which expands and then contracts according to the modulation frequency of the light. The associated pressure variation is a sound wave whose intensity can be measured with a microphone. A disadvantage of this method is that because of its complexity it is not suitable for routine measurements.

In the calorimeter method a volume of air containing absorbing substances is irradiated. The heating rate of this volume is compared to a control volume containing the gas without the particles [53].

In the integrating sphere method a sample of aerosol particles, usually collected in an impactor, is placed in an integrating sphere. Both the light scattered and transmitted through the sample reaches the diffusively reflecting surface of the sphere and is scattered in the direction of the photodetector. The signal of the photodetector is proportional to the transmitted plus the scattered light. The difference between a clean substrate and a substrate loaded with particles gives the amount of light absorbed by the particles [53].

In filter techniques the absorption coefficient of atmospheric aerosols is estimated by measuring the absorption of particles collected on filters, e.g., the integrating plate technique [53–54]. In this method an opal glass plate is located between the filter and the optical detector. Because the opal glass is a diffuse reflector, light scattered by particles in the forward direction is detected with the same efficiency as light that enters the glass directly. If backward scattering is small in comparison to absorption, changes in the filter transmittance before and after the collection may be attributed to particle absorption. This class of methods is cheaper, but less accurate, than the photoacoustic spectroscopy method.

References

1. Gray HA, Cass GR, Huntzicker JJ, Heyerdahl EK, Rau JA (1986) Envir Sci Technol 20:580
2. Finlayson-Pitts JB, Pitts JN (1986) Atmospheric chemistry fundamentals and experimental techniques. Wiley, New York
3. Reid RC, Prausnitz JM, Sherwood TK (1977) The properties of gases and liquids. McGraw Hill, New York
4. Allen MD, Raabe OG (1985) J Aerosol Sci 16:57
5. Fuchs NA (1964) The mechanics of aerosols. Pergamon, Oxford
6. Kelly WP, McMurry PH (1992) Aerosol Sci Technol 17:199
7. Cohen MD, Flagan RC, Seinfeld JH (1987) J Phys Chem 91:4563
8. Richarson CB, Spann JF (1984) J Aeros Sci 15:563
9. Junge C (1952) Ann Met 5:1
10. Pilinis C, Seinfeld JH (1987) Atmos Environ 21:2453
11. Cohen MD, Flagan RC, Seinfeld JH (1987) J Phys Chem 91:4575
12. Pilinis C, Seinfeld JH, Grosjean D (1989) Atmos Environ 23:1601
13. Wexler AS, Seinfeld JH (1991) Atmos Environ 25A:2731
14. Stelson AW, Seinfeld JH (1982) Atmos Environ 16:2507
15. Tang IN, Munkelwitz HR (1993) Atmos Environ 27A:467
16. Wagman DD, Evans WH, Parker VB, Schumm RH, Harlow I, Bailey SM, Churney KL, Nuttall RL (1982) J Phys Chem Ref Data 11 Suppl 2
17. Seinfeld JH (1986) Atmospheric chemistry and physics of air pollution. Wiley, New York

18. Pruppacher HR, Klett JD (1980) Microphysics of clouds and precipitation. D Reidel Publishing Company, Dordrecht, Holland
19. Köhler H (1927) Geofysiske Publ 2, No. 6
20. Köhler H (1927) Geofysiske Publ 5, No. 1
21. Köhler H (1936) Trans Farad Soc 32:1152
22. Pandis SN, Seinfeld JH, Pilinis C (1992) Atmos Environ 26A:2509
23. Ogren JA, Clarkson RJ (1992) Tellus 44B:208
24. Herrera JR, Castro JR (1988) J Appl Meteor 27:1189
25. Radke LF, Hobbs PV (1969) J Atmos Sci 26:281
26. Squires P (1956) Tellus 8:443
27. Hudson JG (1993) J Appl Meteor 32:596
28. Twomey S (1974) Atmos Environ 8:1251
29. Twomey S (1977) J Atmos Sci 34:1149
30. Twomey S, Piepgrass M, Wolfe TL (1984) Tellus 36B:356
31. Wigley TML (1989) Nature 339:356
32. Kaufman YJ, Fraser RS, Mahoney RL (1991) J Climate 4:578
33. Fouquart Y, Isaka H (1992) Ann Geophys 10:462
34. Kotronarou A, Sigg L (1993) Environ Sci Technol 27:2725
35. Middleton P, Kiang CS, Mohnen VA (1980) Atmos Environ 14:463
36. Saxena P, Seigneur C (1987) Atmos Environ 21:807
37. Luria M, Sievering H (1991) Atmos Environ 25A:1489
38. Sievering H, Boatman J, Galloway J, Keene W, Kim Y, Luria M, Ray J (1991) Atmos Environ 25A:1479
39. Kerker M (1969) The Scattering of light and other electromagnetic radiation, Academic, San Diego
40. Ouimette JR, Flagan RC (1982) Atmos Environ 16:2453
41. Sloan CS (1983) Atmos Environ 17:409
42. Sloan CS (1984) Atmos Environ 18:871
43. Larson SM, Cass GR, Hyssey KJ, Luce F (1988) Environ Sci Technol 22:629
44. Pilinis C (1989) J Geoph Res 94:9937
45. Friedlander SK (1977) Smoke, dust and haze. Wiley, New York
46. Van de Hulst HC (1957) Light scattering by small particles. Wiley, New York
47. McCartney EJ (1976) Optics of the atmosphere-scattering by molecules and particles. Wiley, New York
48. Wickramasinghe NC (1973) Light scattering functions for small particles with applications in astronomy. Wiley, New York
49. Cohen A (1975) Appl Opt 14:2878
50. Charlson RJ, Wigley ML (1994) Scientific American 266:48
51. Charlson RJ, Schwartz SE, Hales JM, Cess RD, Coakley JA, Hansen JE, Hofmann DJ (1992) Science 255:423
52. National Research Council (1993) 'Protecting visibility in national parks and wilderness areas'. National Academy Press. Washington DC
53. Horvath H (1993) Atmos Environ 27A:293
54. Lin CI, Baker M, Charlson RJ (1973) Appl Opt 12:1356

Chemical Mass Balance

Jozef M. Pacyna and Eva Selin Lindgren*

Chalmers University of Technology, Department of Physics S-14296, Göteborg, Sweden
*NILU, P.O. Box 100, N-2007, Kjeller, Norway

Introduction 126
Chemical Mass Balance as a Tool to Estimate Emission Rates of Selected Trace Metals 127
Chemical Mass Balance used to Assess the Impact of Local Source Emissions on the Surroundings 131
Origin of Pollutants at Remote Locations Studied by Mass Balance Methods 134
Chemical Mass Balance used to Assess a Global Budget of Selected Pollutants 138
Concluding Remarks 141
References 141

Summary

Various applications of the chemical mass balance technique to study source-receptor relationships for air pollutants deposited on particles are described. Examples of the technique used to study the physico-chemical behavior of trace elements during combustion of fossil fuels in electric power plants and wastes in waste incinerators are presented on the basis of measurements carried out in Poland, the United States, Germany, and the Netherlands. Methods to elaborate emission rates of these elements to the atmosphere are described. The application of the chemical mass balance method to assess the impact of emissions from various point sources on the surrounding environment is discussed. Examples of various chemical mass balances for trace elements in the atmosphere and the terrestrial and aquatic environments are presented in order to illustrate the impact of large point sources on the local scale. Studies on the impact of emissions from coal-fired power plants on the local environment in Poland are reviewed. Chemical mass balance used to study the origin of pollutants at remote locations is also described. The application of the method to study the regional scale of air contamination by various pollutants is discussed on the basis of measurements in the Arctic. Finally, the use of chemical mass balance to assess global budgets of selected pollutants is demonstrated. Global budgets of Hg, Pb, other trace elements and organic compounds are discussed using information on their presence in major compartments of the environment.

Introduction

Various techniques based on chemical mass balance are used to study source-receptor relationships for several air pollutants. The concept of the chemical mass balance is based on the mass conservation of each element considered. It can be simply assumed that in a given reservoir defined by characteristic physical, chemical and biological properties, the sum of incoming flows and the contents of the reservoir of an element, e.g. sulphur, is equal to the sum of outcoming flows and the remaining mass of the element. There are various kinds of reservoirs for which the concept of chemical mass balance is useful, such as the whole environment, its individual compartments or even local regions. The sources and sinks of elements being balanced are compared in these cases as rates of creation or destruction of these elements within a given reservoir per unit of time. The mass balance can also be prepared for elements taking part in various processes, e.g. trace metals, during combustion of fuels.

However, the mass balance technique also has some limitations. One obvious shortcoming relates to the fact that it does not contain any information about the chemical transformations that the specific elements are undergoing. All chemical forms of a certain element are treated on the same footing, and thus no dynamic behaviors can be studied, e.g. important processes like gas-to-particle conversion of sulphur compounds as well as the distribution of sulphur over different particle sizes and compounds cannot be identified.

Information will be given on the application of the mass balance technique to study source-receptor relationships for some major pollutants in local, regional and global scales. Examples of chemical mass balance applied to the assessment of the physico-chemical behavior of selected trace metals during combustion of fossil fuels in electric power plants and wastes in waste incinerators are presented. As a result of the assessment, emission rates of these metals are derived. Finally, the impact of emissions of trace metals from power plants and waste incinerators on local environments is discussed.

Chemical mass balance as a tool is used together with information on physical and chemical characteristics of the aerosol to study the origin of pollutants. Long range transport of trace metals from source regions in Eurasia to receptors in the Arctic will be discussed.

Global budgets of mercury and lead will be discussed on the basis of their content in major compartments of the environment. Historical records on production and use of these metals will be related to time-resolved information on their emission fluxes.

Chemical Mass Balance as a Tool to Estimate Emission Rates of Selected Trace Metals

Emission data, which form the basis for environmental policy strategies and input information for the environmental models, can be obtained either by concentration measurements of a given compound in the flue gas or by calculation. The former method is regarded as more accurate although, not always preferable due to the high cost of some of the measurements or problems in collecting a representative sample. In such cases, emission data are estimated mostly with the help of emission factors or other methods which are based on the chemical mass balance for emission generating processes. An example of such a case is the estimation of emissions of volatile trace metals during combustion of fuels or incineration of wastes.

Coal contains more than 60 trace elements which are brought to the combustion burner as coal impurities. At a typical peak temperature in the burner of about 1550°C the volatile trace elements in the coal ash evaporate [1]. Later, they condense as submicron aerosol particles, or are deposited on the surface of ash particles as the flue gas cools to 370–450°C in the convective heat transfer sections of the power plant. Details on the volatilization-condensation process are available from the literature [2–4]. Most of the elements are present in the residues remaining after the combustion in the same relative proportions as in the feed coal, but are enriched in the ash by a factor equivalent to the material emitted in the flue gas when the coal is burned [5, 6]. An enrichment factor (EF) normalizes the concentration of a trace element in the input and output streams (usually coal and fly ash respectively) to the concentration of an element which is considered non-volatile during combustion (e.g. Si or Al):

$$\mathrm{EF} = [(\mathrm{C})\mathrm{output} \,/\, (\mathrm{E})\mathrm{output}] \times [(\mathrm{E})\mathrm{coal} \,/\, (\mathrm{C})\mathrm{coal}] \,.$$

C is the concentration of the trace element and E is the concentration of the normalizing element in the output streams (bottom ash, fly ash, or stack dust) and the coal input streams.

The elements As, Cd, Ga, Ge, Pb, Sb, Sn, Te, Tl, and Zn have the highest values of enrichment factors (well above 1.0) and this enrichment increases with decreasing particle size in the ash streams. Other elements such as Eu, Hf, La, Mn, Rb, Sc, Sm, Th, and Zr are more evenly distributed between different size fractions of ash. These elements do not volatilise easily at the combustion temperatures and, together with the intermediate elements such as Ba, Be, Bi, Co, Cr, Cs, Cu, Mo, Ni, Sr, Ta, U, V, and W, form bottom ash and fly ash particles (a review in [6]). The enrichment factors of these elements are between 0.7 and 1.0. Hg and halogens together with portions of B and Se may remain in the gas phase during the passage through the plant and their enrichment factors in ashes are well below 1.0. The relative distribution of trace elements between residues and the flue gases for the 2000 MW Turow power plant in Poland is presented in Fig. 1 [1].

Failure to achieve closure in mass balance calculations for some elements in Fig. 1 was an indication that about 95% of Hg and 60% of Se is present in the vapor phase. The two gases are completely released from the stack to the atmosphere because no emission control equipment for gases was installed in this power plant.

Emission factors for enriched trace elements were calculated on the basis of results of their mass balances between residues (fly ash and bottom ash) and the flue gas passing the dust control installation, and their content in coal.

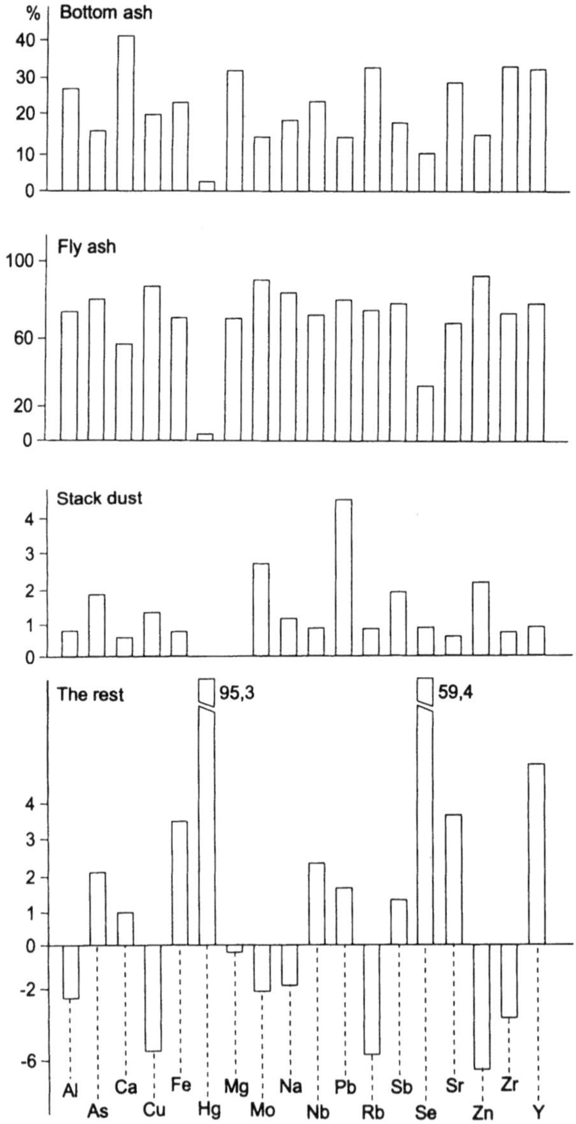

Fig. 1. Percentage distribution of trace elements between the bottom ash, fly ash, and stack dust measured in the 2000 MW Turow power plant in Poland [1]

Similar studies of chemical mass balance for trace elements between residues and the flue gases have been carried out in the Netherlands [7], the United States [8], and Germany [9]. The results of the Dutch study in the Gelderland power station are presented in Fig. 2. Good agreement with the results of earlier studies in Poland can be observed, particularly for Hg.

Chemical mass balance is more complicated when control equipment to reduce contamination of the flue gases is installed. This applies particularly to the reduction of the gaseous portion of trace elements through installation of the flue gas desulfurization (FGD) equipment. Distribution of Hg within various streams of the wet FGD system studied by Gutberlet [9, 10] is presented in Fig. 3. As much as 70% of Hg in flue gases remains in the gas stream after passing through the FGD. In general, the trace element content of the gypsum, a product of desulfurization, and sludge are influenced by the solubility of the elements upstream, the settling behavior of solids in the thickener and the waste water volume. In the waste water cleaning plant, the solubility of the trace elements is decreased by increasing the pH to the range of 8.0 to 9.5 so the waste water contains only a very small amount and the elements are concentrated in the waste sludge at low levels (e.g. 6 ppm for Hg).

Chemical mass balance studies were also performed on solid waste incineration, although information on the input of trace elements with wastes to be incinerated is usually less accurate than the information on the input with coals to be burned in power plants. An example of percentage distribution of some trace elements in incinerator effluent streams in the United Kingdom is shown in Table 1 [11]. The presented values are means of average metal distribution as found by the Warren Spring Laboratory at 13 plants. As they are averages, the

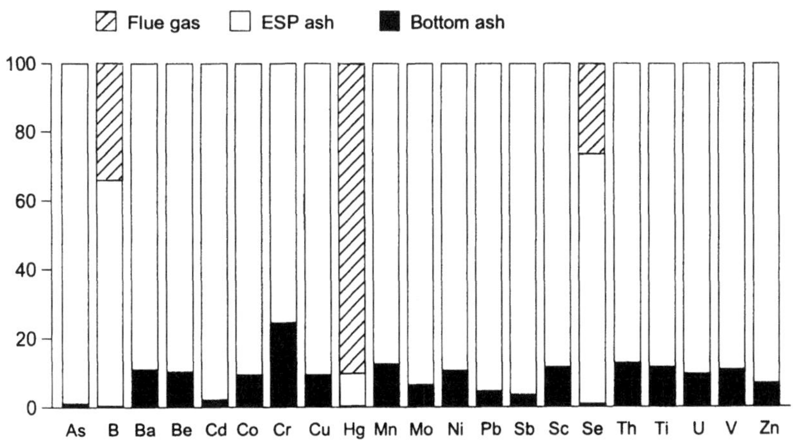

Fig. 2. Percentage distribution of trace elements between the bottom ash, fly ash collected at the electrostatic precipitator (ESP), and flue gas measured in the Gelderland power station in the Netherlands [7]

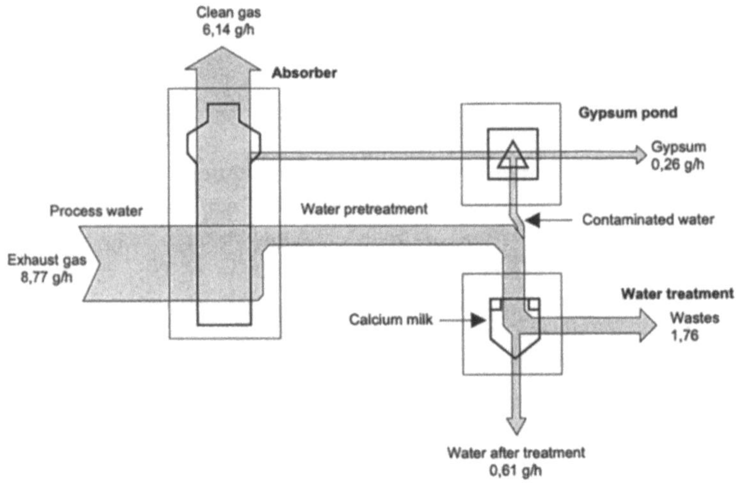

Fig. 3. Distribution of Hg within various streams of the wet flue gas desulfurization system, as presented in [9]

Table 1. Mean percentage distribution of metals in incinerator effluent streams [11].

	Coarse emitted particles	Fine emitted particles	Vapor phase	Emitted subtotal	Arrested fly ash	Grate residue
Pb	3.1	6.3	NA	9.4	19.6	70.0
Cu	1.2	0.6	NA	1.8	4.8	94.1
Cd	5.1	10.6	NA	15.3	34.6	54.8
Cr	2.6	0.5	NA	3.0	9.4	88.8
Ni	2.1	0.6	NA	3.0	8.3	90.3
Hg	1.6	0.0	87.5	88.6	2.6	8.7

percentage distributions may not add up to exactly 100%. Almost 90% of Hg was emitted from the stack, the vast majority of it in the vapor phase. Around 15% of Cd and 10% of Pb was found in the emitted particles, with more in the fine fraction than in the coarse. By contrast, only about 2–3% of Cu, Cr, and Ni were emitted through the stack, with most of this in the coarse fraction.

Methods to calculate emission factors on the basis of results of the chemical mass balances are quite straightforward. Information on the amounts of trace elements in flue gas after the FGD and liquid waste is used, together with the information on the input of trace elements and total amount of coal, to estimate emission factors to the atmosphere and aquatic environment, respectively. Information on the amount of coal burned can be replaced by the information on the amount of electricity produced from a given amount of coal. The results from measurements of trace element concentrations in fly ash, bottom ash, and gypsum are the basis for calculating enrichment factors of these elements in solid wastes.

The major problem in preparing useful chemical mass balances for combustion processes is lack of data. Although the number of available data points for plant emissions of various chemical species is often large, the number of paired data sets – inlet and outlet – on a given control device or part of technology is rather sparse [8].

The chemical mass balances are also used to estimate atmospheric emissions of S compounds from combustion operations and non-methane volatile organic compounds (NMVOCs), particularly from solvent use [12]. It can be concluded that as far as the emission estimation is concerned. the chemical mass balances are the most complete for sulfur compounds, followed by trace elements on particles. The balances for NMVOCs and gaseous trace elements are far less complete due to lack of accuracy of available paired data sets.

Chemical Mass Balance used to Assess the Impact of Local Source Emissions on the Surroundings

Receptor models were developed in the early 1970s in an attempt to identify the source(s) of particulate matter in large urban areas and to quantify the amount of particulate matter emitted from the source(s) [13, 14]. The chemical mass balance method is based upon the premise that the emissions characteristics, in terms of chemical and elemental composition as well as physical size and morphology, of various source types are different enough that one can identify their contributions by measuring the characteristics in samples collected at a receptor site. Thus, an important first step in the application of chemical mass balance model to apportion the sources of air pollutants measured in a specific urban area is to define an emissions inventory of the number and source types of the important sources of air pollution. These models assume that the compositions of all contributing source types are known. This is often not the case either because the sources are not easily characterized or because the source classes have widely varying compositions [15]. This is an important limitation of the chemical mass balance method in that the lack of specific source profile information for the pollutants of interest prevent this approach from being applied. Emissions data have often been sparse in the past but the situation is improving.

An example of the application of the chemical mass balance technique to assess the impact of emissions from a given point source on the quality of the environment in the surroundings is given below. The 2000 MW Turow power plant in Poland is the only major source of several trace elements in the vicinity of several tens of kilometers [1]. The contamination of air, water, soils, and plants by As, Cu, Mo, Pb, Sb, Zn, Hg, and Se in the area of about 30 km radius from the power plant was a subject of extensive measurements. As a result, a model of the environmental migration of the studied elements was developed. The chemical mass balance formed the basis for the model which is presented in Fig. 4 [1]. The following discussion of the model results was offered. The transport of trace elements from the atmosphere to other compartments of the environment occurs through dry and wet deposition. For example, this deposition results in the 50% input of As to water, while the rest comes from soil leaching. Concerning Pb almost 90% of the input to the surface water in the area arrives with atmospheric deposition. The contamination of plants from the air is almost equal to the input from soil. Obviously there are differences in these estimates related to physico-chemical and mechanical properties of soils, type of vegetation, the chemical composition of water, soil, soil solution, and the atmosphere, etc. The results in Fig. 4 should be regarded as one of the first and rough assessments of the trace element migration through the local environment.

Further development of local budgets of pollutants, prepared on the basis of chemical mass balances, was made for selected trace elements or specific features, such as budgets over time. Hg was a particular subject of studies in various regions. As an example, a mass balance model for total Hg in the treatment basin of Little Rock Lake, Wisconsin is presented in Fig. 5 (adapted from [16]). It was concluded [17] that total atmospheric Hg deposition (one third dry and the rest wet) readily accounts for the total mass of Hg in fish, water and accumulation in the sediments. It should be noted that only a relatively small fraction of the input, perhaps below 10% can supply the estimated accumulation of Hg in fish. If so, even modest increases in atmospheric mercury loading could lead directly to elevated levels in the fish stock [17].

The above presented results of studies in the United States and the research on the Hg flow for lakes in southern Sweden [18, 19] were used to suggest that atmospheric deposition dominates the flux of Hg to lacustrine systems and the open ocean. Previous studies, such as the one in Poland around the power plant [1], suggested that at least half of the total Hg in the surface water is removed from the atmosphere.

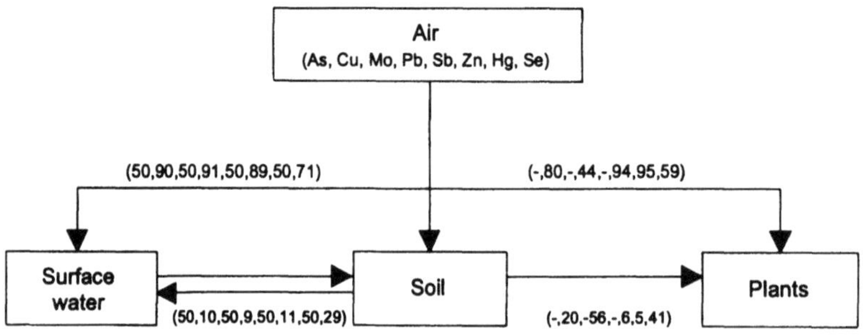

Fig. 4. Results of the migration model for several trace elements emitted from the 2000 MW Turow power plant in Poland [1]. The numbers indicate the percentage contributions of the pollutants from air and soil to surface waters and plants in the surroundings of the power plant. For example, 50% of As in surface waters comes from the air and the remaining 50% from the soil

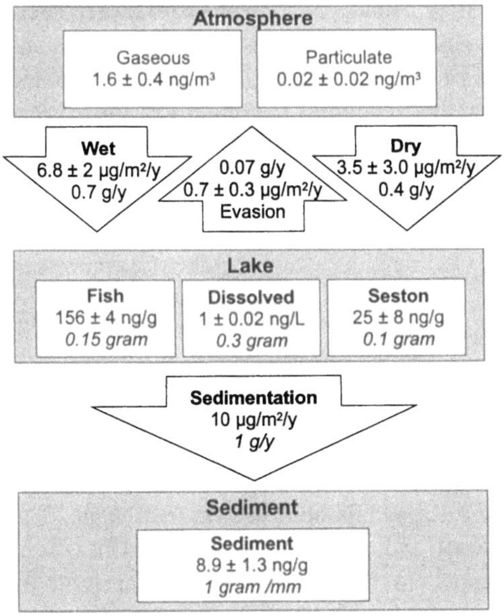

Fig. 5. Mass balance model for Hg in the treatment basin of Little Rock Lake, Wisconsin, adapted from [16]

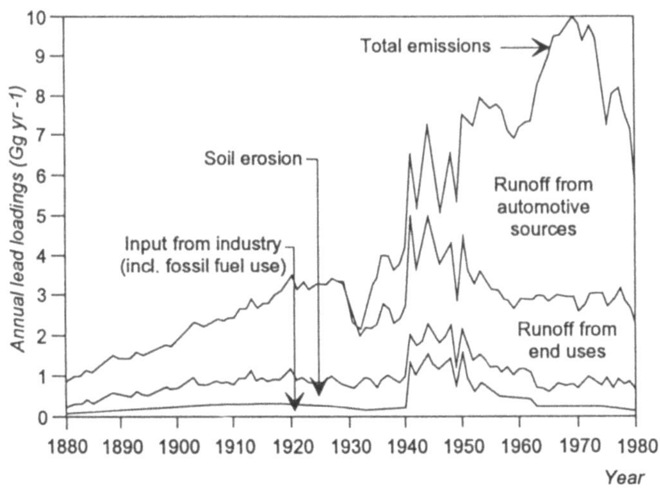

Fig. 6. Annual Pb inputs into the surface waters of the Hudson-Raritan Basin, 1880–1980 [20]

Budgets over time can be regarded as an extension of chemical mass balances constructed for a particular species in a particular regime at a particular time. When such a budget over time is prepared, changes of sources and sinks must be taken into account in addition to current source and sink magnitudes. An example of the budget over time study is mass balance for annual Pb inputs into the surface waters of the Hudson-Raritan basin in the time period from

1880 to 1980 [20]. The spacial scale of this study is somewhere between local and regional scales. The portion of the 1880–1980 Hudson-Raritan Pb emissions ending up in surface waters is indicated in Fig. 6 [20]. The authors also estimated input of atmospheric Pb to soils and then the fraction of that input lost by erosion. A clear decrease of Pb emissions can be noted from the beginning of the 1970s, due to phasing out PbBrCl as an additive to gasoline. Consequently, a decrease of Pb runoff from automotive sources can be observed.

Origin of Pollutants at Remote Locations Studied by Mass Balance Methods

While the chemical mass balance method has primarily applied to urban scale data, Lowenthal and Rahn [21] also applied this technique to their "regional signatures" to apportion the sulfate and trace metals observed on particulate. The application of receptor models to regional and global scale problems has been controversial and has yet to be developed to the level necessary for it to be thought of as definitive in nature. Successful applications of receptor modelling techniques to solve regional scale problems have been documented by Stevens et al. [22] and Dzubay et al. [23]. However, an independent verification for the appropriateness of the trace element ratio approach was performed and indicated that this technique can be quite powerful [24, 25].

The question addressed here is to what extent mass balance for the chemical elements in airborne particles can be used to assess the influence from large source areas or regions on the air quality in remotely located "clean" areas. Many problems can be foreseen in such an approach; the most obvious are differences in dispersion and deposition rates for the different chemical constituents that characterize each element along the route of the air mass from source to receptor. Gas-to-particle conversion as well as growth and evaporation of particles during transport will affect the dry and wet deposition rates. Differences in deposition rates will accordingly change the elemental ratios as the air mass proceeds along its route. Thus it is relevant to ask for how long a period of time the elemental ratios that characterize a certain source will keep their original values. Two examples of this problem will be discussed.

The first example is the V/Ni ratio which was studied in a measurement campaign in 1988 in the vicinity of a refinery [26] located on the west coast of Sweden. The V/Ni ratio in the fine particle fraction (diameter < 3.5 μm) measured at several locations 0.5–10 km from the plant was 2.7 when derived as the slope of a regression line in a V-Ni plot containing all data (70 data points). An earlier study (1987) of the V/Ni ratio in moss around the refinery gave a ratio of 2.8 [27]. In a location some 60 km away from the refinery the V/Ni ratio was 2.7 during the same sampling campaign as the measurements at the refinery (1988) and had the same value (2.7) two years later [28]. Furthermore, in all the plots of the V-Ni data mentioned above there was a linear relationship between the two elements with a high correlation (R >0.8). Thus one can conclude that the V/Ni ratio as a "fingerprint" of that particular refinery stays reasonably constant on a local/regional scale.

The second example concerns the Br/Pb ratio as a fingerprint for leaded petrol. In fresh car exhaust Pb in particular form is emitted as PbBrCl, which gives a mass ratio between Br and Pb equal to 0.39, the "ethyl ratio". Br/Pb ratios have been investigated in many studies [29], and the ratio has shown a great variability. However, from studies made in a relatively unpolluted site on the Swedish west coast in campaigns from 1981 and onwards, it appears that the Br/Pb ratio in winter aerosols transported across the North Sea and Norway is very close to the ethyl ratio [30–32] whereas the summer ratios and the ratios for transport from more polluted regions are lower. Still the Br/Pb ratios from these regions can be used for identification of leaded car exhaust mainly because of the linear relationship between the elements as illustrated in Fig. 7.

A complication in the use of chemical mass balance for source identification is that no receptor site is remote in the sense that all local sources are negligible. Even in non-populated areas, e.g. the Arctic and Antarctic, local sources of sea spray, biogenic and/or soil-derived dust can be identified [33, 34] and are in many cases the main sources of airborne particulates. Thus one may be faced with the problem of identifying weak contributions from distant anthropogenic sources in an aerosol which contains strong local sources of naturally derived particles. This problem is documented in an article by Artaxo et al. [33], in which fine (diameter < 2.0 μm) and coarse (2.0 μm < diameter < 15 μm) particles were sampled

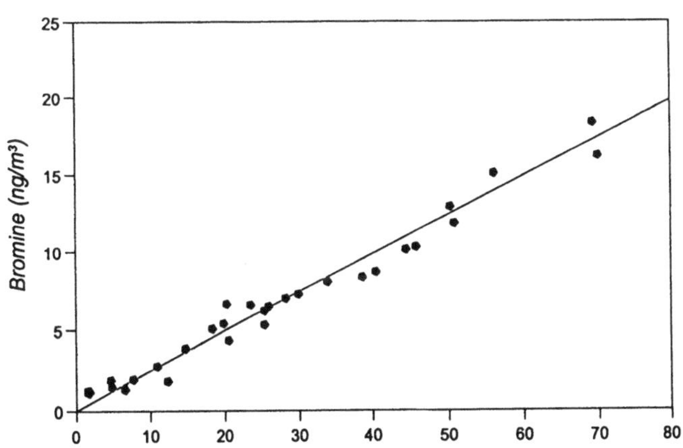

Fig. 7. Scatter diagram for Pb and Br measured on the west coast of Sweden in February 1984. The data include measurements of fine particles at the background and urban sites and coarse particles at the urban sites. Only days with air mass history from the eastern part of Europe were included. The best linear fit to the data gives a Br/Pb ratio of 0.24

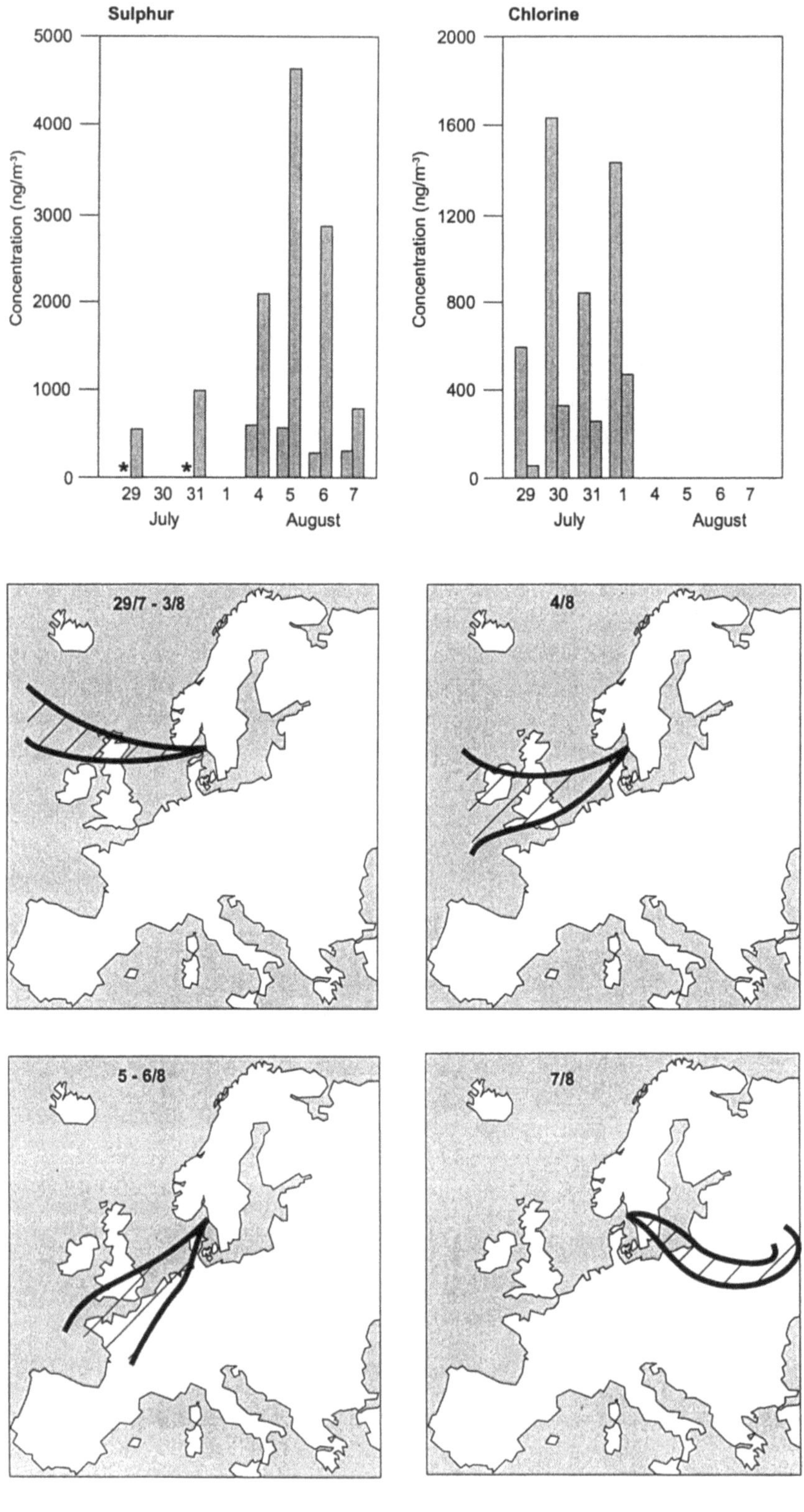

Sulphur
Chlorine
Concentration (ng/m⁻³)
5000
4000
3000
2000
1000
0
2000
1600
1200
800
400
*
29
30
31
1
4
5
6
7
July
August
29/7 - 3/8
4/8
5 - 6/8
7/8

in the Antarctic peninsula and analysed for 23 elements. In this study contributions from sea-salt and soil dust were found in both size fractions, but no anthropogenic sources could be identified.

In the identification process statistical methods, e.g. principal component analysis (PCA), are important tools for establishing local as well as distant sources and source regions [32, 33, 35]. As a general rule it may be stated that the identification process will be facilitated the more data of different kinds that are available, e.g. meteorological, chemical and physical. Thus size segregation plays an important part as a strong indication of whether the particulates in the aerosol are of anthropogenic or natural origin. How many size fractions that are optimal for characterizing a certain aerosol will depend on the specific receptor site and the choice of other measurement parameters, such as sampling time and flow rate for the sampler. Too many size fractions will give less material in every fraction, which might give poor statistics or long sampling times. Considering the bimodal size distribution of the natural aerosol a minimum of two fractions is required.

Another highly relevant piece of information is that of the air mass back trajectories for the aerosol. This is of crucial importance for the identification of the direction from which the aerosol has arrived to the receptor site [31]. This is illustrated in Fig. 8, in which the air mass back trajectories to a rural site on the Swedish west coast were calculated, and the levels of S and Cl in fine (diameter $< 3.5\ \mu m$) and coarse ($3.5\ \mu m <$ diameter $< 18\ \mu m$) particles are exposed. In this study [31] it was shown that fine particle Zn and Pb varied in the same manner as S. In contrast, the sea-salt derived Cl exhibited quite a different pattern (Fig. 8). The covariation of the anthropogenic elements that has often been observed in similar studies indicates that they can be regarded as deriving from the same source region in western and central Europe as indicated on the map in Fig. 8. In more recent measurements [32] the "fingerprint" of this European source region was exposed by PCA analysis of elements in the fine and coarse particle fractions of aerosols arriving at the same site on the Swedish west coast. In this analysis the fine and coarse particles of each element were treated as separate variables. The correlation matrix for this aerosol indicated a high degree of complexity of the source region. It can be noted that in order to resolve such a complex matrix into distinct sources more physical, chemical, meteorological or other information than that available in the study [32] would be needed.

Fig. 8. *The upper part of the figure* shows how the levels of S and Cl change from one day to the next in summer. Note that *the left bar* gives daily concentration of coarse particles and *the right bar* the concentration of fine particles. Thus it is seen that S dominates in the fine and Cl in the coarse particle mode. *The lower part of the figure* shows the region from where the aerosol has arrived to the measuring site on the Swedish west coast. Thus it is observed that as the aerosol transport to Sweden changes from Scotland to south England on the 4th of August, the S levels increase considerably [31]

Chemical Mass Balance used to Assess a Global Budget of Selected Pollutants

Chemical mass balance technique proved to be a good method to construct global budgets of several pollutants and to assess the perturbations of their geochemical cycles caused by various anthropogenic activities. The budgets have been prepared in order to attempt a quantitative understanding of the global flows into, the chemical transformations within, and the removal from the atmosphere or other parts of the environment of C and its compounds [36, 37], Cl ([38]), N species ([38]), and S species ([39] and the modified version by [38]). A global budget for Hg has also been attempted (e.g. Lindqvist, 1991). The overall conclusion can be formulated that the budgets for the above pollutants are out of balance in the positive direction. As a result, it is expected that the concentrations of these pollutants should be growing. The assessment of the imbalance in some of the above mentioned budgets of long-lived atmospheric species is presented in Table 2 [38].

Global budgets for chemical compounds on particles, particularly trace elements, became available only after the first quantitative assessment of worldwide contamination of air, water, and soils by these pollutants had been prepared [40]. The assessment was prepared for the following elements: As, Cd, Cr, Cu, Hg, Mn, Mo, Ni, Pb, Sb, Se, V, and Zn.

As a first step, major sources of the element releases to the air, water, and soil were identified and their fluxes determined. Later on the atmospheric inputs of the elements to water and soil were estimated. The atmospheric inputs and direct releases of the elements to water and soil on global scale are presented in Tables 3, and 4, respectively. Finally, the mass balance technique was used to assess the man-induced mobilization of trace elements into the biosphere, determined as median values in thousand tonnes per year of the terrestrial plus aquatic inputs of a given element minus its atmospheric emissions. It was concluded that the mobilization comes to 120 for As, 30 for Cd, 2150 for Cu, 11 for Hg, 110 for Mo, 470 for Ni, 1160 for Pb, 72 for Sb, 79 for Se, 71 for V, and 2340 for

Table 2. Assessment of imbalance in budgets of long-lived atmospheric species [38].

Species	1990 Concentration	Estimated annual concentration change	Estimated annual percentage change
CO_2	354 ppmv	1.8 ppmv	0.5
CH_4	1.72 ppmv	15 ppbv	0.9
CO	90 ppbv	0.6 ppbv	0.7
N_2O	310 ppbv	0.8 ppbv	0.3
$CFCl_3$	0.28 ppbv	0.01 ppbv	4
CF_2Cl_2	0.48 ppbv	0.02 ppbv	4
CH_3CCl_3	0.14 ppbv	0.01 ppbv	4

Table 3. Global atmospheric inputs and direct releases of various trace elements to aquatic ecosystem [40] (in 1000 tonnes/year)

Input	As	Cd	Cr	Cu	Hg	Mn	Mo	Ni	Pb	Sb	Se	V	Zn
Direct releases	8–55.6	1.1–12.1	37–191	26.1–53	0–6.7	73.8–288	0.6–14.5	27.1–158	7.1–51	3.3–28.4	9.2–67.1	0–7.6	53.4–286
Atmospheric deposition	3.6–7.7	0.9–3.6	2.2–16	6.0–15	0.2–1.8	3.2–20	0.2–1.7	4.6–16	87–113	0.4–1.7	0.5–1.1	1.4–9.1	21–58
Dumping of sewage sludge	0.4–6.7	0.1–1.3	5.8–32	2.9–22	0.1–0.3	32–106	1.0–4.8	1.3–20	2.9–16	0.2–2.9	0.3–3.8	0.7–4.3	2.6–31
TOTAL INPUT	12–70	2.1–17	45–239	35–90	0.3–8.8	109–414	1.8–21	33–194	97–180	3.9–33	10–72	2.1–21	77–375
Median value	41	9.4	142	112	4.6	262	11	113	138	18	41	12	226

Table 4. Global atmospheric inputs and direct releases of various trace elements to the soil [40] (in 1000 tonnes/year)

Input	As	Cd	Cr	Cu	Hg	Mn	Mo	Ni	Pb	Sb	Se	V	Zn
Direct releases	43.6–94	3.4–29.6	479–1271	527–1331	1.0–10.7	698.6–2587	29.4–141	95–507	277–850	3.7–43.1	4.7–73.4	39.8–201	640–1919
Atmospheric deposition	8.4–18	2.2–8.4	5.1–38	14–36	0.6–4.3	7.4–46	0.6–4.0	11–37	202–263	1.0–3.9	1.3–2.6	3.2–21	49–135
TOTAL INPUT	52–112	5.6–38	484–1309	541–1367	1.6–15	706–2633	30–145	106–544	479–1113	4.7–47	6.0–76	43–222	689–2054
Median value	82	22	896	954	8.3	1670	88	325	796	26	41	132	1372

Zn. To illustrate the expected environmental effects of this mobilization, it was calculated that the annual total toxicity of all the elements mobilized exceeds the combined total toxicity of all the radioactive and organic wastes generated each year, as measured by the quantity of water needed to dilute such wastes to drinking water standard [40].

Reconstruction of historical emissions of heavy metals and S has also been studied using the chemical mass balance. The discovery of Pb and its industrial and cultural applications in society were mentioned in pre-Roman times. Widespread application of the element in the Roman Empire and its use for glazing pottery during the period 7000–5000 B.C. are reported. Chinese history also indicates Pb coins in the society around 2000 B.C. Nriagu [41] has used the information on the production of Pb over time to estimate the element emissions to the air. This production – emission balance for Pb has been clearly seen in the historical trend of Pb concentrations in the snow and ice layers deposited in Greenland and Antarctica ([42]). The chemical mass balance method was also used to reconstruct emissions of trace elements during shorter periods of time. For example, the emissions of Cd, Pb and S from the Swedish alum works during the period 1726 through 1840 were related first to the production of the elements and then to their likely deposition in the emission generation region [43].

An interesting concept of a chemical time bomb was introduced recently [44], which implies that there exists a legacy of accumulated chemical pollutants in the environment stored in soils and sediments. These pollutants may cause the delayed and sudden occurrence of harmful effects due to their mobilization in response to slow alterations of the environment. Assessments calculating the risk of chemical time bombs can be determined from estimation of changing vulnerability over time. These changes may occur either because the capacity for storage of toxic materials in the soil and sediments becomes saturated, or because changes in environmental conditions cause the capacity to shrink or expand [44]. The estimation of the storage capacity is necessary in order to evalute the changes resulting either way. The chemical mass balance method can be used to estimate the storage capacities.

Another application of the chemical mass balance technique was made to the study of global distribution of persistent organic pollutants ([45]). A non-steady-state, multi-compartmental mass balance model for these pollutants was developed in which the global environment is represented by several sequentially arranged climatic zones. Information on the air, ocean water, fresh water, fresh water sediments and two soil compartments for each of the zones is needed to run the model together with the information on advective and intermedia transport processes. The mass balances for each of the compartments result in a system of 54 partial differentail equations, which are solved numerically to obtain estimates of concentrations, masses, transport fluxes, and reaction rates as a function of time.

Concluding Remarks

There exists a wide variety of applications of the chemical mass balance method in the science of the total environment. Most of them use the concept of mass budget. In some of these applications the budget imbalance is the key result. For example, the budget imbalance for Hg in power plant ash streams is used to estimate the emission of the gaseous portion of the element to the air. In other cases, budget imbalance is not a central issue, because spatial variability always appears and some regional budgets may be out of balance to the extent and direction needed to balance other regional budgets ([38]). Therefore, it is very important to identify clearly the purpose of study which will utilize the chemical mass balance technique.

The scale of the mass balance technique application depends strongly on the lifetime of studied pollutants. For the elements deposited on particles one should be aware that their lifetime in the atmosphere is rather short and the application of chemical mass balance can be appropriate to study the atmospheric phenomena on local and regional scale rather than on global scale.

References

1. Pacyna JM (1980) Coal-fired power plants as a source of environmental contamination by trace metals and radionuclides. Habilitation thesis. Technical University of Wroclaw, Poland
2. Flagan RG, Friedlander SK (1976). In: Proc Symp on Aerosol Science and Technology of the 82nd National Meeting of the American Institute of Chemical Engineers, Atlantic City, NJ
3. Kaakinen JW, Jorden RM, Lawasani MH, West RE (1975) Environ Sci Technol 9:862–863
4. Ensor DS, Cowen S, Shendrikar A, Markowski G, Waffinden G (1981) Kramer station fabric filter evaluation. EPRI Rep CS-1669. Electric Power Research Institute, Palo Alto, CA
5. Meij R, van der Kooij J, van der Sluys JLG, Seipman F, van der Sloot HA (1984) Characteristics of emitted fly ash and trace elements from utility boilers fired with pulverised coal. KEMA Scientific and Technical Rep 2, (1), 1–8, Arnhem, the Netherlands
6. Clarke LB, Sloss LL (1992) Trace elements – emissions from coal combustion and gasification. Int. Energy Agency, Coal Research, IEACR/49 Report, London, the United Kingdom.
7. Meij R (1989) Tracking trace elements at a coal-fired power plant equipped with a wet flue-gas desulphurisation facility. KEMA Scientific and Technical Reports, 7, (5), 267–355
8. Chow W, Nott B (1992). In: G Vourvopoulos (ed.), Elemental Analysis of Coal and Its By-products, World Scientific, Singapore.
9. Gutberlet H (1984) Measurement of heavy metal removal by a flue gas desulfurization plant working by the lime scrubbing method. Research report ENV-492-D (B). Commission of the European Communities, Luxembourg
10. Gutberlet H (1985) Determination of heavy metal retention by a flue gas desulfurization plant using lime scrubbing. Proc. VGB Conference "Kraftwerke und Umwelt", Essen, Germany
11. Clayton P, Coleman P, Leonard A, Loader A, Marlowe I, Mitchell D, Richardson S, Scott D (1989) Review of municipal solid waste incineration in the UK. Warren Spring Laboratory, LR 776 (PA), Stevenage, the United Kingdom
12. Pacyna JM, Joerss KE (1991) Proceedings of the EMEP Workshop on Emission Inventory Techniques, Regensburg, Germany, 2–5 July, 1991. Norwegian Institute for Air Research, EMEP/CCC-Report 1/91, Lillestrøm, Norway
13. Miller MS, Friedlander SK, Hidy GM (1972) J Colloid Interface Sci., 39:165–178
14. Dzubay TG, Stevens RK, Gordon GE, Olmez I, Sheffield AE (1988) Environ Sci Tech 22:1132–1141
15. Henry RC (1991). In: PK Hopke (ed.), Receptor Modeling for Air Quality Management, Elsevier, New York, NY

16. Fitzgerald WF, Mason RP, Vandal GM (1991) Water, Air, and Soil Pollution, 56:745–767
17. Baker JE, Church TM, Eisenreich SJ, Fitzgerald WF, Scudlark JR (1993) Relative atmospheric loadings of toxic contaminants and nitrogen to the Great Waters. A report for the U.S.EPA, the University of Maryland System, Solomons, MD
18. Johansson K, Aaastrup M, Anderson A, Bringmark L, Iverfeldt Å (1991) Water, Air and Soil Pollution, 56:267–281
19. Lindqvist O, Jernelov A, Hobansson K, Rodhe H (1984) Mercury in the Swedish Environment: Global and Local Sources. National Swedish Environment Protection Board, Solna, Sweden
20. Ayres RU, Ayres LW, Tarr JA, Widgery RC (1988) A historical reconstruction of major pollutant levels in the Hudson-Raritan basin: 1880–1980. NOAA Tech. Memo. NOSOMA43, National Oceanic and Atmospheric Administration, Rockville, MD
21. Lowenthal DH, Rahn KA (1985) Atmos. Environ, 19:2011–24
22. Stevens RK, Dzubay TG, Shaw RW, McClenory WA, Lewis CW, Wilson WE (1980) Environ. Sci. Tech. 14, 14911–14918
23. Dzubay TG, Stevens RK, Haageson PL (1984) Environ. Sci. Tech. 18:873–878
24. Keeler GJ (1987) A hybrid approach for source apportionment of atmospheric pollutants in the Northeastern United States. Ph. D. Thesis, The University of Michigan, Ann Arbor, MI
25. Keeler GJ, Samson PJ (1989) Environ. Sci. Technol. 23, 1358–1364.
26. Djupström M, Selin E, Öblad M (1989) J of Aerosol Science, 20, 8:1453–1459
27. Ruhling A (1987) Nedfallet av kadmium, nickel och vanadin kring Scanraff IVL-Rep, Stockholm, Sweden.
28. Selin E, Mnubi A., Isakson J, Foltescu V, Djupström M (1992) Environment, Lifestyle and Health. 92:3, 956
29. Harrison RM, Sturges W (1983) Atmos Environ, 17, 2, 311
30. Öblad M, Selin E (1985) Physica Scripta, 32, 462
31. Öblad M, Selin E (1986) Atmos Environ 20, 7, 1419
32. Foltescu V, Isakson J, Selin E, Stikans M (1994) Atmos Environ Accepted for publication.
33. Artaxo P, Andrade F, Maenhaut W (1992) Nucl Instrum and Meths in Phys. Res.
34. Barrie L, Gregor D, Hargrave B, Lake R, Muir D, Shearer R, Tracey B, Bidleman T (1992) Sci Tot Environ 122:1–174
35. Echalar F, Artaxo P, Thurston D (1990). In: Aerosols: Science, Industry, Health and Environment Pergamon Press
36. Bolin B (1983). In: B Bolin, RB Cook (eds.), The Major Biogeochemical Cycles and Their Interactions, SCOPE 21, Wiley, Chichester, UK
37. Cicerone RJ (1988). In: FS Rowland, ISA Isaksen (eds.), The Changing Atmosphere, Wiley, New York, NY
38. Graedel TE, Crutzen PJ (1993) Atmospheric Change: An Earth System Perspective, WH Freeman and Co., New York, NY
39. Granat L, Rodhe H, Hallberg RO (1976). In: BH Svensson, R Søderlund (eds.), Nitrogen, Phosphorus, and Sulphur: Global Cycles, Ecological Bulletins, 32, Stockholm, Sweden
40. Nriagu JO, Pacyna JM (1988) Nature, 333:134–138
41. Nriagu JO (1983) Lead and Lead Poisoning in Antiquity. Wiley, New York, NY
42. Boutron CF, Gorlach U, Candelone JP, Bolshov MA, Delmas RJ (1991) Nature, 353:153–156
43. Eklund M, Bergback B, Lohm U (1994) The Science of the Total Environment (accepted for publication)
44. Stigliani WM (ed.) (1991) Chemical time bombs: definition, concepts, and examples. Executive Report 16 (CTB Basic Document 1), International Institute for Applied Systems Analysis, Laxenburg, Austria
45. Wania F, Mackay D (1994) A global distribution model for persistent organic chemicals. The Science of the Total Environment (accepted for publication)

Sampling of Airborne Particulate Matter

Werner Holländer

Fraunhofer-Institut of Toxicology and Aerosol Research Nikolai-Fuchs-Str. 1, D-30625 Hannover, Germany

1 Introduction 145
2 Basic Hydrodynamics 146
2.1 The Carrier Gas 146
2.2 Bernoulli's Principle 147
2.3 Laminar and Turbulent Flow 147
3 Elements of Single Particle Mechanics 147
3.1 Hydrodynamic Force 148
3.1.1 Stokes' Law and Mobility for Large Spheres 148
3.1.2 Small Spheres 148
3.1.3 Particles of Arbitrary Size and Shape 149
3.2 Other Forces 150
3.2.1 Gravity and Acceleration 150
3.2.2 Electrical Forces 150
3.2.3 Van der Waals Forces 152
3.2.4 Stochastic Forces 152
3.3 Relaxation Time and Aerodynamic Diameter 153
4 Inlet Efficiencies 154
4.1 General Considerations 154
4.2 Thin-Walled Inlets 156
4.3 Blunt Inlets 156
5 Special Sampling Conditions 157
5.1 Calm Air Sampling 157
5.2 Isokinetic and Isoaxial Sampling 158
5.3 Nonisokinetic, Nonisoaxial Sampling 158
5.4 Sampling from Duct Flows 159
5.5 Ambient Air 159
5.6 Sampling from Containments 159
6 Aerosol Transport in Tubes 160
6.1 Particle Deposition 160
6.1.1 Directed Deposition 160
6.1.1.1 Laminar Sedimentation 160
6.1.1.2 Deposition in Bends 160
6.1.1.3 Deposition by Phoretic Effects 161
6.1.1.4 Deposition by Electrical Forces 161

6.1.2 Stochastic Deposition . . . 161
6.1.2.1 Turbulent Sedimentation . . . 162
6.1.2.2 Laminar Diffusion . . . 162
6.1.2.3 Turbulent Diffusion and Impaction . . . 162
6.2 Particle Population Dynamics . . . 163
6.2.1 Evaporation and Condensation . . . 163
6.2.2 Coagulation . . . 166
6.3 Particle Resuspension . . . 166
7 Practical Sampling . . . 167
7.1 Auxiliary Devices . . . 168
7.1.1 Flow Measuring Devices and Controllers . . . 168
7.1.2 Inlets . . . 168
7.1.3 Dilution Systems . . . 168
7.1.4 Neutralizers . . . 169
7.2 Active Samplers . . . 169
7.2.1 Stack Samplers . . . 169
7.2.2 Workplace Samplers . . . 170
7.2.3 Ambient Samplers . . . 170
7.3 Passive Samplers . . . 170
8 Concluding Remarks . . . 171
9 References . . . 172

1 Introduction

Airborne particulate matter varies widely in the atmosphere [1] and has impact mainly on human health (e.g. [2,3]), atmospheric visibility and the global radiation budget (e.g. [4,5]). Certainly a complete knowledge of the spatial and temporal distribution of the aerosol in the atmosphere and at the workplace with respect to number and mass concentration, size distribution, chemical composition etc. would be too difficult and expensive to obtain. Therefore, one would like to make a few representative and unbiased measurements of the quantity one is interested in. In principle, there are two possibilities: the measurement can be made

- in situ or
- from a sample.

In the latter case, the total instrument performance is determined by the combined effects of inlet and transport efficiencies for the particles. This is essentially what the present paper is about. The actual measurement is dealt with in the companion paper 'Particle Counting and Particle Size Analysis' chapter 10 of this volume.

Typically, optical measurements (scattering, extinction, spectroscopy) on particle collectives can be done in situ comparatively easily. More often, particles have to be prepared in a special way for the measurement (chemical analysis, electrical charge etc.), so that the only way to do the measurement is by taking a sample with all the risks of making errors.

In the simplest approach, health and environmental effects are assumed to be proportional to the particle mass or number concentration. For gaseous pollutants, typically the mixing ratios ppm or ppb etc. are used for measuring concentrations, but they are unsuitable for particulate pollutants because of the widely varying mass of the individual particulates. As an alternative, one can relate the number to some standard conditions, e.g. standard temperature (0 °C) and pressure (101 325 Pa). However, for measuring convenience one often relates the above particle characteristics to a specified volume, neglecting the actual pressure and temperature conditions.

Aerosol particles cover a size range roughly between 0.003 μm and 100 μm and a concentration range between 10^{-7} cm^{-3} in clean rooms and 10^8 cm^{-3} near sources. At high concentrations particles are close enough to collide with each other so that the size distribution may change even in short times. Occasionally, collection of bioaerosol is required which must remain viable during this process [6].

Often one is not interested in the total particulate population but rather in a certain fraction relevant to specific questions. An example is the particulate fraction smaller than 10 μm aerodynamic diameter (see below), which is called PM10, or the so-called health related sampling.

Also, it is very important to discriminate between inlet efficiency which refers to cross sectional area, and the transport efficiency of the sampling line, where particle deposition may occur.

It is not the objective of this work to give an exhaustive survey on the design and performance of innumerable sampling devices, but to provide the reader with enough physical background in order to allow him to make sound judgements on whether he can rely on the sampling device chosen for particular circumstances.

2 Basic Hydrodynamics

The following short introduction provides all the information necessary for performing calculations with respect to the carrier gas in which the aerosol particles are suspended.

2.1 The Carrier Gas

Due to the fact that condensible species like water vapor are not present in large amounts in the atmosphere, Boyle Mariotte's law can be applied:

$$pV = nRT \tag{1}$$

where p is the pressure, V is the volume, n is the number of moles, $R = 8.31$ J/(Kmol) is the universal gas constant, and T is the temperature. The gas density ρ_g is then

$$\rho_g = \frac{pM}{RT} \tag{2}$$

where M is the molecular weight. The gas molecules have a mean free path λ_g before they collide; its relation to macroscopically measurable quantities is

$$\lambda_g = \frac{2.004\mu_g}{p} \cdot \sqrt{\frac{\pi RT}{8M}}\,. \tag{3}$$

Since the viscosity of gases μ_g (for example $\mu_{air}(T_0{=}293\text{ K}){=}1.83{\cdot}10^{-5}$ [kg/ms]) is $\propto \sqrt{T}$, the mean free path is essentially inversely proportional to the pressure and proportional to the temperature; it has a numerical value of $\lambda_g \approx 65$ nm for ambient pressure and temperature. The kinematic viscosity ν is defined as $\nu = \mu_g/\rho_g$. A more precise temperature dependence of the gas viscosity is given by Sutherland's approximation

$$\mu_g = \mu_0 \cdot \sqrt{\frac{T}{T_0}} \cdot \frac{1 + S/T_0}{1 + S/T} \tag{4}$$

where the Sutherland constant S has the numerical value $S = 123.6$ K for air.

2.2 Bernoulli's Principle

When a gas is accelerated to a velocity v, small with respect to the sonic velocity, this will give rise to a pressure drop Δp according to Bernoulli's principle

$$\Delta p = \frac{\rho_g}{2} v^2 . \tag{5}$$

This principle is important for designing inertial impactors and for flow measurements using calibrated orifices.

2.3 Laminar and Turbulent Flow

When a flow is sufficiently slow that fluid inertia does not play a role with respect to viscous effects, the flow has an ordered structure and is called laminar. This means that streamlines can be identified, which never cross. On the other hand, swift flows have the tendency to break up irregularly with respect to time and space: this is called turbulence. The parameter describing the state of flow is the Reynolds number

$$Re = \frac{vd}{\nu} \tag{6}$$

where v is the average fluid velocity, and d is a characteristic dimension of the flow, for instance a tube diameter. Flow in the atmosphere is mostly turbulent, while inside a tube it will be laminar if the Reynolds number $Re \leq 2400$. The so-called friction velocity u_* which is a measure of exchange velocity perpendicular to the flow with average velocity v is related to the Reynolds number via [7]

$$u_* = 0.14 v Re^{-0.085} . \tag{7}$$

3 Elements of Single Particle Mechanics

In the beginning we will restrict ourselves to spherical particles, because in this case for size-dependent forces the diameter d_p suffices for a complete characterization while for more complicated shapes (prisms, fibers, flocs, fluffy structures etc.) additional parameters describing the shape may be needed for a unique calculation of the force on a particle.

3.1 Hydrodynamic Force

For the interaction between particle and carrier gas, different regimes for large and small particles must be discerned. The presence of small particles does not significantly disturb the velocity distribution of the molecules of the carrier gas: for this reason this regime is called the free molecular regime. The surface of large particles on the other hand is reached only by molecules which have undergone many collisions with other gas molecules, the velocity distribution of which has been modified by the particle presence. Here the molecular gas structure does not play a role any more, and the laws of macroscopic hydrodynamics apply in the so-called continuum regime. The Knudsen number

$$Kn = \frac{2\lambda_g}{d_p} \tag{8}$$

is used to characterize the transition from the free molecular regime $Kn \to \infty$, to the continuum regime $Kn \to 0$ (λ_g is again the mean free path of the carrier gas molecules). In the range $0.1 \leq Kn \leq 10$, which is called transition regime, no simple approximations, as in the free molecular or continuum regimes, can be made.

3.1.1 Stokes' Law and Mobility for Large Spheres

In the continuum regime, the hydrodynamic friction force F_{cont} on a particle moving with uniform velocity $\vec{v} = \vec{v}_p - \vec{v}_g$ relative to the gas

$$\vec{F}_{cont} = -3\pi\mu_g d_p \vec{v} = -\vec{v}/B_{cont} \tag{9}$$

was first calculated by Stokes. The above definition of the mobility B_{cont} can be generalized to stochastic, nonuniform motion as is the case in diffusion, and to arbitrary Knudsen numbers. Then Einstein's well-known relation between particle diffusion coefficient D_p, mobility B and temperature applies:

$$D_p = Bk_b T \tag{10}$$

where $k_b = 1.38 \cdot 10^{-23}$ J/K is Boltzmann's constant. Here B has no index since the above relation is valid for any size, whether or not the relation between hydrodynamic force and velocity given by Eq. (9) applies.

3.1.2 Small Spheres

For smaller particles Stokes' law is still applied for convenience with the so-called Cunningham correction factor

$$C(d_p) = 1 + Kn\left[\alpha + \beta \cdot \exp\left(-\frac{\gamma}{Kn}\right)\right] \tag{11}$$

which is a function of the particle size d_p via the Knudsen number Kn. Here, the empirical parameters α, β and γ were first determined by Millikan and coworkers. Allen and Raabe [8] reevaluated these parameters to $\alpha = 1.142, \beta = 0.558$ and $\gamma = 0.999$.

Hence, the hydrodynamic force $F_{\mathrm{hyd,s}}$ for a spherical particle (index s) of arbitrary size is given by

$$F_{\mathrm{hyd,s}} = F_{\mathrm{cont}}/C = v/B = v/(B_{\mathrm{cont}}C)\,. \tag{12}$$

The mobility, which has the dimension s/kg, is therefore given by

$$B = \frac{C(d_p)}{3\pi\mu_g d_p}\,. \tag{13}$$

3.1.3 Particles of Arbitrary Size and Shape

For nonisometric (nonspherical) particles one can define a shape factor χ as the ratio of the hydrodynamic force acting on the nonspherical (index ns) particle $F_{\mathrm{hyd,ns}}$ to the force on a volume equivalent sphere (index ves) $F_{\mathrm{hyd,ves}}$ [9]:

$$\chi = \frac{F_{\mathrm{hyd,ns}}}{F_{\mathrm{hyd,ves}}} = \frac{B_{\mathrm{ves}}}{B_{\mathrm{ns}}} = \frac{d_{\mathrm{ns}} \cdot C(d_{\mathrm{ves}})}{d_{\mathrm{ves}} \cdot C(d_{\mathrm{ns}})}\,. \tag{14}$$

For nonsymmetrical particles, χ will in general depend on the direction and therefore be a tensor. For a cube $\chi = 1.08$. Even for rather prolate (cigar-shaped) or oblate (disk-like) particles, the largest of the shape factor components is typically below 2; numerical values are given by, e.g., Happel and Brenner [10] and Hinds [11].

During the last decade, the concept of fractals, i.e. of selfsimilarity over a limited range of length scales [12] has also been successfully applied to characterizing the structure of particle aggregates m_N consisting of N primary particles of diameter d_0 and mass m_0. If the particles are arranged (almost) in a line, the linear dimension is $d_{\mathrm{line}} \propto Nd_0$ whereas for a closely packed sphere it will be $d_{\mathrm{cps}} \propto N^{1/3}d_0$. For any particle with an intermediate packing structure, a fractal dimension D_f can be defined which relates the geometric extension d_{fp} of the corresponding fractal particle to the number of primary particles by $d_{\mathrm{fp}} \propto N^{1/D_f}d_0$ where D_f has any noninteger value between 1 and 3. Normally, soot particles have values around 1.9, branched fibers around 1.3.

Rogak et al. [13] give a review of the present knowledge of the mobility of aerosol aggregates. For the continuum regime, they find for the shape factor

$$\chi = 0.9 \cdot \sqrt{\frac{D_f}{D_f + 2}} \cdot N^{1/D_f - 1/3}\,. \tag{15}$$

It is interesting to note that for $D_f \leq 2$ most of the primary particles in the free molecular regime are exposed to momentum exchange with the carrier gas so that the mobility diameter can be expected to be nearly equal to the area diameter $d_a = d_0 \cdot N^{1/2}$ so that one would expect

$$\chi = \frac{d_a}{d_{ves}} = \frac{d_0 N^{1/2}}{d_0 N^{1/3}} \propto N^{1/6} \tag{16}$$

which agrees with the above formula for $D_f = 2$ and also describes well the numerous experimental results [14, 15].

It is important to note that size and shape of the particle determine its hydrodynamic resistance uniquely (if minor effects of the molecule-surface interaction are neglected). The diameter derived from this friction force is called the mobility diameter. It plays a role mostly for small particles, where Brownian diffusion is important and if the hydrodynamic force is in equilibrium with an external force.

3.2 Other Forces

While (internal) hydrodynamic forces tend to slow down particles, other (external) forces try to restore motion until eventually an equilibrium is reached.

3.2.1 Gravity and Acceleration

Under the simplest circumstances the earth's gravitational field provides an acceleration $\vec{g}$ exerting a force $\vec{F}_g = m\vec{g}$. If buoyancy cannot be neglected, the force on a particle with volume V_p and density ρ_p is

$$\vec{F} = V_p(\rho_p - \rho_g)\vec{g}\,. \tag{17}$$

Also, if the particle is accelerated in a nonuniform flow with acceleration $\vec{b}$, the resulting force tends to separate the particle from the carrier gas flow, which may occur at the sampling inlet or in bends in the sampling tube.

3.2.2 Electrical Forces

Another very common force $\vec{F}_e$ is due to the interaction between n elementary electric charges e on particles and a surrounding electric field $\vec{E}$:

$$\vec{F}_e = ne\vec{E} = q\vec{E}\,. \tag{18}$$

When comparing this expression with Eq. (9) one finds for the migration velocity of a particle with charge q in an electric field

$$v = E \cdot qB \tag{19}$$

where $Z = qB$ is the electrical mobility with the dimension $m^2/(Vs)$.

In cases where the forces derive from a potential W, it is very interesting to look at the ratio W/kT which describes the range and particle size dependence of the force in comparison with the thermal energy. So, for instance, the electrostatic energy W_e between two charges separated by a distance r

$$W_e = \frac{e^2}{4\pi\varepsilon_0 r} \tag{20}$$

defines a scaling length

$$l_e = \frac{e^2}{4\pi\varepsilon_0 kT} \tag{21}$$

which describes the effective interaction length between two elementary charges. For ambient temperature, it has the numerical value 57 nm.

In the so-called Boltzmann charge equilibrium distribution f_n gives the probability for n charges of any sign for a particle in a bipolar ion atmosphere. It is approximated by

$$f_n = \frac{1}{\sqrt{2\pi}} \cdot \frac{1}{n_D} \cdot \exp\left(-\frac{n^2}{2n_D^2}\right) \tag{22}$$

and the standard deviation of the charge distribution n_D is given by $n_D = \sqrt{d_p/(2l_e)}$. Although the Boltzmann equilibrium is not necessarily reached during the neutralizing process which is determined by kinetics [16], a reasonable approach to it is warranted by an exposure to a bipolar ion atmosphere so that effects due to the remaining charge can be neglected. Very small particles carry charges according to the theory of Fuchs [17]. Unipolar charging is frequently used for accumulating maximum charges for electrostatic precipitation of particles in cleaning devices [18].

A list of a few important particle properties like Cunningham factor C at Eq. (11), relaxation time τ_p at Eq. (27), diffusion coefficient D_p at Eq. (10) and sedimentation velocity v_{sed} at Eq. (31) is given in Table 1.

Table 1. Cunningham factor C, relaxation time τ_p, diffusion coefficient D_p and sedimentation velocity v_{sed} for different particle sizes d_p in air (with unit particle density i.e. $\rho_p = 1$ g/cm^3). Note that behavior is determined for small particles by $D_p \propto d_p^{-2}$ and for large particles by $v_{sed} \propto d_p^2$

d_p/µm	C	τ_p/s	D_p/cm^2 s^{-1}	v_{sed}/cm s^{-1}
0.002	109	$1.32 \cdot 10^{-9}$	$1.31 \cdot 10^{-2}$	$1.30 \cdot 10^{-6}$
0.005	44.2	$3.33 \cdot 10^{-9}$	$2.11 \cdot 10^{-3}$	$3.27 \cdot 10^{-6}$
0.01	22.4	$6.76 \cdot 10^{-9}$	$5.35 \cdot 10^{-4}$	$6.64 \cdot 10^{-6}$
0.02	11.5	$1.39 \cdot 10^{-8}$	$1.38 \cdot 10^{-4}$	$1.37 \cdot 10^{-5}$
0.05	5.04	$3.81 \cdot 10^{-8}$	$2.41 \cdot 10^{-5}$	$3.73 \cdot 10^{-5}$
0.1	2.91	$8.78 \cdot 10^{-8}$	$6.95 \cdot 10^{-6}$	$8.62 \cdot 10^{-5}$
0.2	1.89	$2.28 \cdot 10^{-7}$	$2.25 \cdot 10^{-6}$	$2.24 \cdot 10^{-4}$
0.5	1.33	$1.00 \cdot 10^{-6}$	$6.35 \cdot 10^{-7}$	$9.84 \cdot 10^{-4}$
1.0	1.16	$3.51 \cdot 10^{-6}$	$2.78 \cdot 10^{-7}$	$3.44 \cdot 10^{-3}$
2.0	1.08	$1.31 \cdot 10^{-5}$	$1.29 \cdot 10^{-7}$	$1.28 \cdot 10^{-2}$
5.0	1.033	$7.79 \cdot 10^{-5}$	$4.93 \cdot 10^{-8}$	$7.65 \cdot 10^{-2}$
10.0	1.016	$3.07 \cdot 10^{-4}$	$2.43 \cdot 10^{-8}$	$3.01 \cdot 10^{-1}$
20.0	1.008	$1.22 \cdot 10^{-3}$	$1.20 \cdot 10^{-8}$	$1.19 \cdot 10^{0}$
50.0	1.003	$7.57 \cdot 10^{-3}$	$4.79 \cdot 10^{-9}$	$7.43 \cdot 10^{0}$

3.2.3 Van der Waals Forces

The interaction energy between induced dipole moments in matter is the cause of Van der Waals forces. If two molecules 1 and 2 are separated by a distance r_{12}, their interaction potential is given by

$$W_{12} = \frac{C'}{r_{12}^6} . \tag{23}$$

The range of Van der Waals forces is much shorter than that of electrical forces and for $r_{12} > 10$ nm the exponent gradually increases to 7. The so-called Hamaker constant for two materials 1 and 2

$$A_{12} = \rho_1 \rho_2 C' \tag{24}$$

(with the number concentrations of atoms ρ_i) describes the potential strength of interaction between materials 1 and 2. The numerical value of A_{12}/kT typically ranges between 0.1 and 10 [19]. The actual strength depends on the geometric arrangement; for two spheres and with radii r_1 and r_2 separated by a distance h the interaction energy is

$$W_{12} = \frac{A_{12} r_1 r_2}{(r_1 + r_2) h} . \tag{25}$$

Van der Waals forces are responsible for particles sticking to the walls of sampling tubes or to each other during coagulation. If particles have a large velocity upon contact, they may escape by rebounding due to internally stored elastic energy even if some energy is dissipated. This makes, e.g., coating of stages in inertial impactors (see Sect. 4.2.2 in the companion paper) with sticky material necessary in order to prevent bouncing losses [20, 21].

3.2.4 Stochastic Forces

All the abovementioned *directed equilibrium forces* displace the particle by a distance Δx which is proportional to the time t. On the other hand, so-called *stochastic* forces do not act continuously in the same direction so that $\Delta x \propto t^\alpha$ where $\alpha = 1/2$ in the case of the most common Brownian motion, but $\alpha = 3/2$ may occur in turbulence [22].

Recently, a novel kind of charging induced diffusion was predicted, based on the stochastic nature of charge acquisition followed by particle migration in an electrical field [23]. In the presence of electric fields and ionizing radiation, particle diffusion and deposition may be significantly enhanced by this process.

3.3 Relaxation Time and Aerodynamic Diameter

According to Newton's law in the absence of forces other than the hydrodynamic,

$$m_{\mathrm{p}}\frac{d\vec{v}}{dt} = \vec{F}_{\mathrm{hyd}} = -\frac{\vec{v}}{B} \tag{26}$$

and the motion will be damped with a relaxation time

$$\tau_{\mathrm{p}} = m_{\mathrm{p}}B \tag{27}$$

which has a contribution depending on the particle mass and an inverse hydrodynamic coupling strength B (the mobility). A particle with velocity v_0 at $t = 0$ will have a velocity v after time t

$$v = v_{\infty} + (v_0 - v_{\infty}) \cdot \exp\left(-\frac{t}{\tau_{\mathrm{p}}}\right) \tag{28}$$

for a gas velocity v_{∞}. The total distance traveled by a particle injected with v_0 into a stagnant gas ($v_{\infty} = 0$) is called the stopping distance:

$$s_{\mathrm{st}} = v_0 \cdot \tau_{\mathrm{p}} \tag{29}$$

and is obtained by integrating Eq. (28). The relaxation time is the critical parameter for the description of the behavior of large particles with considerable inertia.

If buoyancy can be neglected, in equilibrium

$$F_{\mathrm{hyd}} = F_{\mathrm{grav}} = m_{\mathrm{p}}g \tag{30}$$

which leads with Eq. (12) to a settling velocity v_{sed} of particles given by

$$v_{\mathrm{sed}} = m_{\mathrm{p}}gB\ . \tag{31}$$

For large particles in air at normal pressure,

$$v_{\mathrm{sed}} = \zeta d_{\mathrm{p}}^2 \tag{32}$$

with

$$\zeta \approx 3 \cdot 10^7\ (\mathrm{ms})^{-1}\ . \tag{33}$$

So far we have been considering spherical particles only, which are not very frequent in reality. Therefore, the following simplifying convention is introduced: the so-called aerodynamic diameter d_{ae} of a particle with density ρ_{p}, volume equivalent diameter d_{ve} and arbitrary shape factor χ is then defined as diameter of a sphere with unit density and the same settling velocity, i.e.,

$$v_{\mathrm{sed}} = \frac{d_{\mathrm{ve}}^3 \pi}{6} \rho_{\mathrm{p}} g \cdot B_{\mathrm{cont}}(d_{\mathrm{ve}}) C(d_{\mathrm{ve}})/\chi = \frac{d_{\mathrm{ae}}^3 \pi}{6} \rho_1 g \cdot B_{\mathrm{cont}}(d_{\mathrm{ae}}) C(d_{\mathrm{ae}}) \tag{34}$$

which leads to the definition of

$$d_{\mathrm{ae}} = d_{\mathrm{ve}} \sqrt{\frac{\rho_{\mathrm{p}}\, C(d_{\mathrm{ve}})}{\rho_1\, C(d_{\mathrm{ae}})\chi}} \tag{35}$$

where ρ_1 is the unit density. Virtually all atmospheric aerosol data and sampling efficiencies for particles larger than, say 1 μm refer to aerodynamic diameters for convenience. Note the difference between Eqs. (35) and (14) which defines the so-called mobility diameter.

4 Inlet Efficiencies

Sampling methods are prone to errors because one has to extract a certain volume and to extrapolate from the concentration, size distribution etc. of particles within the sampling volume to the true value in the undisturbed environment from which the sample was taken. In addition to statistical effects (see Sects. 2 and 3 of the companion paper), systematic errors are not unlikely.

The sampling inlet efficiency for the number concentration $c(d_{\mathrm{p}})$ of particles with diameter d_{p} is defined as

$$\eta_{\mathrm{inlet}} = c_{\mathrm{sa}}(d_{\mathrm{p}})/c_{\mathrm{true}}(d_{\mathrm{p}}) \; . \tag{36}$$

Since the situations may differ greatly with respect to particle size and concentration between sampling from the turbulent ambient air, high temperature and high pressure lines, clean rooms, occupational environments etc., great care has to be taken to avoid mistakes which could affect the precision and/or reproducibility of the results obtained.

4.1 General Considerations

The general situation is depicted in Fig. 1. The sampling orifice with a flow rate Q_{sam} and an average inlet velocity $\vec{v}_{\mathrm{sam}}$ is placed at a location L in the aerosol stream. The aerosol at location R is assumed to be representative for the aerosol to be studied. Its average mass velocity is $\vec{v}_{\mathrm{ae}}$ which may be different from the average carrier gas velocity $\vec{v}_{\mathrm{g}}$. In general, $\vec{v}_{\mathrm{sam}} \neq \vec{v}_{\mathrm{ae}}$, especially in outdoor sampling so that inertial particle losses can be expected as shown in Fig. 2 even if the intake flow and the ambient flow directions coincide.

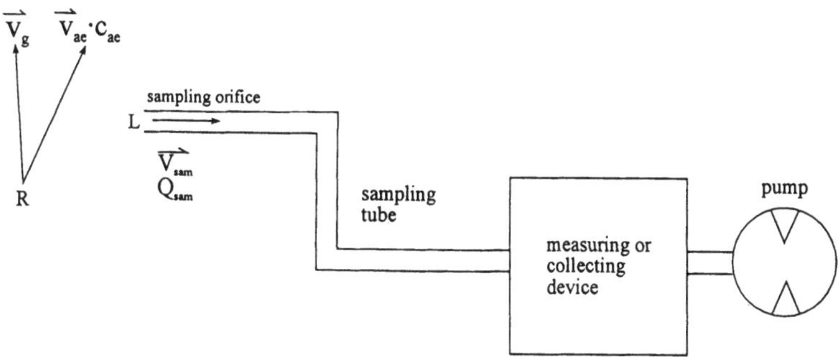

Fig. 1. Typical situation during aerosol sampling. Normally the aerosol flow vector $\vec{v}_{ae}$ and the sampling flow vector $\vec{v}_{sam}$ do not coincide neither in direction nor in magnitude, which may bias the sampling process. Also, transport losses may occur between the sampling point L and the measuring device

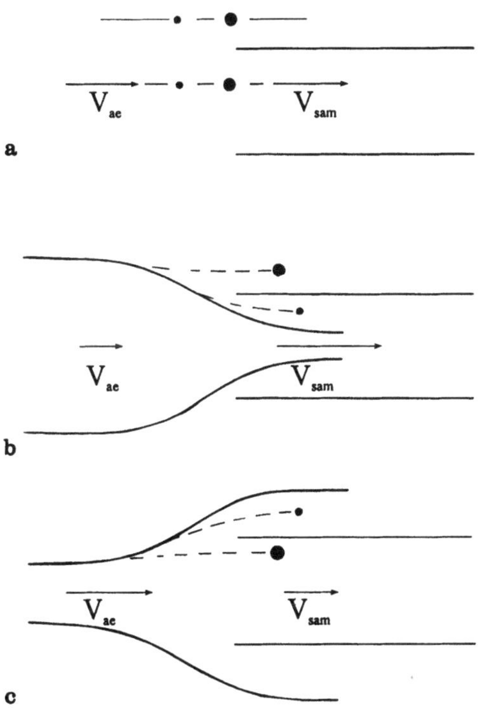

Fig. 2a–c. Isoaxial sampling regimes: **a** isokinetic sampling ($v_{sam} = v_{ae}$); **b** hyperkinetic sampling ($v_{sam} > v_{ae}$) with depletion of large particles; **c** subkinetic sampling ($v_{sam} < v_{ae}$) with enrichment of large particles

The sampling error can be described more quantitatively (see Fig. 3) in terms of the so-called Stokes number which is the ratio of the particle stopping distance s_{st} (see Eq. 29) to a typical sampling tube dimension (often its radius a):

$$Stk = s_{st}/a = v_{rel,\,p} \cdot \tau_p/a \tag{37}$$

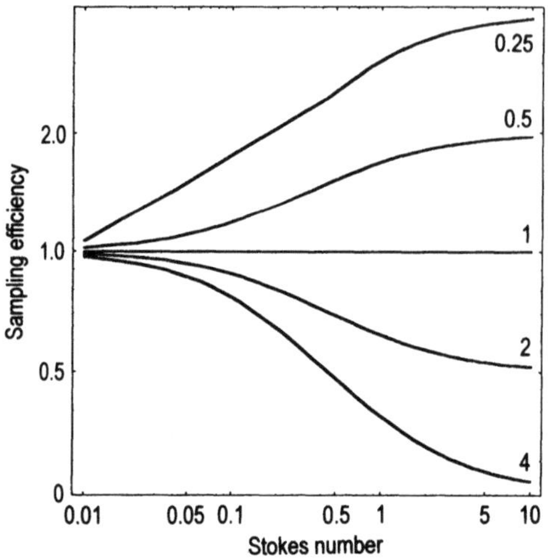

Fig. 3. Inlet sampling efficiency η_{inlet} for isoaxial but anisokinetic sampling with a thin-walled inlet as a function of Stokes number Stk for different aspiration ratios $v_{\text{sam}}/v_{\text{ae}}$

where $v_{\text{rel, p}} = |\vec{v}_{\text{p}} - \vec{v}_{\text{g}}|$ is the relative velocity difference between particle and gas with respect to magnitude as well as orientation. Another important parameter is the so-called aspiration ratio $As = v_{\text{sam}}/v_{\text{ae}}$; the numbers next to the curves in Fig. 3 reflect its influence on sampling efficiency. Inlet efficiencies were calculated according to [24]

$$\eta_{\text{inlet}} = 1 + \left(\frac{v_{\text{ae}}}{v_{\text{sam}}} - 1\right)\left(1 - \frac{1}{1 + Stk(2 + 0.62 \cdot \frac{v_{\text{sam}}}{v_{\text{ae}}})}\right) . \quad (38)$$

4.2 Thin-Walled Inlets

In order to be amenable to mathematical treatment, simplifications have to be made, the most important of which is the assumption of a thin sampling tube which does not disturb the gas flow, at least not if it faces the flow and for isokinetic sampling.

4.3 Blunt Inlets

In reality, all samplers are blunt to a greater or lesser degree. This means that there is a stagnation region with diameter S around the inlet orifice δ from which the sample is actually taken. In general, the ratio between this stagnation region diameter S and the blunt sampler diameter D depends in a complicated way on the aspiration ratio and the ratio δ/D. The literature available on blunt inlets is discussed extensively by Vincent [25]. It should be cautioned, however, that in spite of all theoretical progress in sampling theory, an experimental calibration of sampling inlets is necessary and actually requested by many authorities.

5 Special Sampling Conditions

In the following subsections, several frequently occurring sampling situations are discussed. Although many theoretical and experimental efforts have been made to develop universally applicable sampling theories and devices, one should not forget that the data scatter is very large and the formulas given are meant as rough guidelines only.

5.1 Calm Air Sampling

Here, essentially two errors may arise, both connected to particle inertia. The first involves the velocity ratio v_{sed}/v_{sam}. Obviously, for a sampling orifice forming an angle θ with the vertical $-\theta = 0$ ($\theta = 180°$) means upwards (downwards) facing orifice – particles will have a tendency to fall into (or out of) the orifice while being sampled.

$$\eta = 1 + \cos(\theta)\frac{v_{sed}}{v_{sam}} \,. \tag{39}$$

This is a so-called vertical elutriator characteristic, which was forming the basis for the original ISO convention of the inspirable dust fraction. If one assumes for the human nostrils an average $v_{sam} = 1.03$ m/s, with $\theta = 180°$, (Eqs. 32, 33) the relation

$$\eta = 1 - 2.91 \cdot 10^7 \cdot d_p^2 \geq 0 \tag{40}$$

is obtained, which implies that no particles larger than 185 μm can be inspired (see section on practical sampling).

The second error concerns the so-called dynamic escape introduced by Levin [26]. It is illustrated in Fig. 4.

The mechanism can be explained by a Stokes number based on the particle sedimentation velocity and a sink point flow capture distance L_{cd}. The flow velocity at distance l from a sink point with strength Q is given by

$$Q = 4\pi l^2 \cdot v(l) \,. \tag{41}$$

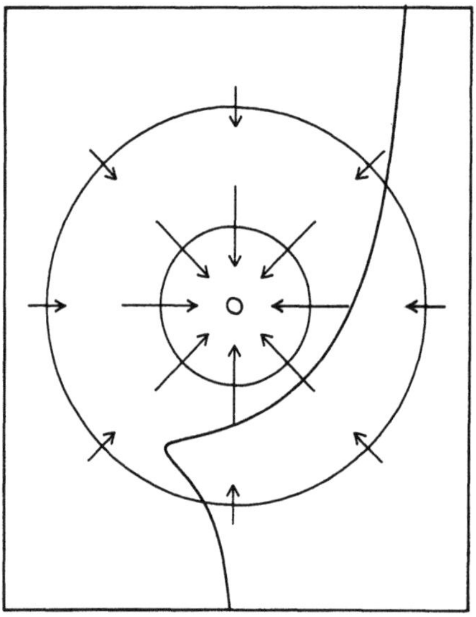

Fig. 4. The flow velocity increases towards a sink as shown schematically by the radially inwards pointing flow vectors. Although a settling particle will be carried towards the sink, it may escape due to its inertia as shown in the example trajectory

With the definition of capture distance, for which $v(L_{cd}) = v_{sed}$, one obtains

$$L_{cd} = \sqrt{\frac{Q}{4\pi v_{sed}}} . \tag{42}$$

According to Levin, the sampling efficiency for this situation is given by

$$\eta = 1 - 0.8K_l + 0.008K_l^2 \tag{43}$$

with the so-called Levin number K_l as parameter:

$$K_l = \frac{\tau_p v_{sed}}{L_{cd}} = \tau_p \sqrt{4\pi v_{sed}^3/Q} . \tag{44}$$

5.2 Isokinetic and Isoaxial Sampling

This is the preferred sampling situation, since it guarantees reliable results. In exhaust ducts with more or less constant velocity this strategy is usually employed, but even in this comparatively simple situation precautions have to be taken as explained in the section on practical sampling.

5.3 Nonisokinetic, Nonisoaxial Sampling

Many different formulae based on theoretical and experimental results have been proposed to cover the whole range of nonisokinetic, nonisoaxial sampling. All are functions of the yaw angle θ between aerosol and sampling flow, the aspiration ratio As, and the Stokes number Stk. Recently, Grinshpun et al. [27, 28] proposed a highly complex equation which seems to represent the wealth of experimental data quite well.

5.4 Sampling from Duct Flows

Although it might seem a simple task to take an isokinetic sample, there are often problems with dust inhomogeneity [29]. Also, if temperature and/or humidity of the stream are high, one has to avoid dew point condensation. Sampling from high pressure lines of ultraclean gases also poses some problems connected with nucleation on trace gases and wall resuspension [30, 31] apart from general sampling orifice design considerations [32].

5.5 Ambient Air

Normally, large particles pose the most serious problems [33, 34], and the large particle fraction may be considerable [35, 36]. In many countries measuring networks have been established, and there is a tremendous number of publications concerning the different measuring strategies (total suspended particulates, PM10, health related particulates) and instruments. Understandably, for commercial reasons, not only scientific arguments have been promoted. Surprisingly, it was found that instruments meeting all criteria and being effectively equivalent in wind tunnel studies, may behave quite differently under field conditions for various reasons [37–48]. As a consequence, in the European Union the certification of instruments will be based on field performance and a study is presently under way to specify the procedure in terms of a CEN standard.

5.6 Sampling from Containments

This case is closest to calm air sampling. The only problem may arise with respect to the choice of a suitable sampling location, if large particles separate by their sedimentation. This will be the case if the velocity ratio of thermal agitation (i.e. the natural convection) to sedimentation v_{th}/v_{sed} is smaller than unity. Usually, for chambers of 1 m height without special temperature control, one would not expect pronounced inhomogeneities for particles smaller than 20 μm.

6 Aerosol Transport in Tubes

Not all particles entering the inlet cross section will actually be carried to the collector or detector by the flow. They may have transverse velocity components, feel forces due to, e.g., gravity, electric fields and turbulent gas flow.

6.1 Particle Deposition

For a circular sampling tube of radius a and length L which is vented by a flow rate Q, the fraction deposited will depend on the transverse distance travelled by the particle, which has a different time dependence for directed and stochastic motion. Also, for laminar flow and directed motion, after a certain time all particles will be deposited, whereas in turbulent flow permanent mixing occurs so that the concentration decays exponentially, which is also true for laminar flows if the particle performs a stochastic motion like Brownian diffusion. In all cases, in the following section the dimensionless deposition parameter μ is not to be confused with the gas viscosity μ_g used in the previous sections.

6.1.1 Directed Deposition

Here the important dimensionless group is the ratio of the particle transverse motion x_{dir} to the tube diameter a

$$\mu' = \frac{x_{dir}}{a} = \frac{v_{dir}t}{a} = \frac{v_{dir}a\pi L}{Q} \ . \tag{45}$$

6.1.1.1 Laminar Sedimentation

This is a special case of directed deposition; Fuchs [9] gives Natanson's result for the penetration through a horizontal tube:

$$P = 1 - \frac{2}{\pi}\left[2\mu\sqrt{1-\mu^{2/3}} + \arcsin(\mu^{1/3}) - \mu^{1/3}\sqrt{1-\mu^{2/3}}\right] \tag{46}$$

where

$$\mu = \frac{3}{8}\cdot\frac{v_{sed}a\pi L}{Q} = \frac{3}{8}\cdot\mu' \ . \tag{47}$$

Here, for $\mu \geq 1$ all particles are deposited.

6.1.1.2 Deposition in Bends

In sampling lines bends often cannot be avoided. The acceleration in the bend eventually separates particles from the flow. The parameter describing deposition is again the Stokes number. There are a number of theories and experiments on this subject listed in [49].

6.1.1.3 Deposition by Phoretic Effects

Phoretic effects arise due to carrier gas inhomogeneities and transport particles along thermal and/or partial pressure gradients.

Diffusiophoresis is caused by carrier gas inhomogeneities and may affect sampling dramatically if the temperature falls below the dew point which has to be avoided under all circumstances since measurements would be uninterpretable.

Another phoretic effect potentially important in sampling is thermophoresis, since temperature differences between the gas stream and the sampling tube occur frequently. Thermophoresis pushes particles contrary to the temperature gradient, and, depending on particle size and heat conductivity, a velocity of $v_{th} \approx 0.05(\mathrm{cm/s})/(\mathrm{K/cm})$ may result. Stratmann and Fißan [50] gave an interpolation formula for thermophoretic tube penetration. Even for temperature differences of as much as 300 K, penetration will usually be above 0.85.

Just to give a few approximate figures, the thermophoretic constant is $\Psi_{th} = 2.5 \cdot 10^{-4} \frac{\mathrm{cm}^2}{\mathrm{Ks}}$ and the diffusiophoretic constant for water in air $\Psi_{dp} = 1.9 \cdot 10^{-6} \frac{\mathrm{cm}^2}{\mathrm{Pa \cdot s}}$

$$v_{th} = \Psi_{th} \cdot \nabla T \tag{48}$$

$$v_{dp} = \Psi_{dp} \cdot \nabla p \tag{49}$$

where ∇T is the temperature gradient and ∇p is the vapor partial pressure gradient.

6.1.1.4 Deposition by Electrical Forces

Electrical charges can also contribute to particle losses. There are essentially three electrical effects leading to particle deposition. Internal pressure due to repulsive forces of particles of the same polarity and image forces by charges induced in the tube walls act without external electric field; both would therefore act in metallic tubes, but these forces would be significant only for very high charges (more than 100 elementary charges).

On the other hand, electric fields might be present in nonconducting tubes and plastic bags, leading to drift of charged particles according to Eq. (19). One should avoid sampling tubes made from Tygon for this reason.

6.1.2 Stochastic Deposition

Here, several cases have to be discerned: For laminar flow particle diffusion may introduce a stochastic element. In the case of turbulent (stochastic) flow the particle motion may be directed or also stochastic: the result will always be an exponentially decaying penetration.

6.1.2.1 Turbulent Sedimentation

As stated above, particle motion is directed, but turbulence permanently mixes which leads to a penetration which is defined as the ratio of outgoing n to incoming particles n_0 [9]:

$$P = \frac{n}{n_0} = \exp\left[-\frac{2aLv_{sed}}{Q}\right] . \tag{50}$$

6.1.2.2 Laminar Diffusion

For one-dimensional Brownian diffusion, the mean square displacement of a particle with diffusion coefficient D_p after time t is $\overline{x^2} = 2D_pt$. Therefore, the dimensionless parameter $\mu' = D_pt/a^2$ will describe diffusional particle deposition in tubes with radius a. Replacing the residence time $t = V_{tube}/Q = a^2\pi L/Q$ in the above expression, one obtains another dimensionless group

$$\mu = \frac{D_pL}{Q} \tag{51}$$

which should describe deposition.

An exact solution for diffusive particle deposition in laminar flow through a circular tube was obtained by Gormley and Kennedy [51].

$$P = \frac{n}{n_0} = \sum_{i=1}^{\infty} a_i \cdot \exp(-b_i\mu) \tag{52}$$

with the first three terms $a_1 = 0.819, b_1 = 11.5, a_2 = 0.0975, b_2 = 70.1, a_3 = 0.0325, b_3 = 179$ valid for $\mu \geq 7 \cdot 10^{-3}$; for smaller values of μ the asymptotic formula $P = 1 - 5.5\mu^{2/3} + 3.77\mu$ should be used. Recently, a formula describing deposition from laminar flow due to sedimentation, diffusion and electrostatic precipitation was proposed [52].

6.1.2.3 Turbulent Diffusion and Impaction

For turbulent tube flows the deposition velocity v_{dep} due to diffusion is given by [53]

$$v_{dep} = 0.058u_*\nu/D_p . \tag{53}$$

If the particle inertia is not negligible, according to Liu and Agarwal [54] the dimensionless particle relaxation time

$$\tau^+ = \frac{\tau_p u_*^2}{\nu} \tag{54}$$

determines the deposition velocity

$$v_{dep} = u_* v^+ \tag{55}$$

where the dimensionless deposition velocity $v^+ = 6 \cdot 10^{-4} \cdot \tau^{+2}$ for $\tau^+ \leq 15.8$ and $v^+ \approx 0.15$ for $\tau^+ > 15.8$.

For long sampling tubes, for instance in nuclear power plants, particle losses may be exceedingly high, and the extrapolation from measured values to actual emission concentrations may not be reliable. As an example, for 1 μm particles a deposition velocity of 5.6 cm/s ($Re = 56\,600$, $u_* = 1.88$ m/s) follows which, according to Eq. (50) would result in a penetration of only $2.4 \cdot 10^{-4}$ for a 100 m tube of 2.5 cm diameter.

6.2 Particle Population Dynamics

For the interpretation of the measuring results, it is of utmost importance to know whether the particle size distribution has changed, which may be due to collisions between particles or to particle growth or shrinking.

6.2.1 Evaporation and Condensation

The saturation vapor pressure $p_s(T)$ of most substances has an almost exponential temperature dependence according to

$$p_s(T) = 10^{A-B/T-C/T^2} \tag{56}$$

where the empirical constants of a few liquids are listed in Table 2 together with other important material constants. Note that A, B, C used in the above equation will yield p_s in Pascals.

Table 2. Vapor pressure constants A, B, C, saturation vapor pressures p_s, saturation densities ρ_s (according to Eq. (2), vapor diffusion coefficients D_v, mass transfer rate factor Φ, nucleation scaling lengths l_n, surface tension σ, molecular weights M, and bulk liquid densities ρ_{liq} of selected liquids (for explanations see text)

substance	water	butanol	glycerol	DEHS
A	9.85373	10.95418	9.131065	12.1998
$B/(K)$	1478.49	2140.028	1144.795	4331.075
$C/(K^2)$	123 815	71 137.70	657 985.9	288 705
$D_v/(cm^2/s)$	0.188	0.094	0.0767	0.0347
p_s(293K)/Pa	2340	670	$3.7 \cdot 10^{-3}$	$1.2 \cdot 10^{-6}$
ρ_s(293K)/(g/m^3)	17.1	20.2	$1.4 \cdot 10^{-4}$	$2 \cdot 10^{-7}$
Φ(293K)(s^{-1})	38 500	28 000	0.1	$7.7 \cdot 10^{-5}$
l_n/(nm)	1.01	1.85	3.82	10.2
σ/(dyn/cm)	73.1	24.6	63.4	32.5
M/(g/mol)	18	74.1	92.1	426.9
ρ_{liq}/(kg/m^3)	1000	810	1260	1120

A consequence of the sensitive temperature dependence is that liquid aerosols are rather unstable unless the actual total vapor density $\rho_{act} \gg \rho_s(T)$. For instance, if the total water concentration is $\rho_{act, H_2O} = 22\ g/m^3$, at a temperature of 20 °C the liquid vapor (here water droplet) concentration would be $\rho_{lv} = 4.7\ g/m^3$, but decreases to $\rho_{lv} = 0.2\ g/m^3$ at a temperature of 24 °C. Also, a decrease of temperature of a humid aerosol may lead to considerable growth of hygroscopic particles or even to formation of new droplets once the dew point is reached. In multicomponent mixtures there may be different dew points, e.g. for water and sulfuric acid.

The kinetics of droplet growth (here in dimensionless form represented by the Knudsen number [55, 56]), and using the saturation ratio

$$S_{act} = \frac{p_{act}}{p_s(T)} \tag{57}$$

(where p_{act} is the actual vapor partial pressure) reads as follows:

$$\frac{dKn}{dt} = -\frac{Kn^3}{1 + Kn \cdot \psi(Kn)} \cdot \Phi \cdot [S_{act} - S_{eq}] \tag{58}$$

where the droplet equilibrium vapor pressure of small droplets is

$$S_{eq} = \exp(\Lambda Kn)\ . \tag{59}$$

The fact that small droplets have a higher equilibrium vapor pressure than a flat bulk liquid is called the Kelvin effect.

Here, $\Lambda = l_n/\lambda_g$ is the ratio of a nucleation scaling length and the mean free path, and

$$l_n = \frac{2\sigma v_0}{k_b T} \tag{60}$$

is the nucleation scaling length.

$$v_0 = \frac{M}{\rho_{liq} N_A} \tag{61}$$

is the volume of a molecule in the bulk liquid, N_A is Avogadro's number, and the mass transfer rate factor Φ is defined by

$$\Phi = \frac{D_v M p_s(T)}{2\rho_{liq} R T \lambda_g^2}\ . \tag{62}$$

D_v is the vapor diffusion coefficient, and ρ_{liq} is the bulk liquid density. Λ has typical values around 0.01–0.03 for volatile vapors so that the Kelvin effect (represented by the exponential) plays a role for $Kn \geq 10$ for S_{act} markedly different from unity.

$$\psi(Kn) = \frac{0.71 + 1.33Kn}{1 + Kn} \; . \tag{63}$$

Equation (58) can only be solved numerically if the Kelvin effect (in the exponential of S_{eq}) is taken into account. Without the Kelvin effect, and for small Knudsen numbers (i.e. $\psi(Kn) \to 0.71 = \alpha_M$, see above) the simplified equation

$$\frac{dKn}{dt} = -\frac{Kn^3}{1 + \alpha_M Kn} \cdot \Phi \cdot [S_{act} - 1] \tag{64}$$

has the analytical solution (which can be applied for $Kn < 0.3$)

$$\Phi(S_{act} - 1)(t - t_0) = \left[\frac{1}{2Kn^2} + \frac{\alpha_M}{Kn}\right] - \left[\frac{1}{2Kn_0^2} + \frac{\alpha_M}{Kn_0}\right] \; . \tag{65}$$

Kn_0 is the droplet Knudsen number at time t_0. Growth and evaporation curves for different Kn_0 are shown in Fig. 5.

Two important features can be seen: first of all, condensational growth leads to monodisperse particles as indicated by the convergence of growth curves in the upper left corner, and secondly, evaporation times of particles in the continuum regime are $\propto d_p^2$ as can be seen from the vertical distance of the evaporation curves on the right side of the graph.

Even at $S_{act} = 1$, droplets evaporate due to the Kelvin effect. If, however, the droplet contains a dissociating substance which lowers the vapor pressure according to Raoult's law, a stable size may result where Kelvin's and Raoult's effects cancel. This is observed when haze is formed at high relative humidities around salt particles.

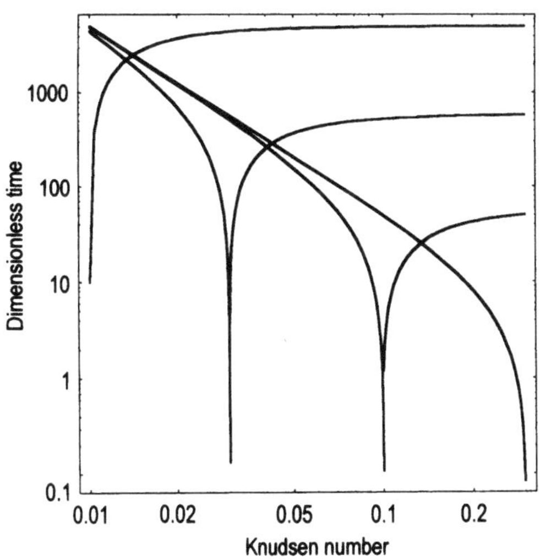

Fig. 5. Growth and evaporation curves for $Kn_0 = 0.01, 0.03, 0.1, 0.3$. The dimensionless time is $t' = \Phi(S_{act} - 1)(t - t_0)$. Using the data of Table 2, actual times t can be read off the graph. Growth (evaporation) curves have a negative (positive) slope

6.2.2 Coagulation

An exact account of coagulation effects is quite difficult and beyond the scope of this paper [57]. On the other hand, the salient point for the importance of coagulation for aerosol sampling is quite easy to fix. Particle coagulation depends on the relative motion of two particles. Let us assume two size fractions, small (index s) and large (index l). Then

$$\frac{dn_s}{dt} = -n_s \cdot n_l \cdot K_{sl} \,. \tag{66}$$

It can be assumed that the large particle concentration remains essentially constant, because their size is not much affected by the addition of small particles, which on the other hand disappear. For the simplest case of continuum mechanics, the coagulation constant K_{sl} is given by

$$K_{sl} = 8\pi(D_s + D_l)\frac{d_s + d_l}{2} \approx 4\pi D_s d_l \tag{67}$$

where D_s and D_l are the diffusion coefficients of the small and the large particles, respectively. Therefore

$$\frac{1}{n_s}\frac{dn_s}{dt} = -4\pi D_s n_l d_l \tag{68}$$

which leads to an exponential concentration decay of small particles with a time constant

$$\tau = 1/(4\pi D_s n_l d_l) \,. \tag{69}$$

As a rule of thumb, one can say that aerosols with particle concentrations $> 10^6$ cm^{-3} need to be checked for whether coagulation might play a role. In this case dilution helps to avoid it. If particle size distributions are determined by diffusion battery measurements, one has to be rather careful because of the long and different residence times in the different batteries. To give an example, a polydisperse submicron aerosol with a diffusion battery penetration of around 0.81 at low concentrations will penetrate with $P = 0.50$ for $c = 2 \cdot 10^7$ cm^{-3} and only $P = 0.20$ for $c = 2 \cdot 10^8$ cm^{-3} [58].

6.3 Particle Resuspension

Normally, particles deposited on the walls would not interfere with the sampled aerosol, although for long sampling lines operated for a long time with concentrated aerosol, reentrainment from the walls will have to be taken into account eventually. The situation is different for ultraclean gas lines in the microelectronics industry. Particles deposited on the walls might eventually come off after some time due to shocks or spontaneously [31].

7 Practical Sampling

In order to achieve the main goal of total suspended particulates (TSP) sampling, an unbiased sample of the aerosol has to be taken from a representative place and be transferred to the measuring device (the sampler) without losses or distortions of the size distribution.

On the other hand, health related sampling should take a sample weighted in the same way that the human lung samples the inspirable dust, which according to present knowledge follows the curve

$$\eta = 0.5[1 + \exp(-0.06 d_{\mathrm{ae}})] \, . \tag{70}$$

This convention is described in ACGIH [59] and has superseded the older ISO Technical Report [60], which more closely resembles what one would expect from sampling theory.

Another popular convention is the PM10 standard as described in the US Federal Standard [61]. Its physiological justification stems from the fact that it is the particles smaller than roughly 10 μm which enter the lung deeply and form the so-called thoracic fraction. Both the thoracic [60] and the inspirable (health related, see formula above) cut-off-curves are shown in Fig. 6.

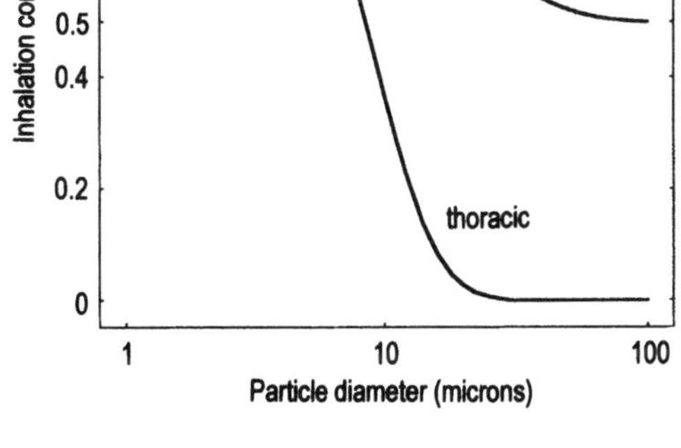

Fig. 6. Sampling efficiency conventions according to the thoracic and the health related conventions

7.1 Auxiliary Devices

In order to meet the specifications of TSP, PM10, or health related sampling, one needs certain auxiliary devices in addition to samplers suitable for the actual application.

7.1.1 Flow Measuring Devices and Controllers

The total volume V sampled during the interval $t*$ to which the measured particulate matter refers is given by

$$V = \int_0^{t*} Q dt \,. \qquad (71)$$

The flow rate Q is therefore of critical importance. Depending on the requirement, it may be defined in terms of the usual volumetric flow rate Q_v or the so-called mass flow rate Q_m which refers to standard temperature and pressure conditions and, hence, implies a certain number of moles.

The most popular measuring (and controlling) devices are:

- turbine gas meters
- bellow gas meters
- orifices
- mass flow controllers based on different principles
- rotameters (including those with electronic display).

While most of the mass flow controllers available commercially use the energy transfer of a heated sensor (for the physical background see [62]), recently another principle was proposed [47]. Here the pressure drop Δp across an orifice is measured, and the mass flow is kept constant for $\Delta p \cdot p/(MRT) = const$, where p, T are the absolute pressure and temperature.

7.1.2 Inlets

Depending on the specific environment, representative sampling inlets vary widely. It is always a good approach to try isokinetic sampling, and several techniques including some suitable for ambient air have been described but it is probably the inlets which still cause most trouble.

7.1.3 Dilution Systems

Dilution systems are used in order to avoid real distortions of size distributions by coagulation during transport from the sampling to the measuring point, or apparent distortions in the measuring device by exceeding the maximum permissible particle concentrations. They simultaneously reduce the risk of dew formation in wet aerosols and of thermophoretic losses since the temperature difference between gas and wall will be smaller.

However, there is a risk of changing volatile aerosol size distributions by diluting. There are different designs of dilution systems. In one type, a small aerosol flow crosses a capillary while the mainstream is filtered [63]. This has the advantage of not changing the overall composition of the carrier gas. In other systems, external particle-free dilution air is added [64].

7.1.4 Neutralizers

Particles produced by (liquid or solid) dispersion processes very often carry high charges, which may lead to losses in sampling lines. In such cases it is advisable to expose the aerosol to high concentrations of ions of both polarities so that excess charges of one sign are significantly reduced. Even if the so-called Boltzmann equilibrium is not reached, charges will be reduced enough to prevent problems with image charge deposition. For this purpose, α- as well as β-sources can be employed, and various types are available commercially. Recently a jet corona ionizer for neutralization has been described [65], and commercial units are available.

7.2 Active Samplers

Active samplers are characterized by the fact that power is needed for the sampling process. Large particles can easily be collected by a rotary impaction surface [66, 67]. The problem with this type of sampler is that the sampling volume is not precisely known due to the secondary flow induced by the rotation, and, for low wind speeds there may even be sampling from a depleted volume. On the other hand, this type of sampler is very suitable for fog water collection [68].

Contrary to rotary samplers, for pump samplers the total volume is well defined by an actively maintained pressure drop which moves air through the sampling line at a known rate. While most of them are quite similar with respect to their flow control and measuring devices, their design philosophies differ considerably. There are 'tunnel' samplers with vanes facing the wind (e.g. [69, 70] some of them even with isokinetic control), and stationary samplers with a vertical symmetry axis, with and without shrouds or shields [71].

7.2.1 Stack Samplers

From the technical point of view, probably the most important sampling is done from stacks. Since the flow direction is constant, and the velocity not very variable, usually isokinetic sampling will be possible. Many well-characterized samplers and impactors are available. Detailed operating instructions are given in VDI guideline 2066 [72]. An extreme case of 'stack' sampling would be sampling from airplanes [73].

7.2.2 Workplace Samplers

It is clearly impossible to list all the different samplers used in the different countries of the world. Because of this wide variety, developing performance standards for those instruments is an important task [74]. There are basically two different types: stationary and personal samplers. While the first type can be designed in the usual way in order to meet all aerosol sampling criteria, personal samplers face very severe restrictions in terms of weight, since they must be carried along by the worker. He may have special clothing, a breathing mask etc. because he is often exposed to a harsh environment since otherwise there would not be the necessity for a personal sampler. The small flow rate of the personal samplers often poses restrictions on the sampling performance. A few techniques used in Germany for workplace sampling are described in VDI guidelines 2265 and 2266 [75].

7.2.3 Ambient Samplers

The most demanding operating conditions prevail, without doubt, when using ambient TSP samplers, because at high wind speeds large particles cannot be easily collected due to their inertia. As the European standard instrument for ambient TSP sampling a modified wide range aerosol classifier is employed [47]. Detailed instructions for specific instruments are given in VDI guideline 2463 [76].

7.3 Passive Samplers

In passive samplers there is neither a pump nor an active flow control, which makes it difficult to define a measuring volume.

Nevertheless, for gas measurements, passive sampling is quite popular, because diffusion sampling is possible. The diffusive flux density $j[g/(\mathrm{cm}^2\mathrm{s})]$ is given by

$$j = D\nabla c = D\frac{c}{\delta_{\mathrm{c}}} \tag{72}$$

where $D = D_{\mathrm{m}} + D_{\mathrm{e}}$ is the total diffusion coefficient, to which molecular (index

m) and eddy (index e) diffusion contribute and ∇c is the concentration gradient at the collection surface which can be described by a concentration boundary layer thickness δ_c. By proper design, eddy diffusion can be made negligible, and the molecular diffusion coefficient is characteristic for the gas species to be sampled. Therefore, reasonable correlation coefficients between passive gas samplers and reference methods can be achieved. The situation is much more difficult for aerosol particles, because

- the diffusion coefficient of particles strongly depends on their size,
- the diffusive deposition is only one of the important deposition mechanisms, and
- shielding for suppressing the influence of turbulence on the deposition is not possible because of size selection effects.

In spite of the abovementioned problems, passive samplers (e.g. [77]) are quite popular and useful for screening purposes. A possible way for size-independent passive sampling of particles would be an approach assuring a sampling flux density

$$j = c \cdot v_{dep} \tag{73}$$

where the deposition velocity v_{dep} is independent of particle size. Unfortunately, this is hardly feasible because of the strong size dependencies of all deposition mechanisms.

If, however, the deposition velocity were made large enough for all particles to be deposited, the aerosol sample would again depend on the volume collected, which is not constant and not known because it depends on the ambient conditions. One way out of this dilemma would be a device which takes samples only at certain times (say every 5 min) and does not collect particles in the meantime. In this case, the collected volume would be given by the sampler volume and the number of collection cycles. Such an instrument has been developed recently [78] and will become commercially available soon.

Recently, in Europe, the decision was made to replace the TSP strategy by the PM10 philosophy. In this context, passive samplers with PM10 characteristics would also have a market.

8 Concluding Remarks

This review article lists the most important principles for successful aerosol sampling under a variety of conditions and gives the design and performance criteria for a large number of commercially available devices. The survey showed that the area of sampling is still – in spite of its relative maturity – an active field of research and development.

9 References

1. Jaenicke R (1987). In: Fischer G (ed) Landolt-Börnstein; New Series V/4b, vol meteorology, London
2. VDI guideline 2310/part19 (1992) Maximum immission concentrations for suspended particulate matter
3. Lodge JP (1986) AST 5: 459
4. Kiehl JT, Briegleb BP (1993) Science 260: 311
5. Charlson RJ, Schwartz SE, Hales JM, Cess RD, Coakley JA, Hansen JE, Hofmann DJ (1992) Science 255: 423
6. Macher JM, Willeke K (1992) J. Aerosol Sci. 23: S647
7. Davies CN (1966). In: Davies CN (ed) Aerosol Science, Academic, Oxford, p 393
8. Allen MD, Raabe OG (1985) AST 4: 269
9. Fuchs NA (1964) The mechanics of aerosols. Dover, New York
10. Happel J, Brenner H (1973) Low Reynolds number hydrodynamics, Noordhoff, Leyden
11. Hinds WC (1982) Aerosol technology, Wiley, New York
12. Kaye BH (1989) A random walk through fractal dimensions, VCH, Weinheim
13. Rogak SN, Flagan RC, Nguyen HV (1993) AST 18: 25
14. Kasper G (1982) AST 1: 201
15. Kops J, Dibbets G, Hermans L, Van de Vate JF (1975) J. Aerosol Sci. 6: 329
16. Mayya YS (1994) J. Aerosol Sci. 25: 617
17. Fuchs NA (1963) Geofisica pura e applicata 56: 185
18. Oglesby S, Nichols GB (1978) Electrostatic precipitation, Marcel Dekker, New York
19. Israelachvili JN (1985) Intermolecular and surface forces, Academic, London
20. Wang HC, John W (1987) AST 17: 285
21. Turner JR, Hering SV (1987) AST 18: 215
22. Shlesinger MF, Zaslavsky GM, Klafter J (1993) Nature 363: 31
23. Mayya YS, Malvankar SV (1993) J. Colloid Interface Sci. 156: 78
24. Durham MD, Lundgren DA (1980) J. Aerosol Sci. 11: 179
25. Vincent JH (1989) Aerosol sampling, Wiley, Chichester
26. Levin LH (1957) Izv. Akad. Nauk SSSR Ser. Geofiz. 7: 914
27. Grinshpun S, Willeke K, Kalatoor S (1993) Atm. Env. 27A: 1459
28. Grinshpun S, Willeke K, Kalatoor S (1994) Atm. Env. 28A: 375
29. Hermann J (1976) Staub in turbulenter Gasströmung, Dissertation TU München
30. Ye Y, Pui DYH, Liu BYH, McMurry PH (1992) J. Aerosol Sci. 23: S173
31. Wen HY, Kasper G (1989) J. Aerosol Sci. 20: 483
32. Lee JK, Rubow KL, Pui DYH, Liu BYH (1993) AST 19: 215
33. Garland JA, Nicholson KW (1991) J. Aerosol Sci. 22: 479
34. May KR, Pomeroy NP, Hibbs S (1976) J. Aerosol Sci. 7: 53
35. Lundgren DA, Hausknecht BJ, Burton RM (1984) AST 3: 467
36. Noll KE, Pontius A, Frey R, Gould M (1985) Atm. Env. 19: 1931
37. Rodes CE, Holland DM, Purdue LJ, Rehme KA (1985) JAPCA 35: 345
38. Hall DJ, Upton SL, Marsland GW, Waters RA (1988) Warren Spring Lab. Report LR669(PA)M; ISBN 0 85624 5291
39. Sweitzer TA (1985) JAPCA 35: 744
40. Wedding JB, Lodge JP, Kim YJ (1985) JAPCA 35: 649
41. Purdue LJ, Rodes CE, Rehme KA, Holland DM, Bond AE (1986) JAPCA 36: 917
42. Shaw RW, Stevens RK, Lewis CW, Chance JH (1983) AST 2: 53
43. John W, Winklmayr W, Wang HC (1991) AST 14: 93 and 165
44. McFarland AR, Ortiz CA, Bertch RW (1984) JAPCA 34: 544
45. Wedding JB, Weigand MA (1985) Atm. Env. 19: 535 and 539
46. Wedding JB, Weigand MA, Ligotke MW, Baumgardner R (1983) Environ. Sci. Technol. 17: 379
47. Holländer W, Dunkhorst W, Pohlmann G (1989) Part. Part. Syst. Charact. 6: 74
48. Holländer W, Morawietz G, Bake D, Laskus L, van Elzakker BG, van der Meulen A, Zierock KH (1990) J. Air Waste Manage. Assoc. 40: 881
49. Pui DYH, Romay-Novas F, Liu BYH (1987) AST 7: 301
50. Stratmann F, Fißan H (1994) J. Aerosol Sci. 25: 1305
51. Gormley PG, Kennedy M (1949) Proc. Roy. Ir. Acad. 52A: 163
52. Chen YK, Yu CP (1993) AST 19: 389

53. Friedlander SK (1977) Smoke, dust and haze, Wiley, New York
54. Liu BYH, Agarwal JK (1974) J. Aerosol Sci. 5: 145
55. Gelbard FM, Seinfeld JH (1979) J. Colloid Interf. Sci. 68: 173
56. Williams MMR, Loyalka SK (1991) Aerosol Science, Pergamon, Oxford
57. Rosner DE, Tassopoulos M (1991) J. Aerosol Sci. 22: 843
58. Holländer W (1994) unpublished
59. American Conference of Governmental Hygienists (ACGIH) (1985) Report of the ACGIH Technical Committee on Air Sampling Procedures, ACGIH, Cincinnati, Ohio
60. ISO Technical Report (1983) ISO/TR/7708-1983(E)
61. Federal Register (1987) 52FR 24729, July 1, 1987
62. Hinze JO (1975) Turbulence, McGraw Hill, New York
63. Remiarz RJ, Johnson EM (1984) TSI Quarterly 10 (1): 7
64. Koch W, Lödding H, Mölter W, Munzinger F (1988) Staub-Reinh. Luft 48: 341
65. Romay FJ, Liu BYH, Pui DYH (1994) AST 20: 31
66. Noll KE (1970) Atm. Env. 4: 9
67. Hameed R, McMurry PH, Whitby KT (1983) AST 2: 69
68. Lesnic D, Elliot L, Ingham DB (1993) J. Aerosol Sci. 24: 163
69. Regtuit HE, de Ruiter CJ, Vrins ELM, Hofschreuder P, Oeseburg F, Benschop FM (1990) J. Aerosol Sci. 21: 919
70. Vawda Y, Colbeck I, Harrison RM, Nicholson KW (1992) J. Aerosol Sci. 23: 233
71. Gong H, Anand NK, McFarland AR (1993) AST 19: 294
72. VDI guideline 2066, parts 1–7 (1975–1990) VDI Handbuch Reinh. Luft
73. Baumgarder D, Huebert B (1993) J. Aerosol Sci. 24: 835
74. Kenny L (1992) J. Aerosol Sci. 23: 773
75. VDI guidelines 2265 and 2266, parts 1–3 (1968–1980) VDI Handbuch Reinh. Luft
76. VDI guideline 2463, parts 1–9 (1974–1987) VDI Handbuch Reinh. Luft
77. Schultz E (1989) Staub Reinh. Luft 48: 113, and VDI guideline 2119, part 4 (1972) VDI Handbuch Reinh. Luft
78. Holländer W (1993). In: VDI-Berichte 1059, p 67

Analysis of Inorganic Particulate Pollutants by Nuclear Methods

Sheldon Landsberger and Steven Biegalski

Department of Nuclear Engineering, University of Illinois, 214 Nuclear Engineering Laboratory, 103 South Goodwin Ave., Urbana, Illinois USA 61801

Introduction 176
Historical Background of Nuclear Methods 178
Literature Review of NAA of Airborne Particles 179
  Remote Area 179
  Rural Area 183
  Urban Area 184
  Biomonitors 186
  Indoor Environment 188
Nuclear Methods 189
  Instrumental Neutron Activation Analysis (INAA) 190
  Fast Neutron Activation Analysis (FNAA) 192
  Instrumental Photon Activation Analysis 192
  Prompt Gamma Activation Analysis (PGAA) 192
  Energy Dispersive X-Ray Fluorescence 193
  Total Reflection X-Ray Fluorescence 194
  Particle Induced X-Ray Emission 195
  Comparison with Other Techniques 196
Filter Choice and Blanks 197
Quality Assurance, Quality Control and Reference Materials 197
Conclusions 198
References 199

Summary

Over the past years the concentrations of many trace metals in the atmosphere have risen dramatically to the point where industrial and automobile emissions now dominate the natural biogeochemical cycling in many urban and to some extent rural and remote environments. From the results of numerous studies it has been suggested that increased urban air concentrations of toxic metals, such as chromium, arsenic, cadmium, and lead, have been associated with various disease risk factors. Airborne particulate matter is a complex pollutant which requires reliable analytical methods to measure accurately the elemental concentrations in air. Besides good trace element analytical techniques, proper sampling methodologies and the use of source-receptor modeling are needed. Nuclear analytical methods such as neutron activation analysis (NAA), X-ray fluorescence (XRF), and particle induced X-ray emission (PIXE) and energy dispersive X-ray fluorescence (EDXRF) are ideally suited to be implemented in research and routine air monitoring programs, particularly in urban environments. An overview of these methods is presented. Special emphasis on NAA is given including its usefulness in several programs conducted in various parts of the world for monitoring and source receptor modeling purposes.

Introduction

Over the past three decades investigation into the characterization and dynamics of air pollution has dramatically increased in size and scope. It is now well known that there are serious health risks associated with air pollution which affect millions of inhabitants living in megacities as well as in rural areas. While the main focus of air pollution studies and monitoring has been the outdoor environment, expanded research has been done on the effects of air pollution in relation to global change in urban, rural and remote areas. More recently, there has been increased awareness of indoor air quality and the role that air pollution has in affecting the lives of people who typically spend 70–80% of their time indoors, either at work or at home. Air pollution can arise from many different industrial and natural sources. Airborne pollutants of concern include oxides of nitrogen and sulphur, carbon monoxide, ozone, photochemical oxidants, synthetic organic compounds, fine particles, and trace metals. There is justifiable concern over the presence of greenhouse gases such as carbon dioxide, methane, nitrous oxide, as well as stratospheric ozone depleting gases such as chlorofluorocarbons, and chlorinated and brominated hydrocarbons. Of genuine fundamental interest in this heterogeneous mixture of air emissions are trace elements, particularly heavy metals such as lead, mercury, cadmium, arsenic, manganese, etc. Once released into the atmosphere they can produce adverse physiological changes in humans when inhaled. The concentrations of many trace metals in the atmosphere have risen dramatically to the point where industrial and automobile emissions now dominate the natural biogeochemical cycling in many urban environments. There have been many studies suggesting that increased urban air concentrations of toxic metals, such as chromium, arsenic, cadmium, and lead, have been associated with cancer risk factors. All these airborne substances can and do undergo short- and long-range transport through the atmosphere. A primary pathway back to earth is by means of precipitation, both wet and dry. Once these heavy metals are integrated into the ecosystems they can have important toxicological effects on aquatic, animal, and plant life. In general terms the study of elemental toxicity can be considered to fall into four main aspects. The first aspect, the most immediately important, is related to the environmental occurrence. A second consideration is the physical and chemical state of the pollutant. It is evident that the toxicity of a specific element can be influenced by its chemical properties and physical characteristics. The next important aspect is the identification of the major routes of entry into the human body, which is dependent on the previous chemical and physical restraints. The final important factor is the metabolic fate and toxicological consequences of the xenobiotics. The initiation of environmental studies can be described in the following supportive and interactive topics:

1. Evaluation of Priorities
   - toxicological implications
   - urgency of toxin
   - availability of analytical methods
2. Assessment of Trends, Concentrations and Distributions
3. Effects on Biological Systems
4. Government Controls

A recent example of major changing priorities is the 1990 United States Clean Air Acts Amendments [1].

Airborne particles, also described as airborne particulate matter or dust, are a combination of solid and liquid fractions in ambient air. It is generally accepted that particles of sizes $< 2.5–5$ μm are considered to be in the fine mode arising from anthropogenic sources. Particles over 5–10 μm, known as the coarse mode, are usually associated with mechanical processes resulting in windblown soil, resuspension of street dust, and dispersion of water/sea spray particles. Years of research have given conclusive proof that particle aerodynamic size is of paramount importance in understanding both the fate of inhalable airborne particles as well as their ability to undergo long-range transportation. Particles of such small sizes usually have very small masses (from several tens of micrograms to tens of milligrams) when collected on most types of air filters. With the advent of source-receptor models there was a need to have multi-elemental characterization of air samples so that a clearer picture of the dynamics of air pollution sources could be readily identified. As a result of the two above criteria, chemical methods which were both very sensitive and multi-elemental were desirable. Airborne particulate matter is a complex pollutant which requires reliable analytical methods to measure accurately the elemental concentrations in air. Besides good trace element analytical techniques, proper sampling methodologies and the use of source-receptor modeling are needed. Nuclear and atomic analytical methods such as neutron activation analysis (NAA), X-ray fluorescence (XRF), and particle induced X-ray emission (PIXE) are ideally suited to be implemented in research and routine air monitoring programs, particularly in urban environments. We will emphasize the importance of nuclear methods, their usefulness in several programs conducted in various parts of the world, and the benefits of using source-receptor models including factor analysis, source apportionment, and chemical mass balances. A complete literature survey of all the methods is not possible. We have concentrated our computer literature search efforts on published works in air monitoring research using NAA.

Historical Background of Nuclear Methods

In the late 1960s and the early 1970s, with the commercial production of high resolution germanium detectors, researchers began to discover the unique features of nuclear methods to investigate and quantify trace elemental concentrations in airborne particles. Neutron activation analysis discovered in 1936 by Hevesy and Levi [2] became the premier method for trace elemental characterization of airborne particles, as well as wet and dry precipitation. Besides being a non-destructive and multi-elemental technique, NAA offered the added unique feature of having detection limits in the nanogram range for many elements of environmental concern. These elements include aluminum, antimony, arsenic, barium, bromine, cadmium, chlorine, chromium, copper, manganese, nickel, selenium, silver, vanadium, and zinc. Besides these trace constituents, NAA could also be effectively used to determine sodium, magnesium, potassium, uranium, thorium and most of the rare-earth elements. The one main disadvantage of conventional NAA is its inability to determine lead and its poor sensitivity for sulphur.

In the early 1970s researchers at Lund University in Sweden [3] developed the technique commonly known as particle induced X-ray emission (PIXE). With the use of a small accelerator and charged particles (usually protons between two to three million electron volts) are employed to bombard air filters to produce X-rays arising from electrons deexciting within atomic shells. These X-rays are characteristic of individual elements and subsequently very low concentrations of many elements can be determined simultaneously. Typically elements from $Z = 11$ (sodium) to $Z = 40$ (zirconium) and $Z = 82$ (lead) are determined routinely in most world-wide PIXE facilities. In principal many more elements can be determined but the PIXE sensitivity of these constituents is very low in airborne particles. The unique advantages of PIXE is its ability to determine some 15–20 elements in air filters in a matter of minutes, analyze very small diameter filters with minute amounts of material, and determine sulphur and lead which are critically important environmental elements. Other competitive methods include X-ray fluorescence and inductively coupled plasma atomic emission spectroscopy (ICP-AES). In general, XRF has limited sensitivity for many elements, particularly for air samples taken in rural and remote areas. However, in polluted cities XRF methods are quite adequate yielding the same elements as PIXE.

Space does not allow for a detailed in depth comparison of these multi-elemental methods. Several review articles, reports, and books have outlined the salient features off these techniques in detail [4–10]. The objectives of the authors is to give an overview of these methods. Special attention is given to the application of neutron activation analysis in air pollution studies in the last several years, to some of the laboratory techniques used, and to some of the newer analytical developments in NAA which have lowered detection limits. An exhaustive computer literature from several data bases was used for this work.

Special attention is also given in NAA methods used in recent global change and indoor air quality studies.

Literature Review of NAA of Airborne Particles

Remote Area

The study of the elemental composition of aerosols in regions far from industrial sources allows us to understand the natural global geochemical cycles. Although industrial and anthropogenic aerosol emissions are large, natural emission predominates in the global aerosol budget. Marine aerosol and soil dust are two major sources of the atmospheric aerosol. Biogenic emission from tropical forests also plays a role in the global aerosol budget [11]. The transport of aerosol particles from deserts to oceanic areas accounts for most of the sediments in some parts of oceans. Receptor modeling can be used with nuclear-related techniques [12] in order to study the global geochemical cycles in areas not greatly affected by anthropogenic pollution. These areas are mainly in developing countries in the southern hemisphere, where knowledge on aerosol elemental concentrations is still rather limited. Even though most of the aerosols found in remote regions of the world are of crustal and marine origin, detectable levels of aerosols from industrial activity may be found. Evaluation of Arctic and Antarctic aerosols helps to determine the global influence of anthropogenic sources. Because aerosol levels in these regions are low, NAA is a desirable analytical technique. Aerosol transport and wind back-trajectory calculations help to determine the origin of aerosols not native to these remote regions. Loureiro et al. [13] used NAA to determine 15 elements in aerosol samples (1986–87) collected on Millipore air filters at the Brazilian Antarctic Station "Comandante Ferraz." Principal factor analysis (PFA) was performed on 10 of the 15 elements measured. The PFA statistical treatment yielded three factors. The first factor represented a sea source and contained Na, Cl, and Br. Aluminum, scandium, vanidium, and iron were included in the second factor and were reported to originate from a crustal component. The third factor contained the elements Zn, Sb, and Au and their source was not identified. Five other elements were measured but were not included in the PFA. These elements included Mn, Ca, K, Th, and La. Bottenheim et al. [14] correlated the depletion of ozone during the Arctic sunrise to Br aerosol levels. NAA was used to analyze hi-vol filters and ion chromatography (IC) to analyze lo-vol filters. The Br values determined by NAA and IC were very well correlated. Ozone measurements were made with a Dasibi Model 1008 analyzer. Results obtained from air sampling at the Alert station during April–May, 1988 showed a definite anti-correlation between the decrease in ozone concentration and the rise in Br. The best explanation for this phenomenon offered by the authors was a "meteorological modulation which alternately brings to the sampling location lower boundary layer air, depleted in ozone and enriched in filterable bromine, and free tropospheric air with abundant ozone and few bromine compounds."

Dick [15] collected aerosol samples at Gipps Ice Rise near the east coast of the Antarctic peninsula during the austral summers of 1984–85 and used NAA as a check for Al as a crustal reference material. Atomic absorption spectrometry (AAS) was the primary means for elemental analysis in this study. The results from the AAS to NAA comparison were inconclusive due to the high Al content in the particular batch of cellulose acetate membrane filters and the low Al content deposited on the filters. The mean value of Al in the blank filters as determined by NAA was 322 $\pm$ 16 ng while the sample levels determined by AAS were between 10 and 35 ng of Al. However, NAA did confirm that there was a low level of Al in the Antarctic aerosol and there was no evidence that the methods and analysis by AAS underestimated the Al concentrations. Pereira et al. [16] used NAA to determine levels of Na, Cl, Al, Mn, V, Ca, K, Fe, Zn, Br, La, Sb, Au, and Sc on aerosol filters collected on the sub-Antarctic King George Island in the South Shetland Islands of the Antarctic peninsula. Aerosols were collected on Millipore membrane filters at flows of 5 l/min for a period of one week. Aluminum and Mn concentrations were higher in the 1986 samples than the 1987 samples, but the two year average values were greater than those at the South Pole as reported by Zoller et al. [17]. Na and Cl values were found to peak from January to April in contrast to Wagenbach et al. [18] findings that Na and UCl peak in November and December on the Ekström ice shelf. Sea-salt derived inorganic components K, Ca, and Br occurred in aerosols with different ratios to sea water when Na and Cl occurred in similar ratios. Trace metals Zn and Sb displayed a seasonal trend with a minimum occurring in the winter. Meeker et al. [19] identified and documented particles of crystalline elemental Au in the emissions of Mount Erebus, Ross Island, Antarctica. This was the first documentation of Au in volcano emissions and the authors indicate that Au is present in the emissions from other volcanoes. Koyama et al. [20] analyzed snow and ice from Antarctica, Mt. Hakuba-Norikura, Japan, and Mt. Naimonanyi, China. The samples were analyzed via NAA and it was found that concentrations of volatile elements like Zn, Cd, As, Sb, Ag, and Se are elevated in snow and ice samples when compared to the non-volatile elements such as Al and Fe. It was found that the ratio of Zn, Cd, As, Sb, Ag, or Se to Al or Fe is 10–1000 times higher in the snow and ice samples than in ordinary soil. Landsberger et al. [21] reported seasonal fluctuations in As, Sb, In, Mn, and V in aerosols from the Alert station in the Canadian Arctic. The analysis covered samples taken from 1987 to 1992. NAA was used to determine the elemental concentrations of these elements on Whatman 41 air filters with the use of epithermal NAA. Enrichment factor analysis indicated that As and Sb were elevated in the wintertime aerosol and potential source contribution function (PSCF) analysis on In, As and Mn showed that these aerosols have a strong probability of origin from the European and Asian areas.

Landsberger et al. [22] analyzed a series of 12 daily samples collected during March 1985 at Alert, Canada. Samples were collected for 24 hour periods on Whatman 41 air filters. NAA was used to determine Al, Br, Ba, Ca, Ce, Co, Cr, Cl, Cu, Hf, Fe, I, K, Na, Sc, Sr, Th, Ti, U, Eu, La, Sm, As, Mn, In, Sb, Se, V, and Zn while IC was used to determine NO_3^-, NH_3^+, and $SO_4^=$. Calculations indicate that the aerosols originate from four sources - crustal, oceanic, anthropogenic, and photochemical reactions of marine bromine compounds. Landsberger [23] discussed improved methodology for the determination of Sb, As, In, Mn, Se, V, and Zn aerosols deposited on Whatman 41 air filters. These seven elements are important in the determination of long-range source signatures. The use of thermal vs epithermal neutrons was also discussed. It was determined that epithermal neutrons yielded the best results for indium. Cheng et al. [24] demonstrated the use of a 3-layer PSCF technique to determine the total PSCF (TPSCF) values for V aerosols observed in the Canadian high Arctic. Meteorological data was collected at three pressure levels (1000, 25, and 850 mb) and was used to determine back-trajectories. The aerosol filters, collected at ground level, were analyzed by NAA. V, which results from fuel burning (especially oil), showed strong probabilities of originating from areas west of Europe Russia including the Ural industrial complex at latitude 50° N to 60° N and longitude from 20° E to 100° E. This analysis covered data collected from January 1984 to December 1987. Cheng et al. [25] examined the distributions of trace elements and ionic species of aerosols collected at the Mould Bay, Igloolik, and Alert stations in the Canadian high Arctic. Ion chromatography, NAA, and inductively coupled plasma (ICP) were used as the analytical techniques to determine the trace elements and ionic species. All of the elemental and ionic species variables illustrated a positive skewness with the mode median arithmetic mean. The positive kurtosis of all the variables, except SO_4 at Alert and Igloolik, indicates that the tails of the distributions are heavier than that of a Gaussian distribution. The values for IC_Cl, IC_NO_3, IC_NH_4, NAA_I, NAA_V, NAA_Al, ICP_Ca, ICP_Mg, ICP_Fe, and ICP_TI fit a log-normal distribution at all three sites while the values for IC_Br, IC_Na, IC_K, NAA_Na, ICP_Ni, ICP_Zn, and ICP_P were found not to be log-normal at all three sites. Landsberger et al. [26] reported on the construction of an epithermal irradiation facility. When compared to NAA via thermal neutrons, this source of epithermal neutrons allows for an eight-fold increase in the sensitivity of In detection on aerosol filters collected in the Canadian high Arctic. In is a by-product of the smelting industry and was used in enrichment factor analysis along with PSCF analysis for this paper. Enrichment factor analysis showed that In levels are elevated in the Arctic atmosphere during the fall, winter and spring months from 1982 to 1987. PSCF analysis indicated strong sources of In to emanate from North America, former central USSR, and parts of East Asia. Pacyna et al. [27] used a trajectory model to investigate the transport of trace elements from European sources to receptors in Norway. NAA was used to analyze 38 elements deposited on a Whatman 40 cellulose filter. Of those elements, As, Sb, V, and Zn were used in the model. Tuncel et al. [28] analyzed samples with NAA which were collected at the Amundsen Scott Station, located at the geographical South Pole, from February 1979 to November 1983. It was found that Antarctic aerosols could be broken down into four components - marine, meteorite, crust I and crust II. It was necessary to break the crustal component into two parts because enrichment factors of some crustal elements, such as Ba, La, Ce, Yb, and Hf were higher in the winter while other crustal element concentrations often double in the summer. Marine element concentrations were highest in the winter. When combined with similar studies at the south pole, no long term elemental concentrations trends could be found in 12 years of sampling. Barrie et al. [29] analyzed samples collected at Alert Canada during April 1986. Samples were collected daily on Whatman 41 air filters and the trace elements were determined by IC and NAA. The elements detected with NAA were placed into three groups: soil (Al, Sc, Fe, Ce, Co, Mn, K, La, Sm), sea-salt (Na, Cl) and anthropogenic (As, Sb). When compared to the work of Lowenthal and Rahn [30], the As/Sb ratios were similar to European and former western Soviet Union. It was found that the V/Sb ratio did not seem to agree with previous documentation. Sheridan and Zoller [31] compared the trace elements present on air filters collected in the troposphere at the North American and Norwegian Arctic, by airplane in the stratosphere, and in the marine boundary layer. The data collected by airplane in the stratosphere correlated well to the ground based data collected in the troposphere. Meteorological and wind trajectory data revealed that relatively unmixed air was traveling from the former central Soviet Union into the sampling areas of interest. The documented As/Sb ratio tracer from the former central Soviet Union was found to be approximately 5–6. It was believed that the V/Sb ratio changed between the time of ground aerosol collection and the time of airplane based collection.

Sturges and Barrie [32] studied the behaviour of Cl, Br, and I in the Canadian Arctic. The samples were collected between 1979 and 1984 and were analyzed by both NAA and IC. The Br values presented were determined by NAA, but the Na and Cl values were determined by IC. Summer Br levels are low and can be explained by equal contributions of marine and automotive exhaust sources. Br concentrations peak after the Arctic dawn (March–April) to levels above what can be explained by marine, automotive and crustal sources. The authors speculated that "this originates from the production of organic gaseous Br by an oceanic bloom, coupled with the onset of photochemically driven processes to convert gaseous Br to particulate." Cl appeared to originate solely from the sea, and I reached peaks in the spring and early autumn at non-marine sites. Barrie and Barrie [33] reviewed the seasonal trends of Arctic aerosols observed at the Alert station from 1980 to 1986. The most polluted period is from December to April with the sea-salt component of the aerosols peaking in December and the anthropogenic component peaking from January to May. A photochemical constituent consisting of Br and I, arises after the polar sunrise in April and May. There exists a second photochemical peak for I, but not for Br, in early fall, possibly due to a secondary bloom in northern oceans occurring in late summer. The soil component to the atmospheric aerosol peaks once in April–May, coinciding with the Asian dust storm, and a second time in September. Barrie et al. [34] characterized the particulate and gaseous fraction of lower tropospheric SO_x, Br, Cl, I and NO_3^- in relation to O_3 depletion events during the Polar Sunrise Experiment at Alert, Canada. An array of denuder and filter measurements were carried out near sea level to characterize these components. Both IC and NAA were used for chemical analysis in this experiment. During the experiment, particulate inorganic Br, gaseous inorganic Br and gaseous organic Br increased during the O_3 depletion events. None of the other measured atmospheric components showed a correlation to the O_3 depletion events.

Rural Area

Toxic aerosols present in rural settings come from both local and long-range sources. The origin of these aerosols needs to be considered as well as the concentrations. Documentation of rural aerosol levels gives a base from which elevated urban concentrations may be compared. Hacisalihoglu et al. [35] collected aerosol samples over the Black Sea in an effort to extensively determine the magnitude and composition of European anthropogenic sources. The influence of natural aerosol components to the region, especially Saharan dust incursions,was also investigated. NAA was used to determine Na, Mg, Al, Cl, K, Ca, Sc, Ti, V, Cr, Mn, Fe, Co, Ni, Cu, Zn, As, Sb, Se, Br, Rb, I, Cs, Ba, La, Hf, W, Au, Ce, Nd, Sm, Eu, Gd, Yb, and Th. SO_4 and NO_3 were determined via IC and Pb analysis was performed using AAS. Of the elements found, Na, Cl, Br, W, Ni, As, Se, Zn, Au, Sb, Cu, Pb, Se, and I were determined to be present at enriched levels. The enrichment of Na, Cl, and Br was attributed to the marine environment. Principal component analysis revealed five source components: crustal, marine, and three anthropogenic ones. One of the anthropogenic sources was identified as originating from a local iron and steel industry. Bichler et al. [36] sampled gasses emitted from Italian volcanoes using an activated carbon filter and later analyzed the samples for Hg, As, Br, Sb, and Se by NAA. The temperature of the fumaroles was between 100 and 300 °C, which made sampling at these locations difficult. Special adaptations had to be made to the sampling equipment to accommodate the elevated temperatures. The elemental concentrations and ratios to Hg were then compared to that of the exhalations from Mt. St. Helens, 1980 and Etna, 1978. The study of toxic metal exhalations from natural sources, such as volcanoes, allows for the evaluation of health effects resulting from active natural sources.

Rusheed et al. [37] analyzed 23 Whatman 41 air filters that were collected in the Niclore, Islamabad (NE Pakistan) area during March to June of 1984. The samples were analyzed for 16 elements with NAA and 9 elements were determined in more than 12 samples. Most of the elements were determined to be crustal in origin with the source of Sb and Zn in question. Sb is believed to have a possible anthropogenic source and Zn could come from sources ranging from local soil enrichment of Zn to paint containing Zn on the building in which the receptor was placed. Hopper and Barrie [38] compared the trace element composition of aerosols collected at Dorset, Ontario, Foret Montmorency, Quebec, and Kejomkujik, Nova Scotia during the autumn of 1984. Aerosols were collected on Whatman 41 air filters and the trace element concentrations were determined with NAA and ICP. Factor analysis for the different sites usually revealed a crustal factor, an urban factor, a marine factor, and a few factors with elements of anthropogenic origin. Back-trajectory analysis of the data showed that In and As were excellent tracers of non-ferrous emissions. As was also found to be enriched in the trajectories from prominent coal-burning areas, but these trajectories were found to have only small amounts of In when compared to the non-ferrous emission back-trajectories. Hacisalihoglu et al. [39] collected samples with Whatman 41 filters over the Black Sea in summer of 1988. The Cl/Na ratio was smaller in the samples collected over the sea than what is observed in sea water. It is suspected that this is due to the evaporation of Cl from sea-salt particles as HCl. The samples were collected with high volume samplers for 24 h. This allowed for deposition of particles at quantities sufficiently above detection limits. The Whatman 41 filters had high blank values for Cr, Au, and I which contributed to statistical uncertainties of those elements. A log-normal particle distribution was observed for most elements, but some anthropogenic constituents like Mn and $SO_4^=$, showed frequency distributions due to mixed source contributions. Ion chromatography was used for $SO_4^=$ and NO_3^- ion analysis.

Urban Area

Yoshizumi [40] determined aerosol concentrations in the Tokyo metropolitan area. This study used NAA to determine the concentrations of 29 elements in the atmosphere: Ag, Al, As, Ba, Br, Ca, Cd, Ce, Co, Cr, Cs, Cu, Fe, Hf, K, La, Lu, Mn, Na, Ni, Sb, Sc, Se, Sm, Th, Ti, V, W, and Zn. Elemental and organic carbon were determined by a total CHN analyzer. The concentrations of $SO_4^=$, NO_3^- Cl^-, and NH_4^+ ions were determined by ion chromatography. The results of the analysis allowed for the resolution of eight separate sources for these aerosols through a source apportionment method. These sources were soil, road dust, marine, fuel oil combustion, iron and steel industry, refuse incineration, auto exhaust, and diesel exhaust. Five of the eight sources were considered to be anthropogenic. The percentage of anthropogenic aerosols gives insight to current and future regulatory needs. Urban aerosol collection and analysis can point out specific pollution sources for toxic aerosols. Ko and Jervis [41] used instrumental NAA in combination with particle induced X-ray emission (PIXE) to study the emissions from hospital incinerators. This study showed that hospital incinerators can be a major anthropogenic contributor to the overall aerosol makeup in urban environments. Air samplers were placed down wind from two separate hospitals to determine the composition of the plume. Concentrations of Ag, Al, As, Br, Ca, Cd, Ce, Cl, Co, Cr, Cs, Cu, Eu, Fe, Hf, I, In, K, La, Mg, Mn, Na, Sb, Sc, Se, Sm, Ti, V, W, and Zn were all determined by NAA. PIXE was employed to add Pb, Ni, Si, S, and Hg to this list of elements. Incorporation of the aerosol, incinerator fly ash, and hospital plastics results showed strong evidence that the hospital incinerators generated elevated levels of Cd, Cl, Cr, Sb, Pb, and Zn to the local environment.

Paiva et al. [42] collected aerosol samples in São Paulo, Brazil with dichotomous samplers for 24 h and then analyzed the samples with energy dispersive X-ray fluorescence (EDXRF) analysis and NAA. Twenty three elements were found by EDXRF and twenty two elements were determined by NAA. Enrichment factor analysis and correlation coefficient calculations were then performed with the data. It was found that soil dust was the major component of fine and coarse particles.

Okamoto and Kobayashi [43] summarizes the Industrial Pollution Control Association of Japan project where samples were collected around the Kashima industrial area in the months of January, April, July–August, and October–November. These samples were then analyzed by NAA, XRF, and IC to determine 33 elements and 3 ions. Crustal components were found to be the major fraction of the aerosol collected throughout the entire year. The marine component was greater in the spring and fall when sea breezes are dominant. It was found that the crustal and marine aerosol sources were larger than any anthropogenic sources generated from the Kashima industrial area.

Qadir et al. [44] used NAA to determine 16 trace elements on aerosol filters collected in the Sheikhupura, Pakistan industrial area. A longitudinal distribution of trace elements was found in order to characterize the environment of the air breathed by people working in the vicinity. Air samples were collected on Whatman 41 filters using low volume air samplers at heights of 4,8 and 18 m. The variation in sampling allowed for determination of source patterns with correlations in aerosol concentrations at the different heights. The concentrations of aerosols found at this location were 20-30 times higher than values determined by the Pakistan Institute of Nuclear Science and Technology at other urban sites. Lobo et al. [45] used NAA, AAS, and IC to determine 23 trace elements which were collected on Whatman 41 and TFA 41 air filters. Atmospheric acid gases were also collected in carbonate buffer solutions and then analyzed for seven acid gases with IC. Major contributors to the degradation of air quality in this data set were vehicle emissions and the aeolian process on stripped soils.

Cassoria et al. [46] analyzed aerosol samples for 15 elements by NAA and for 12 elements with PIXE. The elements determined by PIXE were Al, Si, S, K, Ca,

Ti, Cr, Mn, Fe, Cu, Au, and Pb. NAA was utilized for the determination of Na, Al, Cl, K, Ca, Sc, V, Mn, Fe, Co, Cu, Zn, As, Br, Sb, and Au. Good agreement was found between the two nuclear methods, but a better correlation could have been achieved if the filters had had a larger and more homogeneous deposit. Cassoria stated that the detection limits for NAA are lower than those for PIXE for those elements which could be determined by both methods. Sen [47] determined the concentrations of aerosols collected in Izmir, Turkey on Whatman 41 air filters. NAA determined 22 trace elements including Na, Mg, Al, Cl, K, Sc, Ti, V, Cr, Mn, Fe, Co, Zn, As, Se, Br, Sb, I, Cs, La, Ce, and Th. When combined with meteorological data, it was found that the winter total suspended particulate (TSP) levels were larger with southeasterly winds and decreased with westerly winds. The summer months had higher TSP values due to decreased atmospheric mixing along with long-range transport of the aerosols. Enrichment factor analysis indicated that the levels of Cl, V, Zn, As, Se, Br, Sb, and I in the winter were higher than the 'crustal values.' A comparison between Jungfraujoch (Switzerland), Izmir (Turkey), and Van (Turkey) found the aerosol levels in Izmir to be notably higher.

Sadasivan and Negi [48] reported the results of analysis on aerosol samples collected in Bombay, India with both EDXRF and NAA. The elements determined constitute only about 23% of the TSP matter with associated C, O_2, and N comprising the residual mass. This analysis consisted of the determination of 28 elements on Whatman 41 air filters collected daily for a 10 week period. A varimax rotated factor analysis reveals seven factors: two crustal components, an oil refinery source, a combustion source (probably refuge incineration), sea-salt, a fertilizer industry source, and a mixed source with Th and Cr. EDXRF was recommended as the preferred method for analysis of bulk samples of ambient air while NAA was indicated to be superior for impactor samples or samples with very small volumes.

Biomonitors

In addition to the classical approach to collecting aerosol samples with air filters, an approach through the use of biomonitors has been studied. This method entails taking biological samples to determine aerosol concentrations because certain materials contain a representation of the aerosols from the local environment. These biomonitors seem to be a good reference for the large integrated exposure to which they were subjected. Steinnes et al. [49] demonstrated the feasibility of using the moss *Hylocomium splendens* as a biomonitor in Norway. NAA was the primary analysis method to determine Al, Cl, V, Mn, I, Na, Sc, Cr, Fe, Co, Zn, As, Se, Br, Rb, Mo, Ag, Sb, Cs, La, Sm, and Th. AAS supplemented NAA through the determination of Ni, Cu, Cd, and Pb. These elements were broken down into those resulting from long-range atmospheric transport, local sources, marine environment, local soil dust, and the transfer from soil via higher plants. The geographical plotting of elemental concentrations showed a direct correlation to industry locations in those areas. The results from this study indicate that there is potential for the use of *Hylocomium splendens* moss in the evaluation of environmental aerosols.

Sloof and Wolterbeek [50] carried on a survey of the epiphytic lichen *Parmelia sulcta*, in a Dutch national monitoring survey. NAA was the primary analysis method (AAS was used for the determination of Pb). This survey reported on samples collected in 1986–1987 and compared them to a similar study performed on samples collected in 1982–1983. Target transformation factor analysis (TTFA) for the 1982–1983 data resolved seven separate sources while the 1986–1987 data revealed ten. The results indicated a definite change in the regional concentrations between the two sampling periods with a notable change in the distribution of Cd. This study also resulted in the determination of the time resolution that may be obtained through the use of biomonitoring for the determination of aerosol concentrations. Steinnes [51] determined 28 elements in the moss *Hylocomium splendens* which were collected around Norway. Neutron activation analysis was used to determine 24 elements and AAS was utilized for the determination of Cu, Zn, Cd, and Pb. The results show higher concentrations of Pb, Sb, V, Cr, Cu, Zn, As, Se, Mo, Ag, and Cd to be located in the southernmost parts of Norway. This trend indicates that sources to the south of Scandinavia constitute a large portion of the heavy metal deposition in Norway. The elements which tend to occur as smaller particles ($d < 2\ \mu m$) have the largest difference between northern and southern concentrations.

Markert and Wtorova [52] determined about 50 elemental concentrations in the main vegetation of the Forest Biosphere Reserve near Kalinin, USSR using AAS, ICP-AES, ICP-MS, and NAA. The study included the analysis and comparison of many plant species including *Hypogymna physodes*, *Picea abies* (needles from all years), *Vaccinium vitis-idaea* (leaves), *Vaccinium myrtillus* (leaves), *Polytrichum commune* (above ground parts), and the *Sphagnum* species (entire plant, 1.5 m above ground on the bark of *Picea abies*). A comparison between the plants shows that the lichen *Hypogymna physodes* has the highest trace metal concentrations. Padaki et al. [53] used NAA and ICP for the study of the Spanish Moss *Tillandsia usneoides L.* as an atmospheric biomonitor. The samples were collected in a single location in east Texas and then analyzed for about 35 elements. Samples were collected twice, once in the winter and once in the summer. Contour maps of each element's concentrations were created to display spatial distributions. Elemental concentrations of As, Al, Cs, Co, Cr, Fe, Ga, Hf, Ta, and V correlated well and were able to establish the presence of "hot spots" in certain locations.

Ingrao et al. [54] used NAA to determine Se, Hg, Cr, Cs, Sc, Rb, Zn, Fe, Co, Eu, Sb, Ag, Ce, Ba, and Ni in the head and stem of mushrooms from an area of the Italian Alps. Gamma spectroscopy was used to determine the presence of the radionuclides ^{134}Cs, ^{137}Cs, and ^{40}K. The Cs isotopes were said to originate from the Chernobyl accident. Mushroom heads and stems were used because of their rapid growth rate. A large variability of elemental concentrations was detected in mushroom samples collected in the same location. Thus great caution is recommended for the use of mushrooms as biomonitors. Wyttenbach et al. [55] determined 23 trace elements by NAA in the needles of Norway spruce (*P. abies*) located at 47 sites near Winterthur, Switzerland. The essential elements Ca, Cl, Cu, Fe, K, Mg, Mn, P, and Zn had smaller variations in concentrations at a particular site than the nonessential elements Al, As, Ba, Br, Co, Cs, Hg, La, Na, Rb, Sb, Sc, Sr, and V. It was also found that the elemental variations in concentrations were smaller at particular sites than the variations observed over all the sites. This indicates a measurable difference in elemental concentrations between sites.

Indoor Environment

Toxic element analysis is not limited to the outdoors. Recent works have shown that toxic metal aerosols exist at elevated levels indoors. The air quality of our homes, factories, offices and other public indoor facilities may be determined by NAA methods. Iqbal et al. [56] sampled and analyzed indoor aerosols generated from mild steel welding electrodes. Because welding professionals often complained of weakness after long exposures to fumes, aerosol samples were collected at work stations. Both Whatman 41 and microfibre filters were used for aerosol collection. The filters were then analyzed via NAA. Fifteen trace elements were measured by this method. Elevated levels of Ba, Co, Cr, Sb, Rb, Sc, As, and Fe were found. The authors stated that the results of this experiment will help in the determination and formation of legislation and regulations for the health and welfare of the welders. Kemp [57] used ICP-MS, NAA, and PIXE to analyze airborne dust collected in the Shale Shaker House of an offshore drilling operation in the North Sea. The results obtained by PIXE agreed with the values obtained by NAA. However, the semi-qualitative results by ICP-MS differed by up to a factor of two from the NAA results.

Another area of strong interest is the evaluation of the contributions of environmental tobacco smoke (ETS) to the air quality of public places. The US EPA has declared ETS as a known human carcinogen. In addition to the compounds released to the environment from the burning of a cigarette, such as nicotine, nitrogen oxides, and solanesol, many of the toxic elements are discharged in the form of aerosols (Landsberger et al. [58]). NAA is a good analytical method for the determination of the levels of these toxic elements. Iskander [59] used NAA to determine the elemental makeup of the cigarette tobacco, wrapping paper, cigarette filter before and after smoking, and cigarette ash of a certain brand of Chinese cigarettes. The elemental concentrations found were compared to the concentrations previously reported to exist in other brands, including American ones. The concentrations of Al, Ba, Ce, Fe, La, Mg, Mn, Rb, Sc, and Th were determined to be lower in the Chinese brand, but Zn concentrations were found to be elevated. It was reported that the Chinese filter contained higher concentrations of Br, Cl, K, and Na after the cigarette was smoked. Krivan et al. [60] characterized tobacco smoke condensate with XRF, ICP-MS, NAA, and AAS. A CHN-combustion method determined C, H, and N. S concentrations were found with a precipitation micro-titrimetry method. Of the 57 elements assayed, 35 elements were determined in cigarette smoke condensate resulting from two different cigarette brands. The elements As, Cd, and Pb are most relevant in this determination because they have a relatively high transfer rate from tobacco into smoke.

Nuclear Methods

Nuclear and nuclear-related techniques, i.e. instrumental neutron activation analysis (NAA) and other forms of activation analysis, X-ray fluorescence (XRF), and particle-induced X-ray emission analysis (PIXE), allow a nondestructive, multielement analysis of solid samples, and have therefore found wide applicability in the analysis of atmospheric aerosols. Although the nuclear and nuclear-related techniques (hereafter referred to only as "nuclear techniques") discussed here are all very powerful techniques for such environmental samples, each of them has its own specific advantages and drawbacks [4]. The sensitivities and detection limits in NAA exhibit large fluctuations from element to element, so that the technique is highly sensitive for elements such as Al, As, Co, Cr, Mn, Fe, Sc, Se, V, Zn, halogens and several rare-earths, but cannot generally be used for the determination of Si, P, S, Sn, Tl, Pb or the light elements. ED-XRF and PIXE have the advantage that the detection limits vary in a smooth way with atomic number and that they are, in principle, sensitive to all elements of the periodic table from Na onwards. However, of the elements heavier than Sr, only a rather few are actually observed in atmospheric aerosol samples, and the total number of elements typically detected by the X-ray techniques remains limited to 15–20. By NAA, up to 30–40 elements are commonly observed in atmospheric aerosols. Interesting anthropogenic elements which are easily measured by NAA, but often missed in the X-ray techniques, include V, As, Se, In and Sb. The X-ray techniques, on the other hand, can easily detect Si, S and Pb, which are difficult or impossible to measure by NAA. Although the total number of elements observed by X-ray techniques is far less than in NAA, X-ray techniques offer high speed and the advantage of low cost (in the case of EDXRF) or small sample size requirement (in PIXE), so that they have often been preferred in studies where large sample loads are encountered. However, a significant advantage of NAA over the X-ray techniques (and most other analytical techniques) is that it is virtually free of matrix effects. Consequently, the constraints on aerosol collection substrates and sample types are far less in NAA than in EDXRF and PIXE. From the point of view of sample size, the amount of sample required for NAA depends to a large extent upon the neutron flux available, but is typically of the order of several hundred ug to several mg of particulate material. In EDXRF, the mass requirement is of the order of 100 μg^{-1} mg, whereas in PIXE, it is down to 10–100 μg. The small mass requirement in PIXE makes this truly a microanalytical technique.

Although nuclear techniques generally lend themselves much better to the analysis of solids than of liquids, they may be used for the latter samples too. In such cases, physical or chemical preconcentration techniques are usually applied, so that the liquid is converted to some solid sample which is appropriate for analysis [61]. One variant of EDXRF, i.e. total reflection XRF (TRXRF), requires that the sample be deposited as a thin layer on a suitable reflector substrate. Consequently, solid samples first have to be dissolved, after which a small fraction of the solution is pipetted onto the reflector and dried. Therefore, and in contrast to the other nuclear techniques, TRXRF is more suitable for the analysis of liquid (e.g. rainwater, cloud or fog samples) than of aerosols.

Instrumental Neutron Activation Analysis

NAA with the use of reactor thermal neutrons is one of the most important analytical techniques for the determination of trace elements in environmental samples. The technique has found numerous applications in air pollution and atmospheric trace element studies. Favourable characteristics of NAA are its high sensitivity, the multielement capability, nondestructive character and the virtual absence of matrix effects for airborne particulate matter. The application of NAA to the analysis of airborne particulate matter has been reviewed by Alian and Sansoni [62, 63]. Other relevant information can be found elsewhere [64–67]. The method of NAA takes advantage of the fact that when a sample is irradiated with neutrons some of the nuclei in the target will be converted into radioactive isotopes. The radioisotopes decay at a characteristic rate, in most cases, emitting γ-rays of characteristic energies and intensities. These γ-rays are used to identify the elements present and to determine their concentrations. For sensitive, multi-element NAA, access to a nuclear research reactor is essential. The reactor should preferably be equipped with a pneumatic tube system to provide for the fast transfer of samples to the irradiation position in or near the reactor core. The thermal neutron flux density should be of the order of 10^{12} neutrons $cm^{-2}s^{-1}$ or higher.

Most of the measurements for NAA are carried out by means of a high resolution γ-ray spectrometer with a high-purity (HPGe) coaxial detector (for γ-rays energies between 100 and 3200 keV) or sometimes with low energy photon detector (LEPD) with a planar configuration (energy range 60–200 keV). The detector is normally surrounded by lead shielding and is provided with a sample holder for easy and reproducible positioning of the samples. An automatic sample changer is very useful for routine measurements. The spectrometer further includes a preamplifier, an amplifier (usually with pulse pile-up rejector), an analog-to-digital converter (ADC) and a multichannel analyzer or buffer (MCA) interfaced to micro-(mini) computer or to a personal computer (PC). Spectra are usually stored on disks for further analysis by means of dedicated software.

NAA can be carried out as a relative method by using several multielement standards containing all elements to be determined co-irradiated with the samples. Another possibility is the application of single or multiple comparator techniques using so-called k-factors. This technique has been improved by De Corte and co-workers [65] by introducing more suitable k_0 factors.

Samples of atmospheric particulate matter are usually subjected to two different irradiations in a nuclear reactor, the first for a few minutes to produce radioisotopes with short half-lives of up to several hours. Usually two γ-ray spectra are taken in succession, the first one after 3-7 min decay for measurement of ^{28}Al, ^{49}Ca, ^{51}Ti, ^{52}V and ^{66}Cu and the second one after 15-20 min decay for the determination of ^{24}Na, ^{27}Mg, ^{38}Cl, ^{56}Mn, ^{80}Br, ^{87m}Sr, ^{116m}In, ^{128}I, ^{139}Ba and ^{165}Dy. After a few days decay the samples are irradiated once more for 2-24 h to produce substantial amounts of long-lived activities. After a decay period of 2-5 days the samples are counted to measure ^{24}Na, ^{42}K, ^{64}Cu, ^{69m}Zn, ^{72}Ga, ^{76}As, ^{82}Br, ^{99}Mo, ^{115m}In (from the decay of ^{115}Cd), ^{239}Np (from the decay of ^{239}U) ^{122}Sb, ^{140}La, ^{153}Sm, ^{187}W and ^{198}Au. Finally, after a 15-20 day decay period the radionuclides ^{46}Sc, ^{51}Cr, ^{59}Fe, ^{60}Co, ^{58}Co (formed from ^{58}Ni (n,p) reaction), ^{65}Zn, ^{75}Se, ^{86}Rb, ^{110m}Ag, ^{124}Sb, ^{131}Ba, ^{134}Cs, ^{141}Ce, ^{152}Eu, ^{181}Hf, ^{182}Ta, ^{203}Hg and ^{233}Pa (formed from ^{232}Th) can be measured.

The evaluation of the γ-ray spectra aims at an accurate determination of the net peak areas of the γ-lines of interest and it is invariably done with the aid of a computer. Many γ-ray spectrum evaluation programs also containing multiplet deconvolution procedures have been developed and some of them are commercially available. An up-to-date discussion of the evaluation of γ-ray spectra is given by Erdtmann and Petri [66]. Computer programs are also used for the calculation of final concentration data and for quality assurance purposes. Only a few spectral interferences are encountered in NAA of air particulate matter or wet deposition samples, and they can easily be accounted for. In some practical applications a selective irradiation with epithermal neutrons (energy range 0.4 eV-1 MeV) can be used to advantage Epithermal NAA (ENAA) makes it possible to determine some elements in aerosols that are very difficult to do by thermal NAA, such as Ni, In, Ba, Mo, Sr and U [67]. The determination of silicon is also significantly enhanced using ENAA. Many of these techniques and methodologies are discussed by Landsberger [5]. Landsberger and Wu [68] used epithermal NAA methods in conjunction with Compton suppression counting techniques on samples collected near Lake Ontario to detect trace levels of Sb, As, Cd, In, I, Mo, Si, and U. Detection limits were typically found to be approximately 0.3 ng for Sb, 0.7 ng for As, 8 ng for Cd, 0.2 ng for In, 14 ng for I, 5 ng for Mo, 60 µg for Si, and 2 ng for U. The Si values are especially noteworthy because of the universal problems encountered in the detection of Si through conventional NAA.

Epithermal NAA in conjunction with Compton suppression counting techniques were also used by Landsberger et al. [57] to determine concentrations of Cd in environmental tobacco smoke. Concentrations of Cd were found to be as high as 38 ng/m^3 in certain indoor conditions. This concentration is approximately 30 times higher than in indoor areas where smoking is prohibited. The detection limits in NAA depend upon many factors, including the background in the γ-ray spectrum, and thus the composition of the sample and filter blank. For aerosol research, particularly for samples collected in remote areas, impurity levels in the collection substrates are often the limiting factor.

Fast Neutron Activation Analysis (FNAA)

Although FNAA is used much less extensively than thermal neutron activation analysis, it became a promising technique with the development and availability of high yield 14-MeV neutron generators. Neutrons of about 14-MeV are produced by the $^{3}H(d,n)^{4}He$ reaction. Irradiation of a sample with 14-MeV neutrons gives rise to a variety of reactions, such as (n, 2n), (n, p) etc. This provides certain advantages, such as the production of different radioisotopes. By this technique the concentrations of Si (which cannot conveniently be determined by thermal NAA) and of some other major elements, such as Al and Fe, can be determined in atmospheric particulate material within a few minutes [69].

Instrumental Photon Activation Analysis

IPAA is based on bombarding the samples with high flux γ-rays, which are produced by impacting energetic electrons (typically 35 MeV) on high Z metals, such as W, in a linear accelerator (LINAC). IPAA is an excellent complementary technique to INAA producing isotopes by (γ, n), $(\gamma, 2n)$ and possibly (γ, p) reactions. The most important elements which can easily be determined by IPAA, but not by NAA, are Si, Pb, Cd and Ni. This method has been used for the determination of trace elements in atmospheric particulate material and other environmental samples [70, 71]. Unfortunately, the number of LINACs in operation in the world is declining, and as a result IPAA is not used extensively.

Prompt Gamma Activation Analysis (PGAA)

PGAA is similar to NAA in the sense that thermal neutrons are commonly utilized for irradiation and emitted γ-rays are measured. However, there are some fundamental differences. The determination of the elemental composition of a sample by NAA depends upon the measurement of delayed radiations following neutron capture. Therefore, for an element to be measured by NAA, it is preferable that at least one of its stable isotopes have a fairly large product of its isotopic abundance and thermal neutron cross section. Also the reaction products should emit γ-rays of suitable intensity in their decay. PGAA is independent of the radioactive properties of the product nuclide. It utilizes the large number of γ-rays emitted immediately (within 10^{-12} s) following the decay of the excited nucleus formed by neutron capture. Since the technique does not depend on the measurement of delayed radiation, one can obtain the results immediately after the irradiation. In principle the method can shorten the turn around time of the analysis as well as allow the measurement of some elements, such as H, B, C, N and Si, which are not detectable with conventional NAA [72]. However, long counting times, up to 24 h, may be needed to achieve low detection limits. Since only one sample can be irradiated at a time, and a series of many samples may be needed to be analyzed, there may indeed be a very long time for a turn around for the results. Although PGAA systems can be developed using in-core irradiation, a collimated and thermalized neutron beam guided to the outside of the reactor core is preferable. A recent development is the use of cold neutron sources for PGAA, which offer detection limits that are better by one or two orders of magnitude.

Energy Dispersive X-ray Fluorescence

Energy dispersive X-ray fluorescence analysis (EDXRF) is a powerful non destructive analytical tool [73]. It makes use of the fact that excitation of a sample with energetic photons (up to 60 keV) give rise to X-rays which are characteristic of the elements present. As photon sources, radioisotopes, X-ray tubes or even synchrotrons can be used. The photons impinging on the sample (and substrate) lead to several interaction processes. By the photoelectric effect the emission of characteristic fluorescent radiation is induced. Elastic and inelastic scattering process are mainly responsible for the background which is observed in the spectrum. For the collection of the fluorescent radiation an energy dispersive spectrometer system with a Si(Li) detector is used. The measured spectrum allows both qualitative and quantitative analysis. Water cooled high power X-ray tubes (2–3 kW) are useful for producing an intense flux of photons increasing the sensitivity in standard EDXRF and advanced EDXRF techniques. Radioactive isotopes create a photon flux about 4–5 orders of magnitude lower since the source activity is normally limited to 25 mCi to 1 Ci. As a result of the decaying half-life, intensity and cost, tube excited systems are usually preferred to isotope excited systems. The primary radiation from an X-ray tube is polychromatic, i.e. it consists of a Bremsstrahlung spectrum with the characteristic radiation of the anode superimposed on it. To achieve better detection limits it is advisable to convert the primary spectrum into nearly monochromatic radiation. This can be achieved by using a transmission tube, by placing selective filters in the beam path, by employing the secondary target excitation technique, or by applying the high-energy cut-off effect of a totally reflecting optical flat in the beam path.

Both solids and liquids can be analyzed by EDXRF. However, to increase the sensitivity, liquid samples are almost always converted to solids by using physical and/or chemical preconcentration techniques. When the mass thickness of the sample exceeds a few mg/cm^2, matrix effects (i.e. X-ray attenuation and secondary fluorescence) have to be allowed for and corrected. For this reason, accurate quantitative results are easier to obtain for thin film samples than for bulk samples. Because of the limited resolution of the Si (Li) detector, severe line overlap often occurs in EDXRF spectra (e.g. of K_a and K_β lines of adjacent elements in the periodic table). In addition, evaluation of the background is complicated by the absence of pure background regions between the peaks. For these reasons, sophisticated mathematical treatment of the spectra is generally needed. EDXRF spectrum evaluation software is presently available for personal computers [74]. These programs allow one to perform peak search and identification and to calculate the net peak areas of the various detected X-ray lines. In converting the X-ray intensities into concentrations, matrix effects must be considered. Only in the thin film technique is a linear correlation between X-ray intensity and concentration expected. The detection limits for the standard EDXRF and thin film samples on a thin substrate (a few μm) are in the range of 5–20 ng/cm^2 for most elements. In bulk samples mg/g concentrations are detectable.

Total Reflection X-Ray Fluorescence

In advanced techniques such as TRXRF the detection limits have been improved considerably because the background originating from the sample support is drastically reduced [75]. TRXRF is preferably a thin film technique. Having optimal excitation conditions, absolute detection limits down to the pg level and relative detection limits down to the ng/ml level are achieved for trace elements in aqueous solutions. Dilute solutions can simply be pipetted onto the total reflector and dried without having to rely on chemical preconcentration techniques. TRXRF may also be used for the analysis of atmospheric aerosols but the samples in such cases must first be transferred to a total reflector. Another approach to improve the detection limits by reducing the background is to use linearly polarized X-rays. Samples of intermediate thickness, such as aerosols on filters, can be analyzed without any sample preparation. Detection limits below 100 ng/g are obtainable. If, instead, synchrotron radiation is used as the exciting radiation and a total reflection geometry is employed, even lower detection limits can be achieved. However, the use of synchrotron radiation is still not considered a routine method of analysis.

Particle-induced X-ray Emission

Similar to EDXRF, PIXE is a form of X-ray emission analysis, but while electromagnetic radiation is used in EDXRF to produce the characteristic X-rays of the atoms in the sample, heavy charged particles (typically 1–4 MeV protons) are used for excitation in PIXE. The exciting particle beams are normally produced by a small particle accelerator, such as a Van de Graaff accelerator. For the detection of the X-rays emerging from the sample, an energy-dispersive spectrometer with a Si(Li) detector is almost invariably used. A detailed discussion of the PIXE technique and of its applications can be found in a comprehensive textbook [6]. In most PIXE set-ups incident beam diameters of 1 to 10 μm are used. As it is also advisable to limit the mass density of the sample to less than 1 mg cm^{-2} (in order to keep matrix effects to a strict minimum). PIXE is particularly suitable for the analysis of small, thin samples. In atmospheric aerosol research, where the need to collect sufficient material for analysis often puts severe constraints on sampling devices and pumps, this is a particularly interesting feature. It has led to the development of small, inexpensive and light-weight samplers with high time and/or size resolution to be used in combination with PIXE analysis [76, 77].

With regard to the filters and impaction surfaces to be used in the aerosol samplers, PIXE is subject to the same restrictions as those applicable in ED-XRF. In order to keep the background minimal and to avoid the necessity of accounting for absorption effects in the collection substrate, thin poly carbonate filters (e.g. Nuclepore) or thin organic films (e.g. Mylar), which collect the particulate material on their surfaces, are to be preferred. Although PIXE analysis of wet deposition samples is less straightforward than that of atmospheric aerosols, the technique can also be employed for such samples. The physical and chemical preconcentration methods employed are normally similar to those used in EDXRF [61, 78].

For the determination of the net peak areas of the characteristic X-ray lines in a PIXE spectrum, non-linear or linear least-squares fitting routines, similar to the ones used in EDXRF, are generally employed. A compilation of PIXE spectrum evaluation programs, currently used in many PIXE laboratories, has been reviewed [79]. The matrix effects in PIXE show great similarity to those in EDXRF. However, because of the fact that charged particles instead of photons are used for the excitation in PIXE, the description of that part of the matrix correction, which is related to the incident radiation, is quite different. When dealing with thin aerosol samples (mass thickness less than 1 mg cm^{-2}), the total matrix correction remains typically below 10%, except for the lighter elements (Na–Ca). Compared with standard EDXRF, PIXE offers absolute detection limits which are over two orders of magnitude better. Relative detection limits, expressed in ng cm^{-2} of collection substrate, vary between 0.3 and 5 for many elements in PIXE. Under common conditions, atmospheric trace element concentrations down to 0.05 mg/m^3 can be measured by PIXE.

An advantageous characteristic of PIXE, compared with EDXRF, is that it can be complemented by other nuclear techniques which employ light ions as incident particles, such as particle-induced γ-ray emission analysis (PIGE) and particle elastic scattering analysis (PESA) [80, 81]. These techniques allow one to measure several light elements, including C, O, N, F and Na in atmospheric aerosol samples. Often the PIGE and/or PESA measurements can be done concurrently with the PIXE measurements.

Comparison with Other Techniques

In addition to the nuclear techniques discussed above, various other analytical techniques are available for measuring trace elements. Among these, atomic absorption spectrometry (AAS), inductively-coupled plasma atomic emission spectrometry (ICP–AES) and inductively-coupled plasma mass spectrometry (ICP–MS) are at present the most useful for atmospheric trace element research. When comparing the nuclear techniques with the non-nuclear ones from the viewpoint of sample type, it is evident that NAA, XRF and PIXE lend themselves much better to the analysis of solid samples than of liquids (except for TRXRF), whereas AAS, ICP–AES and ICP–MS are essentially techniques for analyzing solutions. Consequently, wet deposition samples and cloud and fog water samples are ideal samples to be handled by the latter three techniques and TRXRF. However, short-lived NAA can be very successfully employed to determine simultaneously halogens (Br, Cl and I) as well as Al, Mn, Na and V in soluble atmospheric samples [82]. For the analysis of atmospheric aerosols and the insoluble fraction of wet deposition, NAA, standard EDXRF and PIXE offer definitive advantages. The analysis of those samples by TRXRF, AAS, ICP–AES and ICP–MS generally requires decomposition or extraction procedures, which always involve the danger of extraneous additions, losses or incomplete dissolution of the elements of interest. In NAA, standard EDXRF, and PIXE, aerosol samples can be analyzed directly or after very little sample preparation. Another significant advantage of the latter three techniques is their nondestructive character, so that the analyzed sample can still be inspected by other techniques.

The nuclear and non-nuclear techniques discussed here are all subject to spectral interferences and matrix effects. However, in practice, matrix effects are generally entirely negligible in NAA. Matrix effects in XRF and PIXE depend upon the sample thickness and can be very small when, for example, atmospheric particles are collected as a monolayer on the surface of a filter or substrate film. In any case, the two X-ray emission techniques have the advantage over AAS, ICP–AES and ICP–MS that the physical processes responsible for matrix effects are very well understood. Within the AAS techniques, matrix effects are much more pronounced in electrothermal atomization AAS than in flame AAS, and both ICP-based techniques are prone to a variety of matrix effects.

Filter Choice and Blanks

The choice of a filter is critically important when using any of the NAA, XRF or PIXE techniques. For instance, many types of filters undergo neutron damage under irradiation and thus special handling procedures must be used when transferring them to other inert containers. Under PIXE bombardment some filters may actually be burned and even tear. All these methods are also limited by the blanks. For instance, in NAA, relatively high amounts of Na, Br, Cl, and Al may severely compromise low level concentrations of other elements due to the high background levels when the filters are neutron irradiated. This is a common problem in Whatman filters. On the other hand, Teflon filters have much lower inherent contaminants. Special attention must be given to the following criteria for filter choice.

1. Choose a filter which is easy to handle and which can be placed and taken out from irradiation vials. Some filters become highly charged after irradiation and are difficult to handle.
2. Choose a filter which meets your experimental design-size fractionation, PM-10, high volume or low volume sampler, etc.
3. Identify which elements in the filter may add unusually high backgrounds.
4. Analyze the filter for the levels of the blank values and ascertain how reproducible the results are before you begin your sampling.
5. Test the filter under different irradiation conditions to see its integrity.

Quality Assurance, Quality Control and Reference Materials

Reliable analytical measurements of environmental samples are an essential requirement for data interpretation and subsequent applications. Each analytical laboratory has to establish a continuous program to ensure the dependability and validity of the data. This program has to include both the sampling and all other parts of the analytical procedure. Overall objectives should include site selection, network design, meteorology, topography and sampling schedules. The next important phase is the choosing of sampling equipment and proper filter materials. The filters chosen must be appropriate for the technique used. It is recommended that the total mass loading on the filter be determined whenever possible. Special care should be taken with the sampling design so that the appropriate amount of mass will be available for PIXE, EDXRF or NAA. In general 1 mg/cm^2 or less is used in PIXE or EDXRF to assure minimization of matrix effects. It is also recommended that, before any sampling is done, several blank filters be analyzed to establish the level of impurities in the filters. Calibration of the sampling equipment for air flow is essential as well as proper temperature and pressure corrections. To ascertain any contaminant in the procedure, several field blanks should be measured regularly. All samples should be handled with plastic utensils and placed in clean and clearly marked plastic containers. For precipitation samples, contamination during collection can be a serious problem. The utmost care should be devoted to quality assurance of water samples.

It is recommended that calibration of the analytical systems be done by using primary standards (prepared from pure elements or compounds). In the case of PIXE and EDXRF, homogeneity should be tested by irradiating different sections of the filter. If convenient, this procedure can also be used for NAA by dividing the filters in halves or quarters. In general, strict analytical protocols should be followed

To check the accuracy it is highly recommended that certified reference materials be employed whenever available. While several reference materials exist for NAA, only very few are available for PIXE and EDXRF. Laboratories should be encouraged to participate in interlaboratory exercises. These types of exercises will give a good indication of overall laboratory performance including calibration and data evaluation.

Conclusions

Nuclear methods have proven their effectiveness in the evaluation of trace elements collected in airborne particulate matter. Aerosol samples are collected and routinely analyzed for multiple elements. In particular NAA is known to be a reliable analysis technique with low detection limits and samples may be prepared with relative ease when compared to other analytical techniques. NAA has been used to determine the elemental concentrations of aerosols for a variety of samples. Its sensitivity has been effective in studies involving aerosols of the polar regions as well as aerosols collected in urban and rural areas. Aerosols have even been collected indoors and toxic elements such as Cd have been determined as a result of cigarette smoke. Most evaluation of aerosols in the atmosphere originates with the collection of aerosols on filters with a wide variety of sampling devices. Biomonitors have also taken on a role in the collection of aerosols over a longer period of time. Lichens, mosses, and a wide variety of other biological materials act as filters to the environment in which they exist. Variation of metals in the atmosphere are linked to the concentrations of those elements in a particular biological monitor. NAA is very useful for analyzing these types of samples due to the minimal sample preparation. A judicious use of NAA along with meteorological conditions reveals the trace element composition of atmospheric aerosols and their sources. These clues help us understand the condition of our atmosphere and indicate areas of concern.

Acknowledgments. Part of this work was derived from an unpublished International Atomic Energy Agency 1988 report on Nuclear Techniques in Air Pollution Research and Monitoring.

References

1. 1990 United States Clean Air Act Amendments
2. Hevesy G, Levi H (1936) Mathematisk-Fysiske Meddelelser 14:3
3. Johansson JB, Akselsson KR, Johansson SAE (1970) Nucl Inst Meth 84:141
4. Maenhaut W (1992) Int J PIXE 2:605
5. Landsberger S (1992) J Trace and Microprobe Tech 10:1
6. Johansson SAE, Campbell JL (1988) PIXE: a novel technique for elemental analysis. John Wiley, New York
7. Alian A and Sansoni B (1990) in: Alfassi Z (ed) Activation Analysis. CRC Press, Boca Raton, p. 503-565
8. Tölgyessy J and Klehr EH (1987) Nuclear Environmental Chemical Analysis. John Wiley & Sons, New York
9. Parry SJ (1991) Activation Spectrometry in Chemical Analysis. John Wiley and Sons, New York
10. Obrusnik I (1990) Chemia Analitycana. 35:221
11. Artaxo P, Storms H, Bruynseels F, Van Grieken R, Maenhaut W (1988) J Geophys Res 93:1605
12. Artaxo P, Orsini C (1987) Nucl Inst and Meth in Phys Res B22:259
13. Loureiro A, Vasconcellos M, Pereira E (1992), J Radioanal and Nucl Chem 159:21
14. Bottenheim JW, Barrie LA, Atlas E, Heidt LE, Niki H, Rasmussen RA, Shepson PB (1990) J Geophy Res 95:18555
15. Dick A (1991) Geochim Cosmochim Acta 55:1827
16. Pereira EB, Loureiro ALM, Vasconsellos MBV (1992) Atmos Env 26A:1549
17. Zoller Wh, Gladney ES, Duce RA (1974) Science. 183:198
18. Wagenbach D, Gorlach U, Moser K, Munnich KO (1988) Tellus. 40B:426
19. Meeker KA, Chuan RL, Kyle PR, Palais JM (1991) Geophys Res Lett 18:1405
20. Koyama M, Takada J, Kamiyama K, Fujii N, Inoue J, Issiki K, Nakayama E (1988) J Radioanal Nucl Chem 124:235
21. Landsberger S, Vermette VG, Stuenkel D, Hopke PK, Cheng MD, Barrie LA (1992) Env Pollut 75:181
22. Landsberger S, Vermette AJ, Barrie LA (1990) J Geophy Res 95 (D4):3509
23. Landsberger S (1988) Anal Chem 60:1842
24. Cheng Md, Hopke PK, Landsberger S, Barrie LA (1991) in Volume of the Seventh Joint Conference on Applications of Air Pollution Meteorology with AWMA. American Meteorological Society, Boston p. 29
25. Cheng MD, Hopke PK, Landsberger S, Barrie LA (1991) Atmos Environ 25A:2903
26. Landsberger S, Hopke PK, Cheng MD (1992) Nucl Sci Eng 110:79
27. Pacyna JM, Bartonova A, Cornille P, Maenhaut W (1989) Atmos Environ 23:107
28. Tuncel G, Aras NK, Zoller WH (1989) J Geophy Res 94(D10):13025
29. Barrie LA, den Hartog G, Bottenheim JW (1989) J Atmos Chem 9:101
30. Lowenthal DH, Rahn KA (1985) Atmos Environ 19:1995
31. Sheridan PJ, Zoller WH (1989) J Atmos Chem 9:363
32. Sturges WT, Barrie LA (1988) Atmos Env. 22(6):1179
33. Barrie LA, Barrie MJ (1990) J Atmos Chem 11:211
34. Barrie LA, Toom D, Landsberger S, Sturges W (1994), J Geophysical Res, 99D:25435
35. Hacisalihoglu G, Eliyakut F, Olmez I, Balkas TI, Tuncel G (1992) Atmos Environ 26A:3207
36. Bichler M, Platschka R, Poljanc K (1992) J Radioanal Nucl Chem Lett 165:61
37. Rusheed A, Ahmed S, Mannan A, Qureshi IH (1987) Available from the National Technical Information Service, Springfield, Va. 22161 PINSTECH/NCD-109
38. Hopper JF, Barrie LA (1988) Tellus. 40B:446
39. Hacisalihoglu G, Balkas TI, Arami M (1991) Black Sea Oceanography. 351:469
40. Yoshizumi K (1991) Energy and Buildings. 15-16:711
41. Ko MM, Jervis RE (1992) J Radioanal Nucl Chem 161:159
42. Paiva RP, Munita CJS, Cunha IIL, Romano J, Alonso CD (1990) Biol Trace Elem Res 26-27:231
43. Okamoto S, Kobayashi K, (1991) Atmos Env 25B(1):133
44. Qadir MA, Iqbal MZ, Zaidi JH (1992) J Radioanal Nucl Chem Lett 166:465
45. Lobo KK, Tshiashala MD, Lumu BM, Matamba K, Ronnueau C (1990) Biol Trace Elem Res 26-27:295
46. Cassoria V, Rojas X, Gras N, Chuaqui L, Dinator M, Morales JR, Llona F, Romo-Kröger C (1993) Comision Chilena De Energia Nuclear Nucleotecnia. 13:31

47. Sen O (1988) Atmos Environ 22:795
48. Sadasivan S, Negi BS (1990) The Sci. of the Total Env 96:269
49. Steinnes E, Rambæk JP, Hanssen JE (1992) Chemosphere 25(5):735
50. Sloof JE, Wolterbeek HT (1991) Water, Air, and Soil Poll 57-58:785
51. Steinnes E, (1980) J Radioanal Chem 58:387
52. Markert B, Wtorova W (1992) Vegetatio 98:43
53. Padaki PM, McWilliams EL, James WD (1992) J Radioanal Nucl Chem 161:147
54. Ingrao G, Belloni P, Santaroni GP (1992) J Radioanal Nucl Chem 161:113
55. Wyttenbach A, Bajo S, Tobler L (1990) Biol Trace Element Res. 26-27:213
56. Iqbal MZ, Qadir MA, Zaidi JH (1992) J Trace Element Electrolytes Health Dis 6:63
57. Kemp K (1992) J. Trace and Microprobe Tech 10:43
58. Landsberger S, Larson S, Wu D (1993) Anal Chem 65:1506
59. Iskander FY (1992) J Radioanal Nucl Chem 159:105
60. Krivan V, Schneider G, Baumann H, Reus U (1994) J Anal Chem 348:218
61. Landsberger S, Jervis RE, Monaro S (1985) in: Lawerence JF (ed) Trace Analysis. Vol.4, Academic Press, New York, p 237
62. Alian A, Sansoni B (1985) J Radioanal Nucl Chem 89:191
63. Alian A, Sansoni B (1988) Rep. JUL-SPEZ-475 (Kernforschungsanlage Jülich GmbH., Zentral Library, D-5170 Jülich-1, FRG), ISSN 0343-7639
64. Kolthoff IM, Elving PJ, Krivan V (eds) Treatise on Analytical Chemistry, 2nd Ed., Part I, Vol, 14, Sect. K: Nuclear Activation and Radioisotopic Methods of Analysis, Wiley, New York
65. De Corte F, Simonits A, De Wispelare, Hoste J (1987) J Radioanal Nucl Chem 113:145
66. Erdtmann G, Petri H (1986) in: Kolthoff IM et al. (eds) Treatise on Analytical Chemistry, 2nd Ed., Part I, Vol. 14, Wiley, New York
67. Janssens M, Desmet B, Dams R, Hoste J (1975) J Radioanal Nucl Chem 26:303
68. Landsberger S, Wu D (1993) J Radioanal Nucl Chem 167:219
69. Olmez I (1981) Turkish J Nucl Sci 8:15
70. Aras NK, Zoller WH, Gordon GE, Latz G (1973) Anal Chem 45:1981
71. Olmez I, Aras NK, Zoller WH, Gordon GE (1974) Anal Chem 46:935
72. Failey MP, Anderson DL, Zoller WH, Gordon GE, Lindsrom RM (1979) Anal Chem 51:2209
73. Markovic AA, Van Grieken RE (1988) Anal Chem 60:28R
74. Van Espen P, Janssen K, Nobels H, (1986) AXIL-PC software for the analysis of complex X-ray spectra, Chemometrics and Intelligent Laboratory Systems 1:109
75. Wobrauschek P, et al. (1988) J Trace and Microprobe Techniques
76. Akselsson, KR (1984) Nucl Instrum Meth in Phy Res B3:425
77. Annegarn HJ, Cahill TA, Sellschop JPF, Zucchiatti A (1986) Physica Scripta 37:282
78. Hansson H, Ekholm Ak, Ross HB (1988) Environ Sci Technol 22:527
79. Waetjen U (1987) Nucl Instrum Meth in Phys Res B22:29
80. Robaye G, Delbrouck-Habaru JM, Roelandts I, Weber G, Girard-Reydet L, Morelli J, Quisefit JP (1985) Nucl Instrum Meth in Phys Res B6:558
81. Martinsson BG (1986) Nucl Instrum Meth In Phys Res B15:636
82. Landsberger S, Drake JJ, Vermette SJ (1988) Chemosphere. 17:299

Methods for Measuring Atmospheric Acidic Particles and Gases

Constantinos Sioutas and Petros Koutrakis

Harvard University, School of Public Health, 665 Huntington Avenue, Boston, MA 02115, USA

Introduction ... 203
Formation and Physico-Chemical Properties of Acid Aerosols and Gases . 203
Sulfuric Acid ... 203
Nitric, Nitrous and Hydrochloric Acids ... 204
Organic Acids ... 205
Environmental Fate ... 206
Methods to Measure Acidic Particles and Gases ... 206
Filtration ... 207
Diffusion Denuders ... 208
Hollow Denuders ... 209
Denuder Difference Method ... 210
Annular Denuders ... 212
Multiannular Denuders ... 214
Honeycomb Denuders ... 215
Determination of Acid Particle Neutralization due to Interactions of Collected Particles ... 217
Inlet loss of Reactive Gases ... 218
Particle Transmission Efficiency of Denuders ... 219
Sampling of Acidic Aerosols Using Impactors ... 220
Microorifice and Low-Pressure Impactors ... 221
Size Distribution of Atmospheric Sulfates ... 222
Field Method Intercomparison for the Characterization of Acid Aerosols and Gases ... 223
Chemical Analysis of the Collected Material on Denuders, Filters, and Impactor Substrates ... 226
Acidic Aerosol Concentrations: Results from Field Studies ... 226
References ... 228

Summary

This chapter presents an overview of methods to measure atmospheric acidic particles and gases. The main formation mechanisms and physico-chemical properties of inorganic and organic acidic aerosols are discussed first. The second part is focused on the main methods of measuring these pollutants, with special emphasis spent on the shortcomings of using filters, the advantages and disadvantages of diffusion denuders, and the use of conventional impactors to obtain the size distribution of acidic particles in the atmosphere. The final part includes results from field studies conducted in the northern United States, presenting typical levels of ambient acidic pollutants, and discussion of their spatial and temporal concentration variations in urban and rural areas.

Introduction

Atmospheric acidity is defined as acids and their precursors residing in the atmosphere in the gas and aqueous phases. The term aerosol strong acidity refers to the particulate acidity contributed by sulfuric acid or partially neutralized sulfate particles. Weak acids include inorganic species such as nitrous acid, dihydrogen and hydrogen phosphates, and organic species, such as carboxylic acids, phenols, and some pesticide compounds. Oxidation of sulfur dioxide (SO_2), nitrogen oxides (NO_x) and hydrocarbons originating from anthropogenic and natural sources through homogeneous (gas-phase) and heterogeneous (particulate- or water-phase) reactions in the atmosphere results in the formation of acidic gases and particles. Sulfur dioxide is mainly produced by stationary source combustion of coal and residual fuel oil for industrial processes and electric power generation. The major sources of nitrogen oxides are fossil-fuel combustion for electricity and industrial processes and automobile emissions; mobile sources are responsible for roughly half of these emissions [1]. Acid precursors, SO_2, NO_x, and hydrocarbons can dissolve in water or fog droplets and subsequently be oxidized by H_2O_2 to produce acidic particles and gases, as well as react in the gas phase (homogeneous reactions) to produce acid gases. Formation of gases on particles (heterogeneous phase reactions) may also occur, but it is only important near emission sources where the particle concentration is very high [2]. This account consists of three main parts: the first part describes the formation and physico-chemical properties of acid aerosols and gases; in the second part, a description of several integrated methodologies to measure acid particulate and gas phase compounds is given; finally, the third part includes results from recently conducted acid aerosol measurement studies.

Formation and Physico-Chemical Properties of Acid Aerosols and Gases

Sulfuric Acid

Sulfuric acid, H_2SO_4, can be formed through homogeneous gas or aqueous phase reactions. Reaction of SO_2 with OH radicals is the dominant formation mechanism. This reaction occurs during the day, when OH is present. Furthermore, dissolved SO_2 reacts in the aqueous phase with hydrogen peroxide, H_2O_2, and other oxidants to form sulfate droplets. Under usual atmospheric conditions, H_2SO_4 nucleates with water vapor to form aerosol droplets. Oxidation of SO_2 by H_2O_2 is the most important aqueous phase oxidation mechanism, since it can occur either during photochemical (daytime) or dark (nighttime) reactions [3]. In addition, H_2SO_4 can be produced by the oxidation of SO_2 that is adsorbed on the surfaces of solid particles, especially near emission sources where the concentration of these particles is high enough to provide sufficient surface for the reaction to proceed [2].

Hydrogen and bisulfate ions (H^+, HSO_4^-) in a droplet solution react with ammonia (NH_3) to form ammonium ions (NH_4^+). Since the vapor pressure of ammonia over acidic sulfate particles is negligible, ammonia diffuses onto the sulfate particle surface until complete neutralization occurs. Consequently, atmospheric sulfuric acid or partially neutralized sulfuric acid particles cannot co-exist with ammonia [4]. This neutralization may be inhibited by the presence of organic films on the surface of atmospheric aerosols [5].

Depending on the ambient relative humidity and the aerosol droplet composition, the partly or completely neutralized sulfates will remain in the liquid phase or change to the solid phase [5]. Atmospheric sulfates are associated with submicrometer size particles [6, 7]. Results from these studies suggest that aerosol strong acidity (defined as the concentration of the H^+ ion) has a similar size distribution to that of sulfates.

Nitric, Nitrous, and Hydrochloric Acids

During daytime, nitric acid (HNO_3) is formed through the gas phase reaction between NO_2 and OH^-. Since nitrogen oxides are not soluble in water, aqueous phase formation of HNO_3 is negligible. During nighttime, the nitrate radical NO_3, formed by the reaction of NO_2 with ozone (O_3), becomes significant. NO_3 reacts with NO_2 to form nitrogen pentoxide (N_2O_5) which is subsequently hydrolysed to form HNO_3.

Nitrous acid, HONO, is formed through the gas phase reaction of NO with OH. Its daytime concentration is very low due to its photodissociation. In addition, there is some evidence that HONO is emitted from automobile exhaust [8]. HONO is found in higher concentrations indoors [9], where it is produces by the reaction of NO_2 and H_2O. In contrast, HNO_3 is very reactive, and it is adsorbed onto indoor surfaces. Hence its indoor concentration is typically low. HONO is an important source of OH^- and plays a critical role in the formation of HNO_3 and H_2SO_4.

Both HONO and HNO_3 have very high vapor pressures and are present entirely in the gas phase for relative humidities below 98% [10]. HNO_3 can be neutralized by ambient NH_3 to form ammonium nitrate (NH_4NO_3). Ammonia presents a substantial vapor pressure over NH_4NO_3 particles. The dissociation of NH_4NO_3 to NH_3 and HNO_3 has been studied [11], and it was found to be dependent of relative humidity, temperature, and aerosol acidity.

Hydrochloric acid, HCl, is introduced directly into the atmosphere by waste incineration processes. It can also be formed by displacement of chloride from its salts by sulfuric or nitric acids. An example of this formation mechanism is the condensation of freshly generated sulfuric acid vapor on chloride salts, such as NaCl that is found in sea spray. Hydrochloric acid may be neutralized by ammonia, resulting in the formation of NH_4Cl. This is a slightly less stable salt than NH_4NO_3, and its stability increases with relative humidity and decreases with temperature.

Organic Acids

Aliphatic monocarboxylic or dicarboxylic and aromatic organic acids may also be significant contributors to atmospheric acidity. Aliphatic acids with low molecular weights (up to hexanoic or heptanoic acids) are present in the gaseous phase. Formic (HCOOH) and acetic (CH_3COOH) acids are the most abundant acidic gases. There is uncertainty about the sources of organic acids in the atmosphere. Photochemical breakdown of hydrocarbons is a major source of aldehyde precursors, and it is believed to be the source of formic and acetic acid as well [12]. In particular, the photo-oxidation of isoprene, emitted from growing vegetation, appears to be an important source of formic acid [13]. Gas phase ozone-olefin reactions in the atmosphere have been estimated to be a significant source of organic acids [14]. These compounds may also be produced by the reaction of aldehydes and HO_2. Organic acids react with ammonia to form carboxylate ammonium salts.

Carboxylic acids of twelve or more carbons are commonly found in the particulate phase. It has been found [15] that from C_{12} to C_{15} the acids primarily exist in the gas phase; from C_{16} to C_{21} the acids are found primarily in the particulate phase; and above C_{21} the acids are found exclusively in the particulate phase. In reality, volatilization of the lower molecular weight acids during sampling probably skews the ratio of particulate to gas phase organic resulting in higher gas phase, relative to particulate phase concentrations.

Long chain carboxylic acids were found in ambient aerosols in both urban and remote areas [16]. The main sources of these compounds were vascular plant waxes and microbes associated with plant lipids. Other microbial sources of ambient long chain fatty acids include the microbiota of marine and lacustrine environments [17]. Combustion of organic material was also found to be an important source of fatty organic acids.

Dicarboxylic acids, on the other hand, have fewer primary sources. The major primary source has been determined to be motor vehicle emissions [18]. Significant amounts of dicarboxylic acids have also been detected being emitted by frying and charbroiling of meat in residences [19].

Glutaric, adipic and pimelic acids have been found to exist at detectable levels only as particles smaller than 1 μm in diameter, suggesting that photochemical reactions are their primary origin [20]. An additional confirmation of the photochemical origin of dicarboxylic acids lies in the strong correlation between their concentration and the concentration of ozone [21].

Environmental Fate

Deposition of acid particles and gases and their precursors on the earth's surface can have significant adverse effects on the ecosystem and respiratory health [3]. Acid deposition occurs either by wet or dry mechanisms. Soluble particles, such as H_2SO_4, and gases, such as HNO_3, are scavenged by rain droplets; in the absence of rain, acid particles deposit by sedimentation and diffusion, whereas acid gases deposit by adsorption or absorption on the surface of the earth [22]. As mentioned above, most of the acidic ambient particles are found in the size range 0.1–1.0 μm [6]. This size range is also known as the accumulation mode of the atmospheric particle size distribution because it represents the transition between diffusing particles (smaller than 0.1 μm in size) and large particles (larger than approximately 2 μm in size) that have high sedimentation velocities. Consequently, due to their low dry deposition velocities (typically 0.001–0.01 cm/s), acidic particles tend to have long residence times in the atmosphere and can be transported over long ranges [23]. Acid gases are reactive and deposit much faster than acid particles. The deposition velocity of acidic gases is influenced by factors such as atmospheric stability, sunlight, temperature, humidity, and ranges from 0.5–5 cm/s [24].

Ambient ammonia and carbonate particles may react with acidic aerosols and form completely or partly neutralized salts. Ground levels of ammonia are usually enough to neutralize a substantial amount of acid aerosols. Nevertheless, as air layers adjacent to the earth's surface undergo convective mixing during the day, ground NH_3 can be displaced with H_2SO_4-laden air mixing downward [1], resulting in elevated ground-level concentrations of acid aerosols.

Methods to Measure Acidic Particles and Gases

The reactive nature of acidic air pollutants constitutes a special challenge to sampling. Methods for measurement of such compounds include continuous (spectroscopic, chemiluminescent) and time-integrated (filter pack, denuders, impactors) sampling.

Continuous sulfate monitors have been applied in several studies. These instruments utilize flame photometric detection of sulfate species that have been speciated thermally. However, these monitors have limited uses since they cannot distinguish between completely and partially neutralized ammonium sulfate salts [25]. This is a serious limitation since partially neutralized sulfates may be substantial contributors to aerosol acidity [26]. The advantages and shortcomings of continuous methods for measuring acidic particles and gases have been recently reviewed [27] and will not be discussed here. The primary focus will be on integrated methods for measuring atmospheric acid particles and gases, including filter packs, diffusion denuders, and inertial impactors. Precise measurement of acid particles and gases with time-integrated techniques is a complicated process and can lead to sampling artifacts (overestimation or underestimation of the true concentration). What follows is a description of these three basic technologies, and a discussion of the major issues that need to be addressed in acid particle and gas sampling using these methods.

Filtration

The simplest method for simultaneous sampling atmospheric acid particulate and gaseous pollutants employs two filters connected in series; the first filter retains the particulate phase and the second filter, which is coated with a suitable sorbent, retains the gas phase. Nevertheless, several studies in the literature reported both positive and negative artifacts in atmospheric acid aerosol sampling, resulting in an underestimation or overestimation of the particulate phase. This sampling artifact is due to the following phenomena.

1. Acid gas phase compounds, such as SO_2, HNO_3, and to a lesser extent NO_2, can be adsorbed on filter media. The high retention of atmospheric sulfur and nitrogen oxides on filter media led to measurements of atmospheric nitrate and sulfate that were too high [28, 29, 30]. In laboratory studies, glass fiber filters were found to retain 8–21% of the SO_2 as sulfate on the filter at 21 °C and 80% relative humidity, whereas quartz and Teflon retained 5% or less. This retention seems to be related to the filter alkalinity with the most alkaline filters retaining the greatest amount of SO_2 [31].

Adsorption of gas phase HNO_3 onto the filter medium and/or previously collected particles gives a considerable artifact nitrate on certain types of filters, leading to an overestimation of particulate-phase nitrates. In laboratory studies, glass fiber filters retained more than 94% of the gaseous HNO_3 of the sampled air, while quartz filters tested retained 33–99%. Only Teflon filters adsorbed negligible amounts of gaseous HNO_3 [32].

Sulfate and nitrate can also be formed from gaseous SO_2 and HNO_3 on various types of filters. The mechanisms of this artifact formation have already been reviewed [33]. This formation appears to be more severe at higher relative humidities (R.H > 55%).

2. Particle interactions on the filter media during and after sampling, such as reactions of coarse (larger than 2.5 μm in diameter) alkaline particles such as $CaCO_3$, NaCl, or KNO_3, with acidic sulfates can form neutral sulfate salts and acid gases such as CO_2, HCl, or HNO_3. While this interaction between alkaline coarse particles and acidic fine particles can be minimized by using size selective inlets, interactions between fine acidic particles can be very complicated.

Available atmospheric ammonia is first taken up by sulfuric acid, until complete neutralization occurs. Any remaining ammonia is then available to react with nitric acid to form ammonium nitrate [11]. Therefore the coexistence of ammonia and acid aerosol sulfates is thermodynamically impossible. Nevertheless, ammonium nitrate and acid sulfate particles may be present at different times in a sampling period, and consequently they may be collected on the same multihour integrated sample.

3. Volatilization of unstable ammonium particulate salts, such as NH_4NO_3 and NH_4Cl, forms NH_3 and HNO_3 or HCl [10]. The produced ammonia can diffuse to the collected acid sulfate particles and neutralize them, as described in the previous paragraph. The produced acidic gases are collected by the coated filters of the filter pack that are placed downstream of the particle collection filter. Thus, the gas phase concentrations contain the amounts that existed in the atmosphere during sampling plus the amounts that volatilized from the collected particle, resulting in an overestimation of the concentration of acidic gases and an underestimation of acidic particles.

In conclusion, accurate sampling of atmospheric acid particles and gases cannot be achieved using any sampling procedure employing the use of filter packs, since the artifact leads to serious underestimation or overestimation of one of the two phases. It is therefore essential that particle and gases are collected separately. Diffusion denuders, discussed below, have offered an alternative method to sample acid particles and gases avoiding artifact problems.

Diffusion Denuders

During the last decade, diffusion denuders have been used in a variety of atmospheric monitoring studies to collect gaseous atmospheric pollutants [34–39]. In these studies, glass or metallic hollow tubes are coated with an appropriate substance to selectively collect the different gases while allowing other gases and particles to penetrate. The particles are subsequently collected on filter media. Using this sampling technique, particles and gases are separated during sampling and sampling artifacts are minimized.

The performance of diffusion denuders is based on the difference in the deposition rate due to diffusion between gases and particles. Gases diffuse much faster than particles on the walls of a sampler and are adsorbed onto the wall coated with material suitable for collecting the gas. The nature of the material depends on the gas to be collected. Sodium carbonate (Na_2CO_3), sodium hydroxide (NaOH), and potassium hydroxide (KOH) have been extensively used as a coating substrate because they absorb very efficiently acidic gases, such as HNO_3, HCl, HNO_2, SO_2, HCOOH, and CH_3COOH [37, 39, 40, 13]. Citric acid has been used to collect basic gaseous pollutants such as NH_3 [39, 40]. Most of the coating substrates are first mixed with glycerol, then dissolved in water or methanol, and subsequently applied to the surfaces of the denuder. The denuders are dried immediately after coating with clean air [39, 41].

The denuder/filter pack technology has made possible a number of field and laboratory studies which have enhanced our understanding of the physical chemistry of inorganic atmospheric pollutants. Such systems are cost effective because they allow simultaneous sampling of gases and particles using the same pumping system, which can be the most expensive part of the sampling unit. However, this is not the only reason which makes the denuder/filter pack system attractive. The use of a denuder is necessary to prohibit reactions between gases and collected particles on the filter media. For example, a significant portion of particulate acidity can be neutralized on the Teflon filter by ambient ammonia. Therefore, a citric acid-coated denuder is used upstream from the filter to remove ammonia from the air sample. In addition, the coarse alkaline particles are usually removed by inertial methods before the air sample enters the denuders. Typically, either a cyclone [42], or an impactor [43] can be placed upstream the denuder/filter pack to remove coarse alkaline particles. These devices make use of inertial mechanisms to separate particles larger than approximately 2.5 μm in aerodynamic diameter from the rest of the air sample.

The most important issues that need to be addressed in the evaluation of the performance of any denuder are as follows.

1. Collection efficiency, defined as the amount of gas collected on the denuder walls divided by the total amount of gas entering the denuder.
2. Denuder capacity, defined as the maximum amount that can be collected by the denuder without a detectable loss of its efficiency. In practice, capacity is defined as the amount collected before the efficiency of the denuder drops below 95% [41, 40].
3. Sampling loss of reactive gases, which is defined as the amount of gases collected in any part of the sampler other than the denuder (such as inlet or preceeding denuders).
4. Particle losses, defined as the amount of particles lost within the sampler and not collected by the filter pack.

Hollow Denuders

A single glass or Teflon tube has been used to sample gases such as NH_3 and HNO_3 [35, 37]. The dimensions of the tube diameter and length, as well as the flow rate are chosen so that the collection efficiency is 99%. The theoretical efficiency can be calculated with the assumption that once a gaseous pollutant contacts the wall, it reacts with the coated surface and is 100% retained. The theoretical penetration is given for cylindrical tubes by the Gormley-Kennedy equation (1949):

$$\frac{C}{C_0} = 0.82 \exp(-15\mu) + 0.098 \exp(-89\mu) + 0.033 \exp(-228\mu) \qquad (1)$$

where

C: the gas concentration leaving the tube

C_0: the gas concentration entering the tube

$\mu = \pi \; DL/4Q$: a dimensionless number relating the diffusion coefficient to the denuder length and the flow rate
D: the diffusion coefficient of the gas (cm^2/s)
Q: the sampling flow rate (cm^3/s)
L: the length of the tube (cm)
$1–C/C_0$: the collection efficiency of the denuder.
This expression is valid for laminar flows (in which the Reynolds number is less than 2300). The Reynolds number is defined as

$$Re = d \; U \; \rho/\eta \tag{2}$$

where U is the average velocity inside the tube (cm/s), d is the diameter of the tube (cm), ρ is the density of the gas (g/cm^3) and η is the dynamic viscosity of the gas (g/cm.s). In order to meet the laminar flow requirement as well as collect gases efficiently, most of the single tube denuders can operate at low sampling flow rates (less than 3 l/min). For example, a glass tube of 3 mm diameter and 35 cm long would have over 99% collection efficiency for ammonia (whose diffusion coefficient is 0.236 cm^2/s) at 3 l/min flow rate [44, 37].

In sampling practice, however, higher flow rates are often required since they provide a finer resolution of the temporal concentration profile of a pollutant gas. Higher sampling flow rates can be achieved by using parallel tube assemblies [44], such as the one shown in Fig. 1, annular denuders [45, 39], and honeycomb denuders [39, 40].

Denuder Difference Method

The denuder difference method (DDM) has been employed by several researchers to measure accurately particulate and gas phase acidic compounds [37, 46]. The

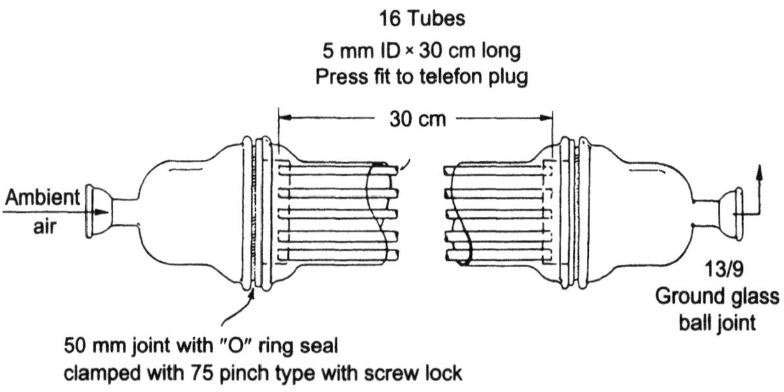

Fig. 1. Schematic diagram of a hollow denuder consisting of 16 glass tubes (From [44]. Reproduced with permission, Copyright American Conference of Industrial Hygienists Inc., 1989)

diagram of this method is shown schematically in Fig. 2. The acidic aerosol is drawn through an inlet that removes coarse alkaline particles (larger than 2.5 μm in diameter). Fine particles and gases enter a manifold where the flow is divided into two parts; one part passes through a denuder coated with MgO, whereas the other part passes through a similar but uncoated denuder. The total sampling flow is 30 l/min. At the exit of the two tubular denuders, the aerosol sample passes through a membrane filter. Since the MgO removes HNO_3, the filter downstream the coated denuder collected the true fine particulate nitrate. The filter downstream the uncoated tube collects both particulate nitrate and gaseous HNO_3, hence the difference between the HNO_3 and nitrate filter and the fine nitrate filter provides an estimate of HNO_3.

Despite its simplicity, the DDM may result in inaccurate estimations of particulate and gas phase acidic compounds for the following reasons [47].

1. The denuders, as well as the membrane filters may retain species other than HNO_3 that form NO_3^-.
2. Ammonium nitrate (NH_4NO_3) may volatilize during passage through the denuder section, resulting in an underestimation of fine particulate nitrate and overestimation of HNO_3.
3. Some HNO_3 can be collected on the uncoated denuder.

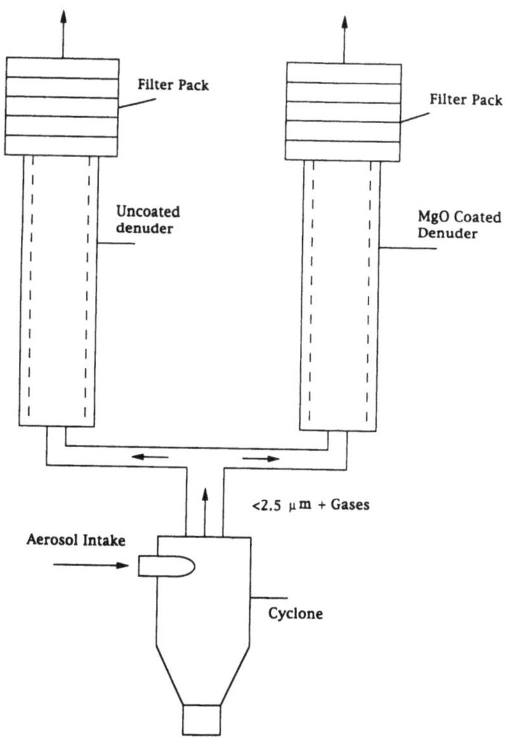

Fig. 2. Schematic diagram of the denuder difference method (DDM)

4. Since measurement of the concentration of gaseous HNO_3 is based on a difference, the limit of detection of the DDM is high [48].

Annular Denuders

Use of annular denuders followed by a filter pack allowed the determination of the concentrations of acid particulate and gaseous pollutants at higher flow rates (10 l/min, or higher). These devices consist of two coaxial cylinders in which the air sample passes through a thin annular space. A denuder was developed [45] that was 30 times more efficient per unit length compared to a tubular denuder. This denuder is shown in Fig. 3, with a cyclone placed upstream the device to remove coarse particles ($d_p > 2.5$ μm). The gas penetration in the case of annular denuders has been derived from a sorption study with SO_2 and is given by the equation [45]:

$$\frac{C}{C_0} = 0.82 \exp(-22.53\mu) \tag{3}$$

where
$\mu = \pi DL(d_1 + d_2)/(4Q(d_2 - d_1))$
d_2: outer diameter (cm)
d_1: inner diameter (cm).
As in the case of single-tube denuders, this equation is valid in laminar flows. In this case the Reynolds number is defined as:

$$Re = \rho U(d_2 - d_1)/\eta \,. \tag{4}$$

Comparing the equations describing the performance of tubular and annular denuders (Eqs. 1 and 3), it can be shown that for the same collection efficiency:

$$(Q/L)_{\text{annular}} = 31.5(Q/L)_{\text{cylindrical}} \,. \tag{5}$$

The main implication of this relationship is that, for the same denuder length, the annular denuder can operate at approximately 30 times higher flow rate than the cylindrical denuder.

The Harvard-EPA Annular Denuder System (HEADS), shown in Fig. 4, was developed [39] in which annular denuders are coupled to a glass impactor to remove coarse alkaline particles, and a filter pack. More specifically, the HEADS system consists of a borosilicate glass impactor, two glass annular denuders, and a FEP Teflon filter pack. The impactor has been calibrated, and at a flow rate of 10 l/min its 50% cutpoint is 2.1 μm [43]. The design of the annular denuders is similar to that of Possanzini et al. [45]. The denuder length is 26.5 cm, the outer diameter of the inner cylinder is 2.20 cm, and the thickness of the annulus is 0.10 cm. The first denuder is coated with Na_2CO_3/glycerol to collect SO_2, HNO_3,

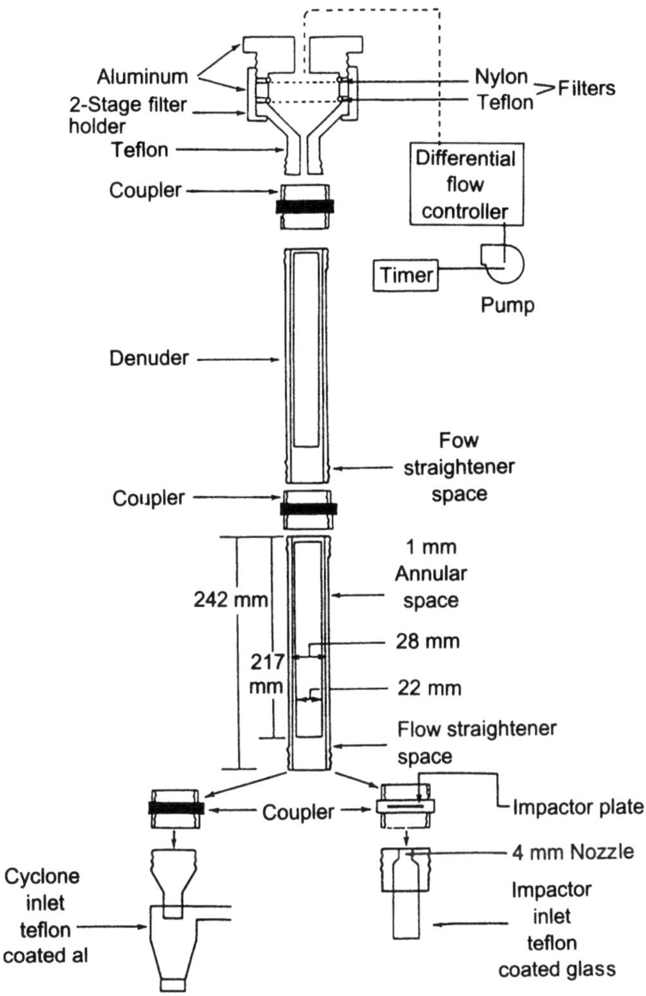

Fig. 3. Glass annular denuder system preceded by a cyclone that removes particles larger than 2.5 μm in diameter (From [53]. Reproduced with permission, Copyright 1991 Elsevier Science Inc)

and HNO_2. The second denuder is coated with 2% citric acid: 1% glycerol in a water: methanol solution to collect ammonia. Following the second denuder is a PTFE Teflon filter pack containing a Teflon filter to collect fine particles for fine particle mass, sulfate, and nitrate concentration measurements. The second filter is a 47-mm diameter glass fiber filter, coated with Na_2CO_3 to trap HNO_3. The origin of HNO_3 will be described in more detail in the section discussing the determination of acidity neutralization due to interaction between collected particles.

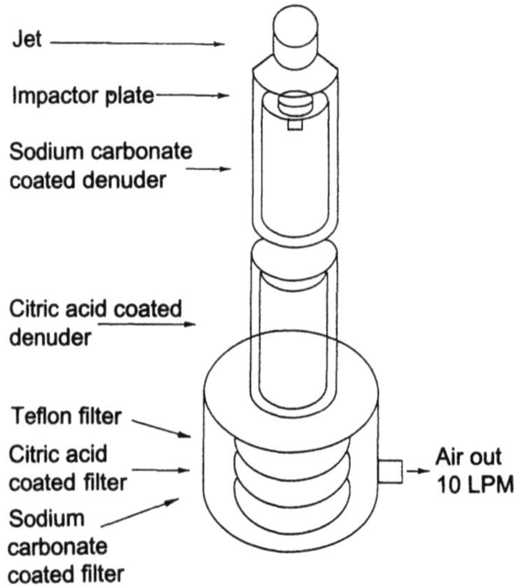

Fig. 4. The Harvard/EPA Annular Denuder System (HEADS). (From [39]. Reproduced with permission, Copyright American Chemical Society, 1988)

Multiannular Denuders

The collection efficiency and capacity of annular denuders can be further increased by using multiannular configurations as collection surfaces [49, 50]. These denuders combine operation at high flow rates (up to 200 l/min) with more economic use of space. In addition, since the collection surface is largely increased for the same length, compared to an annular denuder, these devices enable personal sampling.

The device shown in Fig. 5 is a miniaturized version of a multiannular denuder operating at 4 l/min with a total length of 20 cm. It simultaneously collects acidic gases, ammonia and particulate. It consists of two impactors to remove coarse particles, two coated multiannular denuders, and a three-stage filter pack connected in series. The multiannular denuders consist of four concentric cylinders, each separated by a 1-mm annulus. Total denuder length is 5.1 cm [50]. This miniaturized denuder was used as a personal monitor in a field study to assess exposure to acidic aerosols of children living in Pennsylvania, USA during the summer of 1990. The entire personal monitor, which consists of a backpack containing a battery-operated pump and the miniaturized denuder, weighs 5 lbs.

Another example of a multiannular denuder is a high-volume compound annular denuder (HVCAD) [49]. It consists of 12 annular denuders, the overall length is 20.3 cm and the outer diameter is 8.25 cm. The core is 6.4 mm aluminum rod, and the spacings between the tubes are maintained by three-armed radial rake assemblies attached to the core. This sampler operates at rates up to 200 l/min with an average gas collection efficiency of 94%. Although this sampler has been

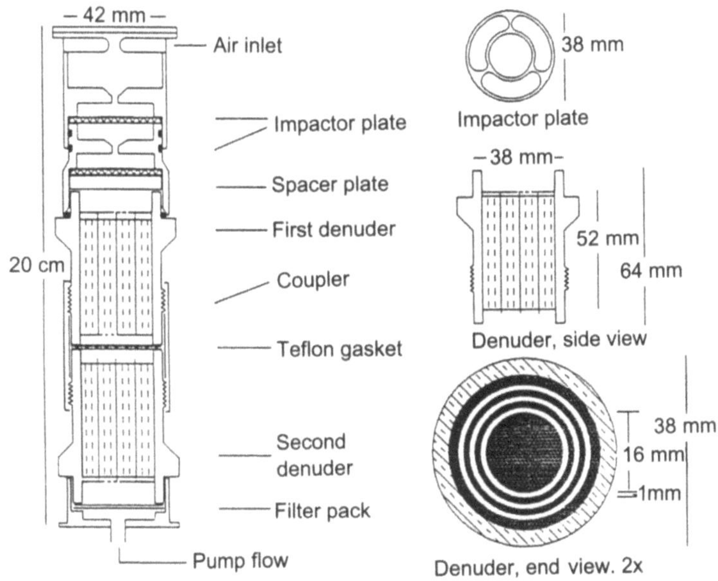

Fig. 5. Personal Multi-annular Denuder System. (From [75]. Reproduced with permission, Pergamon Press, 1994)

developed to collect low diffusion coefficient semivolatile organic compounds, its design illustrates how the efficiency of a denuder can be drastically increased by increasing its collection surface for the same length.

Honeycomb Denuders

In a recent study [40], the development and evaluation of an impactor/honeycomb denuder/filter pack system was presented. The honeycomb denuder/filter pack system, shown in Fig. 6, has three essential features: a) an impactor to remove coarse particles (d_p > 2.1 μm) from the air sample; b) two glass honeycomb denuders; and c) a two-stage filter pack to collect fine particles.

The inlet has a circular array of 10 nozzles which accelerates the sampled aerosol so that particles larger than 2.1 μm impact on the impactor plate, a ring of sintered stainless steel. coated with mineral oil. A transition section allows the sample air to flow smoothly and have uniform flow through the individual honeycomb tubes of the first denuder. A Teflon coating minimizes internal surfaces from interaction with reactive gases.

The main feature of the sampler is the glass honeycomb denuder, also shown in Fig. 6, which is a cylinder 2.54 cm high and 4.7 cm in diameter, containing

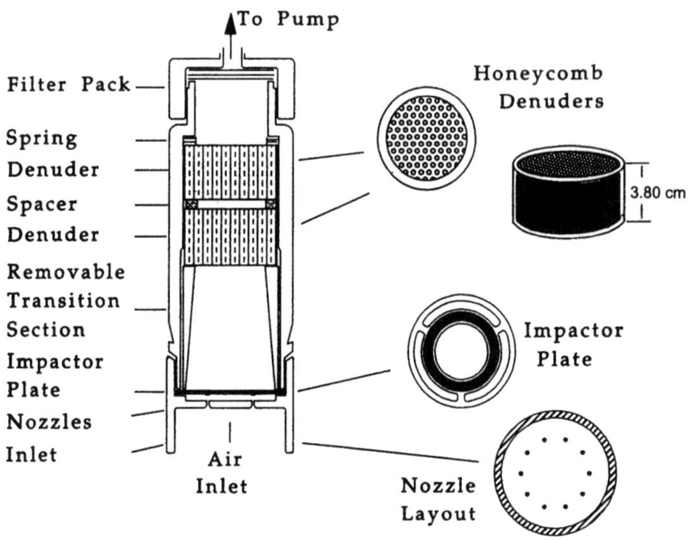

Fig. 6. Schematic diagram of the honeycomb denuder system. (From [40]. Reproduced with permission, Copyright American Chemical Society, 1993)

212 hexagonal glass tubes 2-mm in side. The tubes are sealed inside a thick-walled (4.7 cm OD) glass tube. The first honeycomb denuder (coated with sodium carbonate/glycerol to collect acid gases) is separated by an inert spacer from the second denuder (coated with citric acid/glycerol to collect basic gases). A stainless steel spring keeps the components in place. A third denuder can also be placed, if desired. The two-stage filter pack has a Teflon filter to collect fine particles (below 2.1 μm) and a sodium carbonate-coated glass fiber filter to collect acid gases produced from the reaction of acidic fine particles with particulate ammonium salts of these acid gases. A third stage can be added in the filter pack if it is needed.

The theoretical collection efficiencies for both ammonia and nitric acid for the honeycomb denuder, calculated using the Gormeley-Kennedy equation, are 99.9%, assuming that the 10 l/min flow is divided equally among all the tubes. Correspondingly, the collection efficiencies for the annular denuder are also 99.9%, calculated using the empirical equation of Possanzini, et al. [45]. The collection efficiency and capacity of the honeycomb denuder has been compared to those of the annular denuders under simulated atmospheric conditions in laboratory experiments [40]. Due to its larger collection surface (576 cm^2) compared to the annular denuder (297 cm^2), the honeycomb denuder appears to have larger capacity for both nitric acid and ammonia (Figs. 7 and 8).

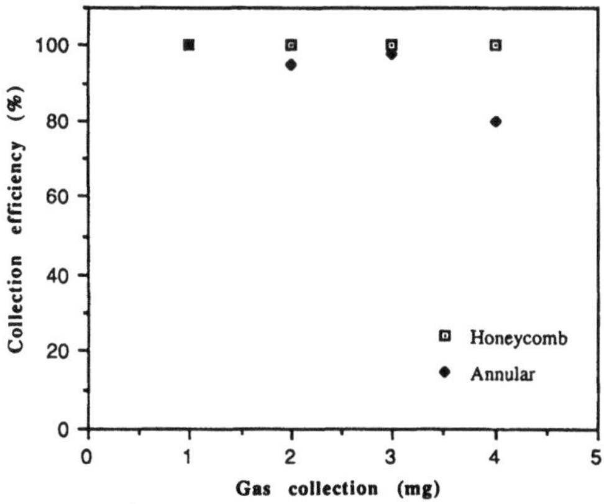

Fig. 7. Collection efficiency for nitric acid. (From [40]. Reproduced with permission, Copyright American Chemical Society, 1993)

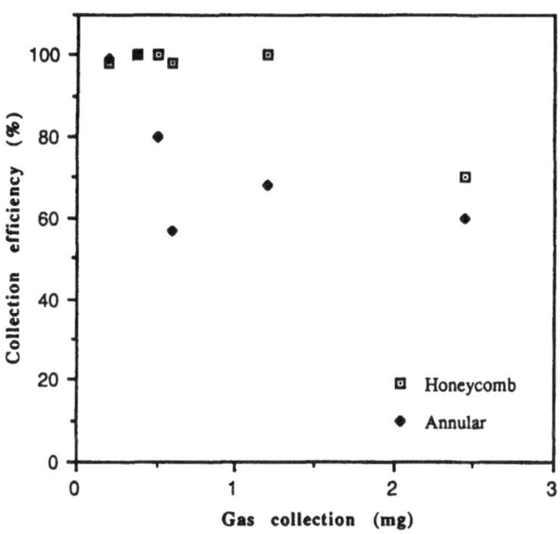

Fig. 8. Collection efficiency for ammonia. (From [40]. Reproduced with permission, Copyright American Chemical Society, 1993)

Determination of Acid Particle and Neutralization due to Interactions of Collected Particles

Ammonium nitrate and acidic sulfate are two of the most abundant aerosol species under typical atmospheres. Although the presence of acidic particles in ambient air implies that no ammonium nitrate exists, ammonium nitrate and acidic sulfate particles may be collected on the same filter medium during prolonged sampling. The interaction of these particles on the filter media is a significant issue in sampling ambient inorganic aerosols. Filter pack techniques use a Teflon filter to collect particles and a nylon filter (or a sodium carbonate-coated glass fiber filter) to collect nitrates (in nitric acid form) downstream from the Teflon filter.

However, the nitrate on the second filter can originate from vapor phase nitric acid or from the volatilization of ammonium nitrate collected on the Teflon filter. Therefore it is impossible to determine ammonium nitrate concentrations unless a nitric acid denuder is used upstream from the filter pack. Note, in addition, that the volatilization of nitrate from the Teflon filter can be associated with two different processes: (a) the dissociation of ammonium nitrate:

$$NH_4NO_3\ (p) \leftrightarrow NH_3(g) + HNO_3\ (g) \tag{6}$$

and (b) the neutralization of acid aerosol sulfate by ammonium nitrate:

$$\begin{aligned} &NH_4NO_3\ (p) + H^+\ (p)\ [NH_4HSO_4,\ (NH_4)_3H(SO_4)_2,\ \text{or}\ H_2SO_4] \rightarrow \\ &NH_4{}^+\ (p)\ [NH_4)_2SO_4, NH_4HSO_4,\ (NH_4)_3H(SO_4)_2] + HNO_3\ (g)\,. \end{aligned} \tag{7}$$

Recently the relative importance of these two mechanisms was determined using a denuder/filter pack system [10]. Annular denuders were used to remove nitric acid and ammonia from the air sample. The filter pack contained three filters: a Teflon filter to collect fine particles, a sodium carbonate-coated glass fiber filter to collect nitrate (which can originate from both ammonium nitrate dissociation and acid sulfate neutralization reaction, as shown in the above reactions), and a citric acid-coated glass fiber filter to collect ammonium (which originates from the dissociation of ammonium nitrate). By subtracting the moles of ammonium collected on the third filter from the moles of nitrate collected on the second filter, the amount of lost particulate acidity from the Teflon filter can be calculated, as indicated by the following relationships:

$$H^+\ (\text{total}) = H^+\ (\text{first filter}) + H^+\ (\text{neutralized}) \tag{8}$$

$$H^+\ (\text{neutralized}) = NO_3^-\ (\text{second filter}) - NH_4^+\ (\text{third filter})\,. \tag{9}$$

Laboratory and field results from this acid aerosol study suggested that the reactions described in Eq. (7) account for 30–80% of neutralization of the observed aerosols acidity. The degree of neutralization depends on the relative amounts of H_2SO_4 and NH_4NO_3 particles and the relative humidity. For high acid particle concentrations ($> 2\ \mu g/m^3$) neutralization becomes negligible, but for lower concentrations the correction factor should be taken into account.

Inlet Loss of Reactive Gases

An important issue in the evaluation of any gas-sampling instrument is the nonspecific collection of gases on the surfaces of the sampler, in particular the sampling inlet. HNO_3 was reported to be deposited on the inlets of various samplers, especially at high relative humidities [48, 41]. This loss is more pronounced at low HNO_3 ambient concentrations, and varies inversely with the residence time at the inlet, as well as the total amount sampled [47].

Annular and honeycomb denuders have been characterized for inlet losses of nitric acid and ammonia [41, 40]. The assessment of inlet losses was done by letting samplers with and without inlets sample from the same air stream, and comparing the concentrations determined by the denuders in both cases. To minimize this loss, all inlet component surfaces of denuders are coated with Teflon, which reportedly decreases such losses [48]. The results of these studies indicated that inlet losses for HNO_3 and NH_3 were less than 5% for inlets coated with Teflon, whereas for uncoated inlets losses of HNO_3 and NH_3 were as high as 10%.

HNO_3 also has a tendency to adsorb onto the surfaces of tubing walls. A study [51] has reported minimum HNO_3 retention for Teflon and Pyrex glass tubing. In addition, these losses decrease with decreasing porosity of the tubing material. Losses on tubing made of polytetrafluoroethylene (PTFE), which has a substantial void content, have been reported to be severe, compared to losses on tubing made of fluorinated ethylene-propylene (FEP), which is made by extrusion molding of polymers, and thus it has a low void content [47].

Particle Transmission Efficiency of Denuders

Little attention has been paid to characterizing particle transmission through denuder systems for fine particle ($d_p < 2.5$ μm). Papers have been published [45, 52] discussing the particle transmission efficiency of annular denuders, measuring particle losses in annular denuders by generating polydisperse aerosols and using optical counters to measure the number concentration upstream and downstream of the denuders. Monodisperse fluorescent particles in the size range 1–2.5 μm have been used [43] to investigate transmission losses through a pair of annular denuders. All these investigations reported particle losses in the order of a few percent. Transmission measurements in annular denuders by controlling the aerosol charge state have been reported [53]. Particles in the size range 0.01–1 μm were used as a test aerosol and losses due to diffusion and electrostatic forces were separately measured. The results of this study suggested that electrostatic effects may increase significantly the particle losses. Losses as high as 65% were reported for singly charged particles 0.03 μm in diameter when three annular denuders in series were used.

Finally, particle losses through 2.54 cm long glass honeycomb denuders were examined [54] for particles in the size range 0.05 to 2.0 μm. This size range was chosen since it represents the majority of the ambient particle mass associated with particles smaller than 2.5 μm in diameter [55, 56]. Losses have been determined using fluorescent monodisperse latex particles at flow rates in the range 1.5–10 l/min. In addition, the effects of relative humidity and the denuder surface coating on the transmission characteristics were examined. The results of this study suggested that deposition of atmospheric fine particles on honeycomb denuder surfaces is small (approximately 5% or less).

Sampling of Acidic Aerosols Using Impactors

Inertial impactors have been used in several studies to determine the size distribution of sulfates [57, 56, 7, 39, 40]. In such devices, collection of the particulate phase occurs prior to the collection of the gas phase. These devices base separation of the particulate and gas phases on the much higher inertia that particles have compared to gases. In a conventional flat plate impactor (Fig. 9), particle-laden air passes through a nozzle where particles are accelerated. At the nozzle exit, the air streamlines are deflected sharply, so that particles larger than a certain size (the impactor's cut-off size) impinge on the collection surface, whereas smaller particles escape and follow the deflected streamlines [58]. By employing more than one such stage, the aerosol sample can be fractionated in several size intervals. This type of impactors is called cascade impactors (Fig. 10). When sampling an acidic aerosol, a diffusion denuder is placed upstream the impactor to remove ammonia, and therefore protect the ambient samples from neutralization. The collection medium used as impaction surface is typically a Teflon filter, chosen for its low background acid and sulfate concentrations.

Due to the extensive theoretical work [59, 60], their performance has become well understood and their characteristics can be predicted quite accurately. The principal parameter determining an impactor's particle capture characteristics is the Stokes number, St, defined as the ratio of the product of the jet velocity, U, and the particle relaxation time, τ, vs the impactor's nozzle diameter, W, [61].

$$St = \frac{\tau U}{W} = \frac{\rho_p C_c d_p^2 U}{9 \mu W} \tag{8}$$

where
W: acceleration nozzle diameter (cm)
U: average velocity of the jet (cm/s)

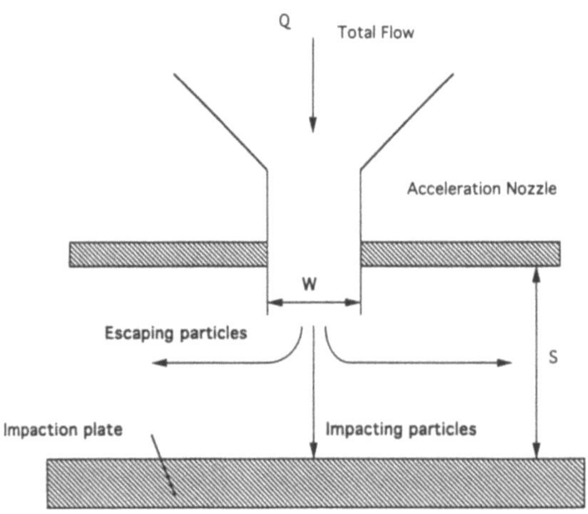

Fig. 9. Schematic diagram of a conventional impactor

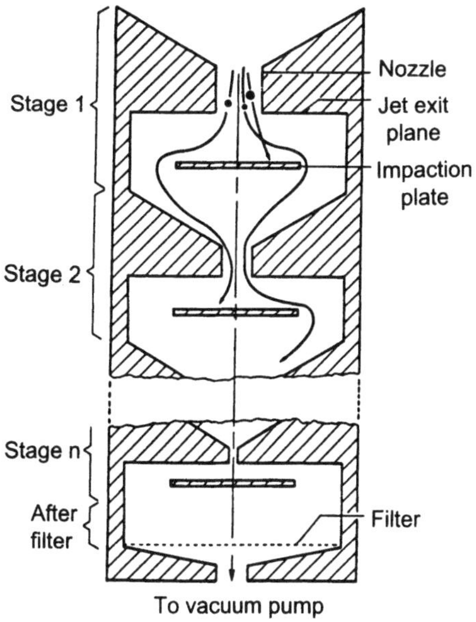

Fig. 10. Schematic diagram of a cascade impactor. (From [76]. Reproduced with permission, Copyright American Conference of Industrial Hygienists Inc., 1989)

ρ_p: the particle density (g/cm^3)
μ: dynamic viscosity of the air (g/cm s)
C_c: Cunningham slip correction factor .
The slip correction factor is given by the equation [58]:

$$C_c = 1 + \frac{2}{Pd_p}[6.32 + 2.01\exp(-0.1095Pd_p)] \tag{9}$$

where P is the absolute pressure in cm Hg and d_p is the particle diameter in µm. Equation (9) underscores the dependence of the slip correction factor on the product Pd_p. C_c increases as the pressure decreases because the mean free path increases. Since St is a dimensionless parameter, Eq. (8), known as the Stokes number equation, indicates that the square root of St provides a measure of the dimensionless particle diameter.

Microorifice and Low-Pressure Impactors

Impactors can typically classify particles larger than a few tenths of a micrometer in diameter, operating at atmospheric pressure and having jets fabricated by conventional machining, as predicted by the Stokes number equation.

To classify smaller acid aerosol particles, two types of impactors have been developed, namely low-pressure impactors and microorifice impactors. Low-pressure impactors resemble ordinary impactors but operate at reduced pressures

(0.05 to 0.4 atmospheres). These devices take advantage of the decreased aerodynamic drag on particles that occurs when the mean free path in air is as large or larger than that of the particle diameter. Therefore, low pressure impactors operate at increased values of the slip correction factor, C_c [62]. All these types of impactors operate at sampling flow rates up to 30 l/min.

Microorifice impactors (Fig. 11) reduce the impactor's cut-off size by accelerating the aerosol to be sampled through small diameter jet orifices [63]. As can be seen from the Stokes number equation, they increase the ratio U/W. Microorifice impactors operate at pressures closer to atmospheric (0.7 to 0.9 atmospheres), and achieve cutpoints on the order of 0.1 µm by employing orifices 0.004 to 0.02 cm in diameter. Operation with a low pressure drop is very important, since volatilization of sulfate-associated water inside the impactor can result in a severe distortion of the sulfate size distribution (Biswas et al., 1987). Water loss could cause a decrease in the particle size. In addition, aerodynamic cooling within the high-speed jet can cause particle size to increase.

Size Distribution of Atmospheric Sulfates

The equilibrium size of atmospheric sulfates has been determined in field studies using microorifice impactors [65, 7, 6]. By combining in situ atmospheric measurements of the sulfate size with thermodynamic calculations, the equilibrium size of sulfates was modelled as a function of particle acidity and humidity [66]. These studies indicated that there is a strong relationship between the mass median diameter of the acidic sulfate particles and both the particle

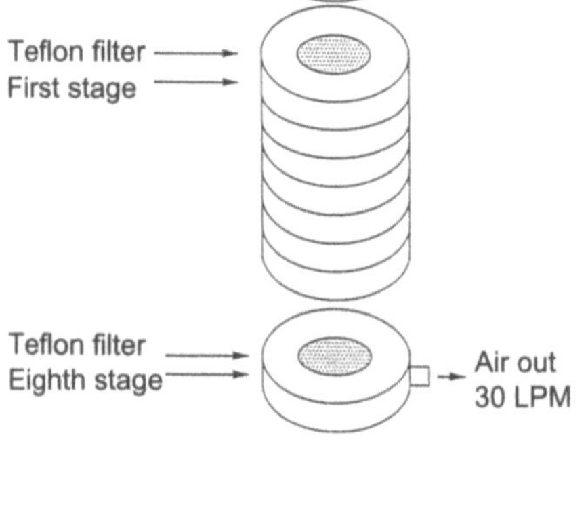

Fig. 11. Microorifice Impactor (MOUDI). (From [67]. Reproduced with permission, Pergamon Press, 1994)

acid content (expressed in terms of the ratio of the hydrogen ion to sulfate ion concentration, H^+/SO_4^{2-}) and the ambient relative humidity. The results of these studies, shown in Fig. 12, indicate that the mass median sulfate diameter varies from 0.2–0.8 μm (with a typical geometric standard deviation of 1.5), and it increases with increasing relative humidity, and particle acid content. Depending on the molar ratio H^+/SO_4^{2-}, the solid phase of the aerosol particles consists of either ammonium sulfate (NH_4NO_3) for $0 < H^+/SO_4^{2-} < 0.5$, letovicite ($(NH_4)_3H(SO_4)_2$) for $0.5 < H^+/SO_4^{2-} < 1.0$, and ammonium bisulfate (NH_4HSO_4) for $1.0 < H^+ < SO_4^{2-} < 1.8$ (Figs. 12 and 13).

Field Method Intercomparison for the Characterization of Acid Aerosols and Gases

A field study was conducted in Western Pennsylvania, USA, during the summers of 1990 and 1991, to compare the performance of five of the most commonly used acid aerosol and gas measurement systems [67]. Acid aerosol and gas concentrations (NH_3, HNO_3, HNO_2, SO_2, H^+, NH_4^+, NO_3^-, SO_4^{2-}) were measured using five sampling systems: the Harvard/EPA Annular Denuder System (HEADS), developed by Koutrakis et al. [39], the Personal Annular Denuder System (PADS),

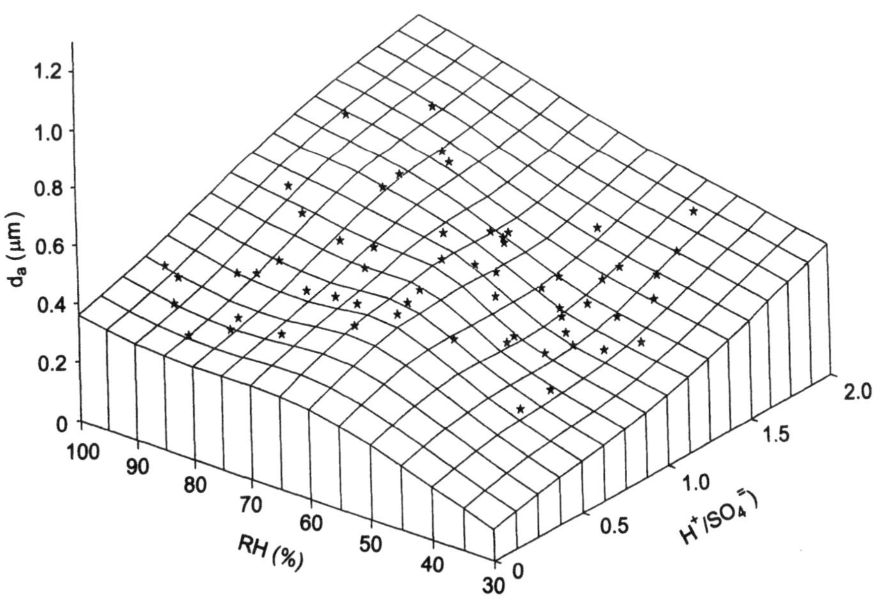

Fig. 12. Aerodynamic diameter of sulfates as a function of relative humidity and H^+/SO_4^{2-} ratio. (From [66]. Reproduced with permission, Copyright American Geophysical Union, 1993)

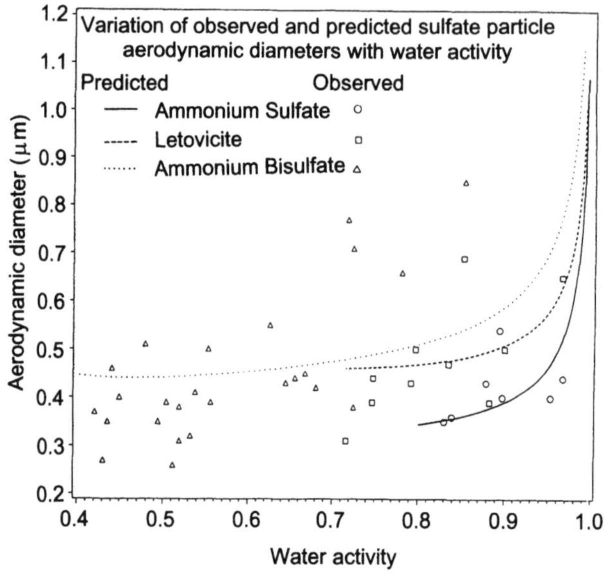

Fig. 13. Observed and predicted sulfate particle aerodynamic diameter as a function of water activity. (From [66]. Reproduced with permission, Copyright American Geophysical Union, 1993)

the Microorifice Impactor (MOI) developed by Marple et al. [63], the Harvard-Marple Impactor (HI), developed by Marple et al. [68], and the Continuous Sulfate/Thermal Speciation System (CSTS), developed by Allen et al. [69]. The first four samplers have already been described in the previous sections covering denuders and impactors; the CSTS provides semi-continuous measurements (10 min) of particulate SO_4^{2-} concentrations. The analyzer is composed of a modified Meloy/CSI (Austin, TX) model 285 flame photometric detector (FPD) preceded by a litharge sulfur gas denuder. Limited speciation of SO_4^{2-} particles is obtained through selective volatilization of the SO_4^{2-} species prior to entering the sulfur gas denuder and the particle sulfur detector. In each measuring cycle (lasting 10 min), a measurement is made of total SO_4^{2-}, sulfuric acid (H_2SO_4), and partially neutralized SO_4^{2-} (ammonium sulfate and ammonium bisulfate) concentrations through thermal volatilization. The analytical method is described in detail elsewhere [69].

Concentrations were measured over 3-, 12-, and 24-h intervals. Univariate regression was used to compare the performance of the various samplers, and the results are shown in Table 1; the correlation coefficient (r^2) indicates the degree of linear relationship between the methods, whereas the slope and intercept of the regression provide estimates of the proportional and constant differences, respectively, between the sampling methods to be compared. The results from this study demonstrate excellent agreement among the particle measurement system.

Table 1. Univariate regression results for SO_2, HNO_3, SO_4^{2-}, H^+, and NH_4^+. 3-h HEADS, MOI, and CSTS samples were averaged over 12-h periods to correspond to 12-h HEADS sample durations. Similarly, 12-h HEADS were averaged over 24-h periods to correspond to HI sample durations. "NS" denotes "non-significant", with a single asterisk indicating a p-value less than 0.05. (From [67]. Reproduced with permission, Copyright Pergamon Press, 1994)

Pollutant species	r^2	Slope	Intercept
SO_2:			
Regression of PADS on HEADS	0.95	0.80 ± 0.04	NS
Regression of 3- on 12-h HEADS	0.97	1.01 ± 0.04	NS
HNO_3:			
Regression of PADS on HEADS	0.79	0.22 ± 0.02	NS
Regression of 3- on 12-h HEADS	0.84	0.96 ± 0.09	NS
SO_4^{2-}:			
Regression of PADS on HEADS	0.97	0.98 ± 0.04	NS
Regression of 3- on 12-h HEADS	0.96	0.98 ± 0.04	NS
Regression of MOI on HEADS	0.95	1.00 ± 0.08	NS
Regression of HI on HEADS	1.00	1.04 ± 0.01	NS
Regression of CSTS on HEADS	0.97	0.94 ± 0.01	24.87 ± 2.22*
H^+:			
Regression of PADS on HEADS	0.96	1.03 ± 0.05	NS
Regression of 3- on 12-h HEADS	0.94	1.00 ± 0.06	NS
Regression of MOI on HEADS	0.96	1.01 ± 0.08	NS
Regression of HI on HEADS	0.96	0.94 ± 0.03	NS
NH_4^+:			
Regression of PADS on HEADS	0.98	0.99 ± 0.03	NS
Regression of 3- on 12-h HEADS	0.88	1.01 ± 0.08	NS
Regression of MOI on HEADS	0.98	1.08 ± 0.06	NS
Regression of HI on HEADS	0.98	1.15 ± 0.03	NS

The sulfate (SO_4^{2-}), aerosol strong acidity (H^+), and ammonium (NH_4^+) concentrations measured by all systems were highly correlated. In addition, 3- and 12-h particle measurements were comparable for HEADS, PADS, and MOI samplers. Although significantly different, mean relative differences between HEADS, CSTS, and HI measurements were small. For the gas nitric acid (HNO_3) and sulfur dioxide (SO_2), the performance of the HEADS was found to be independent of the sampling duration, since the 3- and 12-h SO_2 measurements are comparable. The PADS sampler, however, was found to measure both HNO_3 and SO_2 concentrations poorly, as indicated by the low slope numbers (0.22 and 0.80, respectively) of the regression of its performance on the performance of the HEADS. The lower values were attributed to probable HNO_3 deposition on the inlet surfaces of the PADS.

In conclusion, acid particle measurement conducted with either one of the above samplers gives identical results. For integrated acid gas measurements, the HEADS system appears to be the best currently available method, since its performance is independent of the sampling duration.

Chemical Analysis of the Collected Material on Denuders, Filters, and Impactor Substrates

Regardless of the type of sampler used for measuring acidic particulate and gas phase compounds, the chemical analysis for the determination of acidic particulate and gaseous pollutants is generally the same. The denuders are extracted with deionized distilled water, and the amount of the water primarily depends on the bulk volume of the denuder [39]. The aqueous extracts from the denuders are then analyzed by ion chromatography to determine sulfate, nitrate, formate, acetate and ammonium ions and deduce the concentrations of $SO_2, HNO_3, HCOOH, CH_3COOH$, and NH_3, respectively. An inherent disadvantage of using long, large-volume denuders to increase the gas residence time at higher flow rates, and therefore the collection efficiency, is that the larger amounts of water needed lead to a reduction in the concentration of the sample to be analyzed. Hence, the purpose of using large volume denuders to increase the amount collected per unit time is defeated by the large amount of water needed to extract the denuder. Thus, increase of the collection efficiency at high flow rates should be sought through proper designs that optimize the efficiency while maintaining the small overall denuder volume.

The filters in the filter pack, as well as the substrates of the impaction plates, are analyzed according to the kind of information that is sought. The total amount of particulates with size smaller than 2.5 μm can be determined by simply weighing the filters and the impactor plates on a high sensitivity balance (for example Cahn 21 electrobalance). To determine, however, the concentrations of inorganic and organic ions (sulfate, nitrate, formate, acetate, and ammonium), as well as the amount of atmospheric acidity (H^+) that has been neutralized during sampling, the filters and the substrates need to be extracted in a similar manner to the denuders, using deionized distilled water. One aliquot of the extracted solution is then analyzed by ion chromatography to determine particulate ion concentrations (NO_3^-, SO_4^{2-}, $NH_4^+, COOH^-$). Another aliquot of the extracted filters and impactor substrates is analyzed by a pH-meter with a semi-microelectrode to determine aerosol strong acidity, H^+ [39].

Acidic Aerosol Concentrations: Results from Field Studies

In this last section we give a summary of the current state of acid aerosol research, including results from previously conducted acid aerosol measurement studies. The studies discussed have been conducted in the northern part of the American continent, but the conclusions regarding temporal and spatial distribution of acidic compounds are generalizable.

The midwest and northeastern United States, as well as southeastern Canada, experience high sulfate concentration levels, reflecting the spatial distribution of

SO_2 emission sources and the regional meteorological conditions. The highest concentrations were found in communities located near the Ohio River Valley [1]. In communities with elevated acid aerosol levels, outdoor SO_4^{2-} and H^+ concentrations are highest during the summer months (May through September), when approximately two-thirds of the total yearly acidity is present [70].

The highest SO_4^{2-} and H^+ concentrations have been measured in Uniontown, Pennsylvania, a small semi-rural community located in the southwest corner of the state. In Uniontown, mean summertime SO_4^{2-} and H^+ levels measured over 6-h (10 am–4 pm) periods have been found to equal 150 and 175 nmol m^{-3}, respectively (EPRI report, 1991). For 6-h measurements made during the same summer in Pittsburgh, a metropolitan area located approximately 60 Km northwest from Uniontown, outdoor SO_4^{2-} concentrations were found to be comparable to those in Uniontown, with a summertime average of 125 nmol m^{-3} [71]. For H^+, however, outdoor concentrations measured in Pittsburgh were substantially lower, averaging only 80 nmoles m^{-3} over 6-hr periods [71].

Peak SO_4^{2-} and H^+ concentrations generally occur between noon and 6 pm, but the peak acid aerosol period will vary with radiative heat input and wind speed [72]. Diurnal variation in SO_4^{2-} and H^+ levels is probably due to the effects of atmospheric mixing, where air containing high concentrations of SO_4^{2-} and H^+ is mixed downwards during the daytime hours [72]. For H^+, this effect is enhanced by ammonia (NH_3), where at night H^+ is neutralized below the inversion layer by NH_3 emitted from the ground [72].

Outdoor SO_4^{2-} exhibits little spatial variability, as outdoor concentrations tend to be uniform across large areas. In a study [26] conducted in metropolitan Toronto, outdoor SO_4^{2-} concentrations were similar in three sites, located 33 km apart and with different population densities. Similarly to SO_4^{2-}, outdoor H^+ exhibits little spatial variation within non-urban communities. Spatial variation, however, has been observed within urban communities [26, 73] and across communities with different population densities [71, 74]. In all cases, this spatial variation was due to differences in local NH_3 concentrations, where higher NH_3 concentrations resulted in increased H^+ neutralization and thus lower H^+ concentrations [74]. Further evidence to support this hypothesis was given by the observed inverse relationship between outdoor H^+ and NH_3 concentrations (Fig. 14).

Acid aerosol studies also have demonstrated that indoor SO_4^{2-} and H^+ concentrations measured over 12-h (8 am–8 pm and 8 pm–8 am) periods are typically lower than corresponding outdoor levels and may vary substantially from home-to-home [9]. For SO_4^{2-}, indoor concentrations were found to comprise a substantial fraction of those outdoors, with indoor concentrations highest inside non-air conditioned homes. Indoor H^+ concentrations were also higher in non-air conditioned homes as compared to air conditioned homes; however, indoor levels generally were low. Indoor SO_4^{2-} and H^+ concentrations were highly correlated to outdoor levels, suggesting that outdoor air is the primary source for SO_4^{2-} and H^+ indoors.

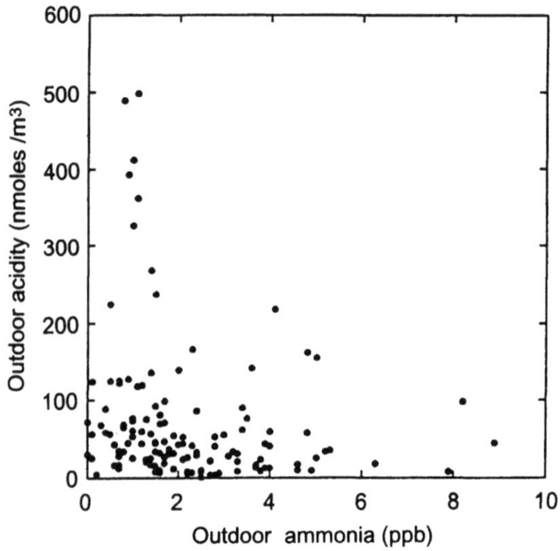

Fig. 14. Outdoor H^+ vs outdoor NH_3. Results from field studies in State College, PA, during summer of 1991. Concentrations were integrated over 24-h (8am–8am) periods, and were measured at the home sites. (From [67]. Reproduced with permission, Copyright Pergamon Press, 1994)

References

1. Spengler JD, Keeler GJ, Koutrakis P, Ryan PB, Raizenne M, Franklin CA (1989) Exposures to acidic aerosols. Environ Health Perspect 79:43
2. Finlayson-Pitts BJ, Pitts JN Jr (1986) Atmospheric chemistry John Wiley & Sons, New York
3. Koutrakis P, Mueller PK (1989) Atmospheric acidity: chemical and physical factors. Proc of Air & Waste Management Association, Anaheim, CA, June 25–30
4. Cadle RD (1972) Formation and chemical reaction of atmospheric particles. J Colloid Interface Sci 13:25–31.
5. Tang H, Eatough DJ, Lewis EA, Hansen LD, Gunther K, Benlap D, Crawford J (1986) The generation and decay of environmental tobacco smoke constituents in an indoor environment. Proc of the EPA/APCA Symposium on the Determination of Toxic and Related Air Pollutants, May 1989, Air and Waste Management Association, Raleigh, NC:596–605
6. Koutrakis P, Wolfson JM, Spengler JD, Stern B, Franklin CA (1989) Equilibrium size of atmospheric sulfates as a function of relative humidity. J Geophys Res 94:6442–6448
7. Pierson WR, Brachaczek WW, Truex TJ, Butler JW, Korniski TJ (1989) Ambient sulfate measurements on Allegheny Mountain and the question of atmospheric sulfate in the northeastern United States. Ann NY Acad Sci 338, 145–173
8. Pitts JN (1984) Atmospheric implications of simultaneous nighttime measurements of NO_3 radicals and HONO. Geophys Res Lett 11:557.
9. Brauer M, Koutrakis P, Keeler GJ, Spengler JD (1991) Indoor and outdoor concentrations of inorganic acidic aerosols and gases. J Air Waste Manage Assoc 41, 171
10. Koutrakis P, Thompson KM, Wolfson JM, Spengler JD, Keeler J, Slater J (1992) Determination of aerosol strong acidity losses due to interactions of collected particles: results from laboratory and field studies. Atmos Environ 26 A:987–995
11. Stelson AW, Seinfeld JH (1982) Thermodynamic prediction of the water activity of NH_4NO_3 dissociation constant, density, and refractive index for the NH_4NO_3-$[NH_4]_2SO_4$-H_2O system at 25°C. Atmos Environ 16:2507–2514
12. Jacob DJ, Wofsy SC (1988) Photochemistry of biogenic emissions over the Amazon forests. J Geophys Res 93:1477–1486
13. Lawrence JE, Koutrakis P (1994) Measurement of formic and acetic acids: methods evaluation and results from field studies. Environ Sci & Technol, in press
14. Grosjean D (1983) Polycyclic aromatic hydrocarbons in Los Angeles air from samples collected on Teflon, glass and quartz filters. Atmos Environ 17:2565–2573

15. Cautreels W, Van Cauwenberghe K, Guzman LA (1977) Comparison between the organic fraction of suspended matter at a background and an urban area. The Science and the Total Environment 8:79–88
16. Simoneit BR, Mazurek MA (1982) Organic matter of the troposphere-II. Natural background of biogenic lipid matter in aerosols from urban areas of China. Atmos Environ 16:2139–2159
17. Stephanou EG (1991) Analysis of anthropogenic and biogenic lipids in the coastal area in East Mediterranean Sea. Fresenius J Anal Chem 170:133–138
18. Kawamura K, Kaplan IR (1987) Motor exhaust emissions as a primary source for dicarboxylic acids in Los Angeles ambient air. Environ Sci & Technol 21:105–110
19. Rogge WF, Hildemann LM, Mazurek MA, Cass GR, Simoneit RT (1991) Sources of fine organic aerosols. 1. Charbroilers and meat cooking operations. Environ Sci & Technol 25:1112–11125
20. Schueltze D (1975) Analysis of complex mixtures by computer controlled high resolution mass spectrometry. I-Application to atmospheric aerosol composition. Biom Mass Spectrom 2:288–298
21. Cronn DR, Charlson RJ, Knights RL, Crittenden AL, Appel BR (1977) A survey of the molecular nature of primary and secondary components of particles in urban air by high-resolution mass spectrometry. Atmos Environ 11:929–937
22. McMahon TA, Denison PJ (1979) Empirical atmospheric deposition parameters-A survey. Atmos Environ 13:571–585
23. Seinfeld JH (1989) Urban air pollution: state of the science. Science 243:745–752
24. Shannon JD (1981) A model of regional long-term average sulfur atmospheric pollution, surface removal and net horizontal flux. Atmos Environ 15:689–701.
25. Hutzincker JJ, Hoffmann RS, Cary RA (1984) Aerosol sulfur episodes in St. Louis, Missouri. Environ Sci and Technol 18:962–967
26. Waldman JM, Lioy PJ, Thurston GD, Lippmann M (1990) Spatial and temporal patterns in summertime sulfate aerosol acidity and neutralization within a metropolitan area. Atmos Environ 24B, 115
27. Tanner RG (1993) Measuring the strong acid content of atmospheric aerosol particles. In: Measurement Challenges in Atmospheric Chemistry, edited by L. Newman, American Chemical Society, Washington, DC
28. Coutant RW (1977) effect of environmental variables on collection of atmospheric sulfate. Environ Sci and Technol 11:873–878.
29. Witz S and McPhee RD (1977) Effect of different types of glass filters on total suspended particulates and their chemical composition. J Air Pollut Contr Assoc 27:239–241.
30. Appel BR, Povard V, Kothny EL (1984) Artifact particulate sulfate and nitrate formation on filter media. Atmos Environ 18:409.
31. Witz S (1985) Effect of environmental factors on filter alkalinity and artifact formation. Environ Sci & Technol 19:831.
32. Shaw RW, Stevens RK, Bowermaster J, Tesch JW, Tew E (1982) Measurements of atmospheric nitrate and nitric acid: the denuder difference experiment. Atmos Environ 16:845
33. Appel BR, Tokiwa Y, Haik M, Kothny EL (1984) Artifact particulate sulfate, and nitrate formation on filter media. Atmos Environ 18:409
34. Durham JL, Wilson WE, Bailey EB (1978) Application of an SO_2 denuder for continuous measurement of sulfur in submicrometric aerosols. Atmos Environ 12:883–886
35. Ferm M (1979) Method for determination of atmospheric ammonia. Atmos Environ 13:1385–1393
36. Shaw RW, Stevens RK, Bowermaster J, Tesch JW, Tew E (1979) Atmos Environ 16:845–853.
37. Forrest J, Spandau DJ, Tanner RL, Newman L (1982) Atmos Environ 16:1473–1485
38. Braman RS, Shelley TJ, McClenney WA (1982) Tungstic acid for the preconcentration and determination of gaseous and particulate ammonia and nitric acid in ambient air. Analytical Chem 54:356–364
39. Koutrakis P, Wolfson JM, Slater JL, Brauer M, Spengler JD (1988) Evaluation of an annular denuder/filter pack system to collect acidic aerosols and gases. Environ Sci and Technol 22(12): 1463–1468
40. Koutrakis P, Sioutas C, Ferguson S, Wolfson JM, Mulik JD, Burton RM (1993) Development and evaluation of a glass honeycomb denuder/filter pack system to collect atmospheric particles and gases. Environ Sci & Technol 27:2497–2501
41. Brauer M, Koutrakis P, Wolfson JM, Spengler JD (1989) Evaluation of an annular denuder system under simulated atmospheric conditions. Atmos Environ 23:1981–1986
42. Appel BR, Povard V, Kothny EL (1988) Loss of nitric acid within inlet devices intended to exclude coarse particles during atmospheric sampling. Atmos Environ 11:2535–2540.

43. Koutrakis P, Wolfson JM, Brauer M, Spengler JD (1990) Design of a glass impactor for an annular denuder/filter pack system. Aerosol Sci and Technol 12:607–613
44. Stevens RK, Dzubay TG, Russwurm G, Rickel D (1978) Sampling and analysis of atmospheric sulfates and related species. Atmos Environ 21:589
45. Possanzini M, Febo A, Liberti A (1983) New design of high performance denuder for the sampling of atmospheric pollutants. Atmos Environ 17:2605–2610
46. Stevens RK, Paur JR, Allegrini I, DeSantis F, Febo A, Perrino C, Possanzini M (1985) Measurement of HNO_3, SO_2, NH_3, and particulate nitrate with an annular denuder system. Proc 5th Annual National Symposium on Recent Advances in the Measurement of Air Pollution, US EPA, Rayleigh, NC
47. Appel BR (1993) Sampling of selected labile atmospheric pollutants. In: Measurement Challenges in Atmospheric Chemistry, edited by L. Newman, American Chemical Society, Washington, DC
48. Appel BR, Povard V, Kohny EL (1988) Loss of nitric acid within inlet devices intended to exclude coarse particles during atmospheric sampling. Atmos Environ 11:2535–2540
49. Coutant RW, Callahan PJ, Kuhlman MR, Lewis RG (1989) Design and performance of a high-volume compound annular denuder. Atmos Environ 23(10):2205–2211
50. Suh HH, Spengler JD, Koutrakis P (1992) Personal exposures to acid aerosols and ammonia. Environ Sci Technol 26: 2507
51. Bowermaster J, Shaw RW (1981) J Air Pollut Contr Assoc 31:787.
52. Rosenberg C, Winiwarter W, Gregori M, Pech G, Casensky V, Puxbaum H (1988) Fresenius Z. Anal Chem 331:1–7
53. Ye Y, Tsai CJ, Pui DYH, Lewis CW (1991) Aerosol Sci and Technol 14:102–111
54. Sioutas C, Koutrakis P, Wolfson JM (1994) Particle losses in glass honeycomb denuder samplers. Aerosol Sci and Technol, in press
55. Whitby KT, Husar RB, Liu BYH (1972) J Colloid Interface Sci 39:177–204
56. Wang HC, John W (1988) Characteristics of the Berner impactor for sampling inorganic ions. Aerosol Sci & Technol 8(2):157–172
57. Ahlberg MS, Winchester JW (1978) Dependence of aerosol sulfur particle size on relative humidity. Atmos Environ 12:1631–1632
58. Hinds WC (1982) Aerosol technology: Properties, behavior and measurement of airborne particles John Wiley & Sons, New York
59. Marple VA, Liu BYH (1974) Characteristics of laminar jet impactors. Environ Sci & Technol 7:648–654
60. Marple VA, Willeke K (1976) In: Fine particles: Aerosol generation, measurement, sampling, and analysis. (B.Y.H Liu ed.)
61. Fuchs NA (1964) The Mechanics of Aerosols Pergammon Press, New York.
62. Hering SV, Flagan RC, Friedlander SK (1978) Design and evaluation of a new low pressure impactor-1. Environ Sci Technol 12:667–673
63. Marple VA, Rubow KL, Behm SM (1991) A micro-orifice uniform deposit impactor (MOUDI). Aerosol Sci & Technol 14: 434–446
64. Biswas P, Jones CL, Flagan RC (1987) Distortion of size distributions by condensation and evaporation in aerosol instruments. Aerosol Sci and Technol 7:231–246.
65. Alberg MS, Winchester JW (1979) Dependence of aerosol sulfur particle size on relative humidity. Atmos Environ 12:1631–1632.
66. Koutrakis P, Kelly BP (1993) Equilibrium size of atmospheric aerosol sulfates as a function of particle acidity and ambient relative humidity. J Geophys Res 98:7141–7147
67. Suh HH, Allen GA, Aurian-Blajeni D, Koutrakis P (1994) Field method intercomparison for the characterization of acid aerosol and gases. Atmos Environ in press
68. Marple VA, Rubow KL, Turner W, Spengler JD (1987) Low flow rate sharp cut impactors for indoor air sampling: design and calibration. J Air Pollut Contr Assoc 37:1303–1307.
69. Allen GA, Turner WA, Wolfson JM, Spengler JD (1984) Description of a continuous sulfuric acid/sulfate monitor. In: Proc National Symposium on Recent Advances in Pollutant Monitoring of Ambient Air and Stationary Sources, US EPA, #EPA-600/9-84-019, 140
70. Thompson KM, Koutrakis P, Brauer M, Spengler JD, Wilson WE, Burton RM (1991) Measurements of aerosol acidity: sampling frequency, seasonal variability, and spatial variation. Proceedings of the AWMA Conference, Paper No. 91–89.5
71. Burton RM, Wilson WE, Koutrakis P, Liu LS (1992) Comparison of aerosol acidity in urban and semi-rural environments. Proceedings of the EPA-AWMA Conference
72. Wilson WE, Koutrakis P, Spengler JD (1991) Diurnal variations of aerosol acidity, sulfate, and ammonia in the atmosphere. Proc of the AWMA Conference, Paper No. 91–89.9

73. Allen GA, Koutrakis PA (1992) Development and validation of a model for predicting short-term acid aerosol concentrations from the Harvard School of Public Health Continuous Sulfate/Thermal Speciation Monitor. Presented at the EPA/Air and Waste management Association Conference on Measurement of Toxic and Related Air Pollutants, Durham, NC, May.
74. Suh HH, Koutrakis P, Spengler JD (1993) Validation of personal exposure models for sulfate and aerosol strong acidity. JAWMA 43:845
75. Suh HH, Koutrakis P, Spengler JD (1994) Housing factors and their influence on indoor acid aerosol and gas concentrations. J Expos Anal and Environ Epidemiol (in press)
76. Mitchell RI, Pilcher JM (1959) Improved cascade impactor for measuring aerosol particle sizes. Ind Eng Chem 51: 1039

Appel BR, Tokiwa Y, Kothny EL (1983) Sampling of carbonaceous particles in the atmosphere. Atmos Environ 17: 1787–1796

Bjorseth O, Krohn C, Togersen S, Fjelstadt PE (1980) The working environment in the aluminum industry. Sampling for PNA by an alumina fluidized bed sampler. Electrochemica Acta 25: 117–124

Brosset C (1979) Possible changes in aerosol composition due to departure from equilibrium conditions during sampling. Am Chem Soc Conference, Hawaii

Friedlander SK (1976) Smoke, dust and haze. John Wiley & Sons

McMurry PH, Zhang XQ (1989) Size distribution of ambient organic and elemental carbon. Aerosol Sci & Technol 10: 430–437

Mercer TT, Staffrord RG (1969) Impaction from round jets. Ann Occup Hyg 12:41

Mercer TT, Tillery MI, Chow HY (1968) Operating characteristics of some compressed air nebulizers. Am Ind Hyg Assoc J 29:66–78

Seinfeld JH (1986) Atmospheric physics and chemistry of air pollution. John Wiley & Sons, New York

Analysis of Organic Particulate Matter

Constantini Samara

Environmental Pollution Control Laboratory, Department of Chemistry, Aristotle University of Thessaloniki, 540 06 Thessaloniki, Greece

Introduction 234
Sampling and Sample Preparation Procedures 235
Sampling 235
Extraction 238
Clean-Up and Fractionation 239
Analytical Methods 241
Polynuclear Aromatic Hydrocarbons (PAHs) 241
Organochlorine Pesticides and Polychlorinated Biphenyls (PCBs) 244
Polychlorinated Dibenzo-*p*-dioxins (PCDDs) and Polychlorinated Dibenzo-*p*-furans (PCDFs) 246
Analytical Method Validation 247
References 249

Summary

The organic fraction of airborne particulate matter contains hundreds or even up to thousands of chemical components of natural and anthropogenic origin. Many of them have received considerable attention in recent years due to their adverse health effects.

To understand the extent of human exposure to such contaminants reliable methods for sampling, sample preparation and analysis of airborne particulate matter have to be employed. Reliable techniques require detailed method validation and careful evaluation regarding efficiency. Nowadays, the analysis of most particulate organics of environmental significance has developed to high standards. The complexity of the samples, however, continues to promote research interest, often directed towards achieving more convenient and cost-effective methods.

Introduction

Particulate matter in urban areas results not only from direct particulate emissions (primary aerosols) but also from gas-phase reactions followed by condensation of the products, or by reactions taking place in the existing atmospheric aerosol phase.

The carbonaceous fraction of the particulate matter usually consists of both elemental carbon (EC) and organic carbon (OC). Elemental carbon, also called black carbon, graphitic carbon, or soot, has a chemical structure similar to impure graphite and is emitted directly into the atmosphere predominantly during combustion. Organic carbon is either emitted directly (primary OC) or can be formed in-situ by condensation of low volatility products of the photo-oxidation of hydrocarbons (secondary OC) [1–5].

The organic fraction of atmospheric aerosol contains hundreds to thousands of chemical components such as biogenic long chain *n*-alkanes, fatty alcohols and fatty acids, *n*-fatty acid salts of marine origin and, of course, anthropogenic organic compounds many of which are of great concern due to their toxic and carcinogenic properties. These include the polynuclear aromatic hydrocarbons (PAHs) and their derivatives, chlorinated pesticides, polychlorinated biphenyls (PCBs), as well as polychlorinated dibenzo-*p*-dioxins (PCDDs) and dibenzofurans (PCDFs). Identifying and determining the concentration of individual components in mixtures of this complexity is extremely difficult, particularly if composition as a function of particle size is required.

Most of the environmentally significant organics contained in airborne particulate matter (PAHs, organochlorines, PCBs, PCDDs/PCDFs) are semivolatile compounds (SOCs), e.g. they are mainly associated with suspended particulate matter, but also exist in the atmosphere as vapours, the vapour-to-particle partitioning being governed by the vapour pressures and ambient temperature.

The vapour-to-particle partitioning has an important influence on atmospheric removal by rain and dry deposition. Vapour scavenging is governed by Henry's law constant, the ratio of vapour pressure to water solubility. Some lower molecular weight SOCs are rained out by vapour dissolution in raindrops (e.g. phenols, two- to four-ring PAHs, hexachloro-cyclohexane, etc). Other SOCs that have substantial fractions in the particle phase and/or Henry's law constants unfavourable for vapour scavenging are washed out of the atmosphere mainly on particles. Some compounds of the latter category are PCBs, DDT, PAHs of five or more rings, the hexa- through octachloro homologues of PCDDs/PCDFs and *n*-alkanes [6–9]. Conversely, scavenging of the particles and ambient temperature considerably modifies the gas/particle ratio [10–12].

Since knowledge of the physical state of an atmospheric contaminant is vital for the understanding of its operative physical and chemical removal processes, and since the vapour pressures of many trace organics fall in the range where substantial amounts of a given compound are expected in both aerosol and vapour phases, a comprehensive organic air sampling system must include both aerosol and vapour measurement capabilities.

Analysis of trace substances in complex matrices such as airborne particulate matter is fraught with hazard. Reliable techniques require detailed method validation and careful evaluation for analytical efficiency by means of recovery tests. In order to define quantitatively the accuracy and precision of the procedure for each determinand, it is necessary to estimate statistically random and systematic errors. In addition, sampling and sample preparation procedures should be considered integrally with the characterization of an analytical procedure, an area too often neglected.

Sampling and Sample Preparation Procedures

Most methods for the analysis of organic particulates in ambient air involve filter collection, extraction and some form of chromatographic separation. Clean-up and prefractionation of the extract before the chromatographic separation is often necessary to separate the compounds of interest and to remove those substances that might interfere with the analysis.

A typical scheme for the analysis of organic constituents of airborne particulate matter is illustrated in Fig. 1.

Sampling

Sampling procedures for organic particulates depend mainly on filter collection. Vapour-to-particle partitioning is an important consideration when semivolatile organic compounds are to be quantitatively sampled.

The most popular configuration for sampling airborne particulates includes a glass-fibre filter and a high volume sampler. The use of high volume samplers for the collection of SOCs has been seriously criticised due to the almost inevitable pressure drop across the filter that accelerates evaporation of the more volatile compounds [13, 14].

The difficulty of quantitatively sampling airborne SOCs, such as PAHs and organochlorine pesticides, on account to their individual varying volatilities has long been known. Losses can occur both by evaporation and sublimation as air is sucked through the organic particles already entrapped by the filter.

The "blow-off" effect was particularly investigated for PAHs [15–19]. An estimate of the distribution of a PAH between vapour and particulate phases can be obtained from its equilibrium vapour concentration (EVC). If the EVC of an individual PAH exceeds the maximum airborne concentration, no significant amounts of this compound will be found on the filter. The "blow-off" effect is of particular importance in the case of 3- and 4-ring compounds with high vapour pressures at ambient temperature and boiling point below 400 °C. However, losses up to 20% can occur even from particle-bound PAHs with low vapour pressures

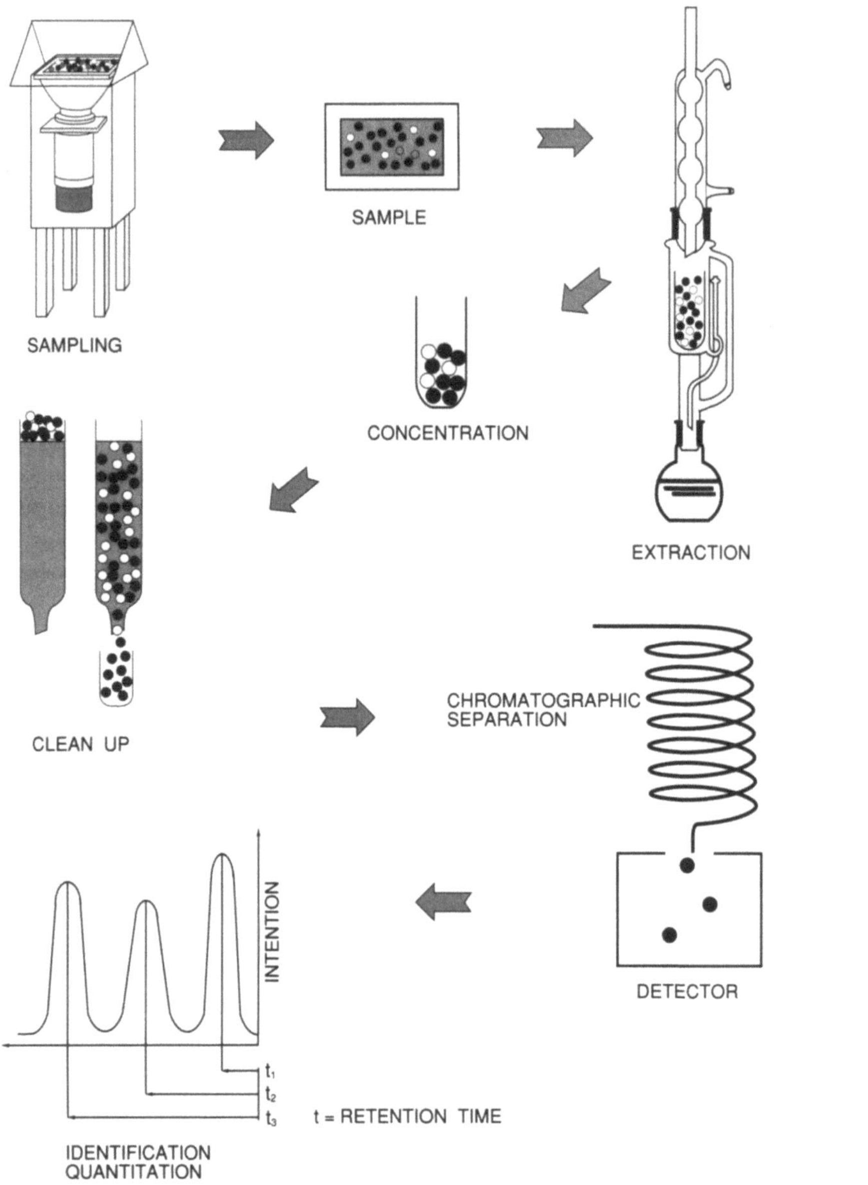

Fig. 1. A typical scheme for the analysis of organic constituents of airborne particulate matter

(such as the benzo[*a*]pyrene), as a result of evaporation if the flow rate through the filter is high and the sampling duration extended [18, 20].

Evaporation problems can be overcome by employing an adsorbent as back-up to trap all volatile fractions that passes through the filter in the vapour state or that sublime during sampling. Adsorbents widely used are the non-polar synthetic polymers Tenax (poly-2,6-diphenyl-*p*-phenylene oxide) and XAD-2 (polystyrene-divinylbenzene) in the form of porous beads, as well as polyurethane foam (PUF) in the form of plugs [21–26]. Other sorbents are XAD-4 (with a collection efficiency for PAHs comparable to that of XAD-2), Florisil, Chromosorb 102 and Porapak [27, 30]. Cryogenic back-up traps to achieve gas-phase enrichment and high volume filters coated with glycerol-tricaprylate have also been tested for quantitative sampling of PAHs [29, 31, 32].

Apart from glass fibre, various other filter materials are used, such as Teflon-coated glass fibre, quartz, polycarbonate, Teflon and cellulose esters [33, 34]. Teflon appears to be the superior filtering medium, because of its reduced surface activity and the very low blank values that make filter precleaning unnecessary. However, owing to its high cost, Teflon finds relatively limited use in comparison to other more cost-effective materials such as fibreglass.

The macroreticular resins XAD-2 and Tenax with their virtually 100% recovery efficiencies have practical application as collection phases for vapour-phase PAHs and pesticides. Polyurethane foam exhibits lower collection efficiency than XAD-2 for volatile PAHs, but it has demonstrated high capabilities in sampling organochlorine pesticides, poly-chlorinated biphenyls and polychlorinated dibenzo-*p*-dioxines. Polyurethane foam also has several other advantages (easy to handle, low cost, good recovery of adsorbed compounds when applying solvent extraction) especially for high-volume sampling due to the low pressure drop of plugs that can be achieved with appropriate dimensions [35, 43]. Although Tenax-GC shows less breakthrough than PUF, small cartridges have a high flow resistance, and large adsorbent beds may be prohibitively expensive for many applications. In addition, Tenax-GC may degrade during sampling in the presence of reactive gases such as O_3 and NO_2 [36].

Adsorbents and filters usually need to be pretreated before use to remove impurities. Usual pretreatment procedures involve reflux or Soxhlet extraction with an organic solvent. The cleaning procedure for PUF is a particularly laborious and time-consuming operation [12, 38, 40], although the development of more time-saving procedures has been attempted [35].

Low volume sampling and/or sampling periods shorter than 24 h have also been employed particularly when sampling is conducted in a polluted region or diurnal variabilities of airborne organic pollutants are to be investigated [41, 42]. The low concentration of target compounds collected in this way requires the employment of an extremely sensitive analytical technique.

High volume sampling over a sampling period of 24 or at least 12 h is more convenient from the point of view that adequate amounts of airborne particulate matter can be aquired so that concentration of organic pollutants is well above the detection threshold of the analytical procedure even when sampling in remote areas. In addition to this, results can be directly related to other parameters such as total suspended particles (TSP), or inhalable particulate matter (IPM), heavy metal concentrations, etc., that are normally sampled on a 24-h basis.

The U.S. EPA recommended method for the determination of PAHs in ambient air depends on a 24-h high volume sampling and the utilisation of XAD-2 or

PUF as back-up adsorbents [35]. Likewise, the corresponding method for the determination of organochlorine pesticides and polychlorinated biphenyls involves 24-h high volume sampling through a glass fibre filter and a PUF adsorbent cartridge [43].

Finally, high volume sampling by using modified commercial high volume samplers in combination with selected retention materials can offer efficient collection for most semivolatile organic pollutants in ambient air.

Extraction

The extraction process itself is a critical step in the analysis and deserves the most careful scrutiny since there are many variables to be considered, such as the nature of the sample, the choice of the appropriate solvent, the agitation technique, etc.

Various procedures have been cited for the extraction of airborne particulate matter after collection, the most prominent being Soxhlet extraction with an appropriate organic solvent.

The aim of the extraction procedure is to recover quantitatively the collected organic pollutants. For this purpose, several factors should be taken into consideration, such as the extraction efficiency of the organic solvent, the ratio of the particulate amount to the solvent volume, the presence of extraneous materials, etc.

Unfortunately, there are minimal standard procedures for the extraction of a single solute or a class of organic pollutants from airborne particulates. Furthermore, only few procedures cited in the literature are described with adequate method validation and evaluation of the extraction step efficiency.

PAHs are usually extracted in a Soxhlet apparatus with organic solvents of low to medium polarity such as cyclohexane, toluene, methylene chloride, tetrahydrofuran, benzene, methanol, acetone etc. [21, 44–49]. Extraction times vary between 1 and 48 h. Among these, acetone, benzene, and cyclohexane are all said to be nearly 100% efficient in the extraction of benzo[*a*]pyrene [50]. Methanol has also been recommended as an efficient solvent for the extraction of PAHs, but has also been found to extract inorganic and organic material other than PAHs. Toluene, as an aromatic solvent, is especially suitable for the extraction of PAHs from diesel soot [41, 42], whereas toluene and dichloromethane have been said to be the solvents of choice if the amount of the heavier PAHs is used as a criterion [47].

Benzene, although having very good extraction efficiency for PAHs, finds only limited use due to its great health hazard to laboratory workers. Since cyclohexane extracts fewer extraneous materials than does benzene and other solvents, and is less hazardous, it seems to be very suitable, and its use has been endorsed by official bodies such as the World Health Organization [49–52].

The length of time of Soxhlet extraction and the possibility of decomposition of some compounds during it makes the use of alternative extraction procedures very attractive. A particularly promising extraction procedure is the use of ultrasonic vibration at ambient temperature since it combines high extraction efficiencies with simplicity and time saving [53–58]. Thus, PAHs and other organics can be completely extracted from air particulates after only 15 or 30 min. The procedure can be refined by sonicating in the presence of silica powder to adsorb polar materials [57]. Ultrasonic extraction has been adopted by the National Institute of Occupational Safety and Health [58].

Similar extraction techniques as that described above for PAHs are also used for the extraction of organochlorine pesticides, PCBs and PCDDs/PCDFs from airborne particulate matter [28, 59–63].

Another extraction technique, although not widely used so far for the extraction of organic solutes from airborne particulate matter, seems to be supercritical fluid extraction (SFE) due to its potential advantages over other extraction technologies. The use of SFE can improve productivity with regard to both time and extraction efficiencies. In addition, thermally labile components can be safely extracted, whereas different extraction selectivities are possible for specific solutes or classes of compounds [64].

The recovery of the volatile fraction of SOCs trapped on back-up adsorbents is usually done by solvent extraction procedures similar to those used for filters [27, 33, 35, 65, 66]. Besides, thermal desorption from some adsorbents, such as Tenax-GC, is often performed when gas chromatographic analysis is subsequently to be employed. Direct thermal desorption into the analytical gas chromatography column is very attractive since high preconcentration factors and method sensitivity can be achieved [66–68]. Ligocki and Pankow [37], found that both adsorption/solvent extraction with PUF plugs and adsorption thermal desorption with Tenax-GC cartridges are valid methods for the analysis of many ambient organic vapours at the ng m^{-3} level.

The EPA method for the determination of organochlorine pesticides and PCBs in ambient air recommends 14 to 24 h. Soxhlet extraction of quartz filters and PUF plugs with 5% diethyl ether in hexane, whereas the EPA Method 610 for PAHs uses 2-h Soxhlet extraction with methylene chloride [43]. The German guidelines for the measurement of PCDDs/PCDFs in ambient air recommends 20-h Soxhlet extraction of the glass fibre filters and PUF plugs with toluene [69].

Clean-Up and Fractionation

Having extracted the organic matter from the particulate sample, the fraction of interest has to be separated so that a simplified mixture of organic species is isolated from the complex arrays of organics present in the extract. Clean-up involves one or a series of steps, and therefore it is often very time consuming.

Clean-up and fractionation of the extracted material is usually done by liquid-liquid partitioning based on the variation of partition coefficients of different classes of organic compounds for a variety of solvents.

Acidic (e.g. phenols) and basic constituents can be separated from the neutral fraction by successive washings with dilute alkali and acid [52, 70, 71]. These steps may be omitted for air particulate samples that contain predominantly neutral materials. Hydrophilic and polar compounds are next removed by partitioning between cyclohexane and either methanol/water [72] or acetone/water [73]. Further enrichment of the analytes that are contained in the neutral non-polar fraction is then possible by extraction from cyclohexane into nitromethane, dimethyl sulphoxide (DMSO), or dimethyl formamide (DMF) and subsequent evaporation or back extraction into an alkane solvent [35, 38, 44, 70, 73–75].

Open column chromatography has been widely used for the clean-up of the extracted organic matter. Silica gel, basic alumina and other adsorbents, such as Florisil, are used for this purpose [35, 76, 77]. Batch methods are less frequently used than column separations, because the latter, although somewhat more tedious, provide more efficient separation.

Besides the clean-up, column chromatography is also used for the separation of the extracted organic fraction into sub-fractions containing compounds of, for example, similar molecular weight [78]. Keller and Bidleman [79] divided air particulate samples into non-polar and polar fractions by using a column of silicic acid overlaid by alumina, thus separating aliphatic (*n*-alkanes) and aromatic components (PAHs).

However, classical adsorbents suffer the disadvantages of high adsorptivity, which can result in losses and in peak tailing with consequent less sharp fractionation, and poor reproducibility because of adsorption of traces of water. Lipophilic gels such as Sephadex LH-20 or the Bio-Beads series have thus been used in adsorption/partition mode to yield fractions containing compounds in sequence of increasing ring number [71, 80, 81].

Disposable minicolumns containing silica or silica bonded organic phases (e.g. octadecyl-, aminopropyl-, cyano-phases) are increasingly used for sample clean-up. With a polar phase, one loads the column with a small volume of sample extract. Polar compounds that have been co-extracted with the analytes are strongly adsorbed onto the column, whilst analytes are flushed with a solvent of low or intermediate polarity. Although the clean-up principles of these small columns are no different from those for other open-column techniques, the convenience of prepacked columns and their disposability can be cost effective in many cases, when large numbers of similar samples have to be cleaned-up.

Since clean-up by column chromatography is less time consuming and less tedious than the liquid–liquid partitioning, it is officially recommended for the analysis of PCDDs/PCDFs, PAHs, PCBs and organochlorine pesticides [43, 69].

Low resolution liquid chromatography provides a useful fractionation technique for separating organic components according to their polarity. Ciccioli et al. [39] have separated four fractions from airborne particulate matter containing aliphatic hydrocarbons, PAHs, PAH derivatives and long-chain polar compounds.

Thrane et al. [38] have used high performance liquid chromatography on silica column to separate airborne PAHs from polar compounds. The fractions collected from the high performance liquid chromatographic column are suitable for subsequent analysis by other chromatographic and spectroscopic methods [38, 39].

Finally, TLC prefractionation has also been used for air particulate extracts [49, 55]. Moriske and Ruden [82] separated the neutral fraction of the ether-extractable organic matter of air particulates into aliphatic component PAHs and polar neutral organic components by TLC on silica gel.

Analytical Methods

The low concentrations of organic pollutants in airborne particulate matter present a great challenge to analytical chemists. Besides the separation of the determinands from extraneous materials and their enrichment during the extraction and clean-up steps, the use of an accurate, precise, sensitive and, if possible, selective analytical technique is crucial for obtaining reliable results.

We will discuss the analytical methods primarily employed for the most significant classes of organic pollutants present within airborne particulate matter: PAHs, organochlorine pesticides, PCBs and PCDDs/PCDFs.

Polynuclear Aromatic Hydrocarbons (PAHs)

PAHs have been extensively analysed by gas chromatography (GC) with flame ionization detection, although the low volatility of these compounds presents a problem [83]. The use of packed columns has achieved only partial separation and resolution of individual PAHs, most success being obtained with non-polar liquid stationary phases, such as OV-17 and high-temperature phases e.g. Dexsil 300 [19]. By far the greatest success has been obtained with the use of glass, and, more recently, fused silica capillary columns coated with various chemically or adsorptive bounded phases like dimethyl-polysiloxane, dimethyl-diphenyl-polysiloxane and dimethyl-cyanopropylenenylpolysiloxane. High efficiency separation and resolution of the common PAHs from naphthalene to coronene are possible with only the benzofluoranthenes often presenting particular separation problems [84–86].

Qualitative analysis is done by comparison of the retention times with those of the reference substances. Quantification is based on the measurement of the peak areas of identified PAHs relative to the peak area or height of an internal standard. Determinand peak area or height can also be measured as analytical response relative to an external standard calibration curve.

Several compounds have been used as internal standards for PAH analysis by GC/FID (e.g. benzo[*b*]fluorene, indeno[1,2,3-*cd*]fluoranthene, 3,6-dimethylphenanthene, β,β′-binaphthyl, 1,3,5-triphenylbenzene, etc) [46, 82, 85, 86].

To use the internal standard calibration procedure, the analyst must select one or most internal standards that are similar in analytical behaviour to the compounds of interest. The analyst must further demonstrate that the measurement of the internal standard is not affected by method or matrix interferences. Due to these limitations, no internal standard applicable to all samples can be suggested. It is recommended that the internal standard approach be used only when the GC/MS procedure is employed due to co-eluting species [43].

The typical detection limits for individual PAHs by capillary column GC with flame ionization detection are of the order of 0.05–0.5 ng. Figure 2 presents a capillary GC/FID chromatogram of PAHs in airborne particulate matter [42].

The most powerful tool for positive compound identification is gas chromatography with mass spectra detection (GC/MS). The use of this technique for the identification and quantitation of PAHs in airborne particulate matter constitutes quite a sensitive and reliable analytical method, particularly where capillary columns are employed.

Identification in GC/MS is achieved by matching the mass spectra of sample components to those of references. Under normal electron impact ionization (EI) conditions, mass spectra of isomeric PAH pairs are virtually identical (even in the cases of isomers with very different structures, such as fluoranthene and pyrene, the mass spectra are mostly indistinguishable), and therefore confirmatory retention time data are required to facilitate compounds identification. On combination of the data of mass spectra and retention index, many isomers can be distinguished. Since intense molecular ions are produced under EI conditions, selected or multiple ion monitoring techniques may be employed to good advantage and detection limits in the picogram range may be achieved [22, 81, 83, 87]. Quantification is made either by rationing the integrated peak area of the molecular ion of the determinand to that of the corresponding internal standard or by external calibration. Isotopically labelled PAHs, such as perylene-d_{12}, chrysene-d_{12}, acenaphthene-d_{10}, naphthalene-d_8 and phenanthrene-d_{10} are suggested as internal standards [43, 22].

However, GC has often been considered inappropriate to the analysis of PAHs because of the thermal instability of that class of compounds and the low volatility of the higher molecular weight species.

Although high pressure liquid chromatography (HPLC) does not approach the high separation efficiency of capillary gas chromatography at present, it does offer several advantages for the determination of PAHs [83, 49, 52]. First, HPLC offers a variety of stationary phases capable of providing unique selectivity for the separation of isomeric PAHs (e.g. benzo[*e*]pyrene and benzo[*a*]pyrene) that are often difficult to separate by GC. Selectivity in HPLC is achieved because of interactions of the solute with both the stationary and the mobile phase rather than with only the stationary phase as in GC. Secondly, PAHs can be separated at ambient temperatures, eliminating the risk of decomposition at higher temperatures. HPLC in combination with UV absorption and fluorescence spectroscopy provides extremely sensitive and, more important, selective detection of PAHs, since it eliminates or reduces interference from compounds which neither absorb ultraviolet light nor fluoresce.

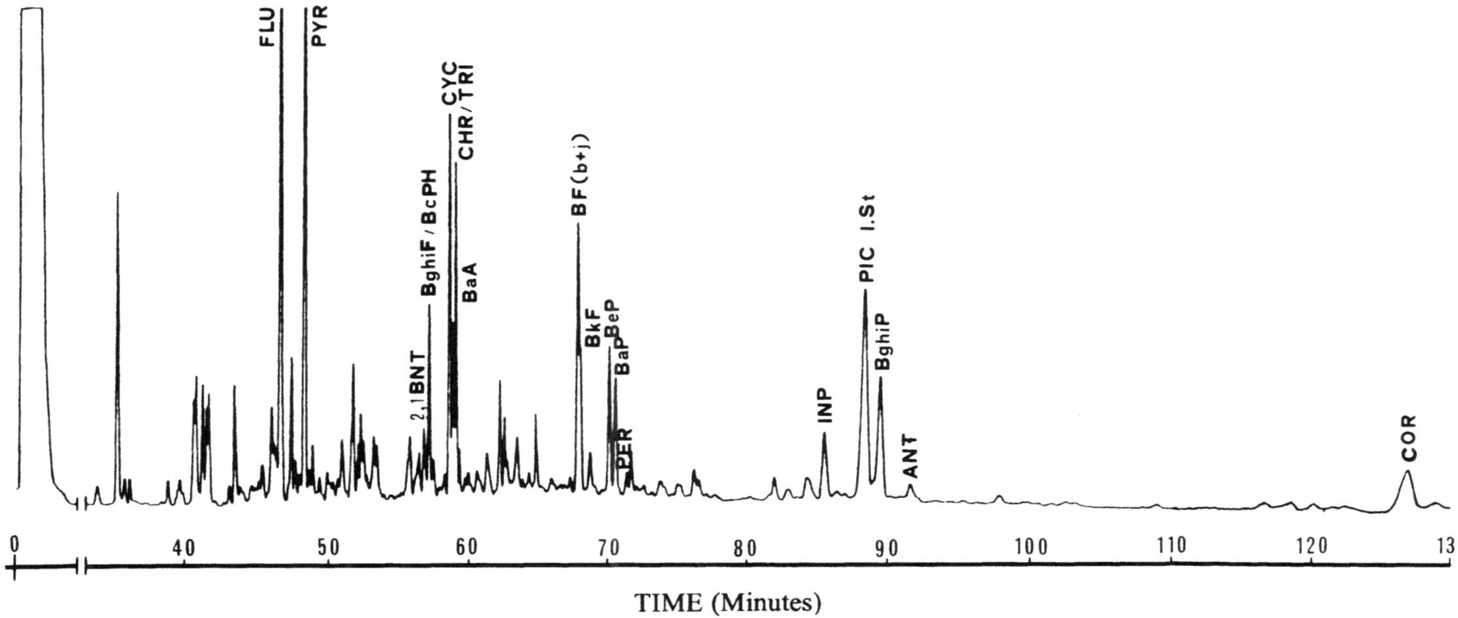

Fig. 2. Glass capillary GC/FID analysis of airborne particulate matter from an area polluted by automobile exhaust gases [42] FLU: Fluoranthene, PYR: Pyrene, 2,1-BNT: Benzo(b)naphtho(2,1-d)thiophene, BghiF: Benzo(ghi)fluoranthene, BcPH: Benzo(c)phenanthrene, CYC: Cyclopenta(cd)-pyrene, CHR: Chrysene, TRI: Triphenylene, BaA: Benz(a)anthracene, BF(b+j+k): Benzofluoranthenes (b+j+k), B(e)P: Benzo(e)pyrene, B(a)P: Benzo(a)pyrene, PER: Perylene, PIC: Picene (Benzo(a)chrysene) INP: Indeno-(1,2,3-cd)pyrene, BghiP: Benzo(ghi)perylene, ANT: Anthanthrene, COR: Coronene

The UV detector is universal for PAHs and accurate quantification is generally difficult, since many PAHs and alkyl-substituted PAHs co-elute. Several workers have described the use of variable-wavelength UV detection to achieve higher degrees of selectivity, identification being based on UV absorption spectra of PAHs or absorbance ratios at several wavelengths. The UV diode-array detector (DAD), recently available, seems to be valuable for positive identification and for quantification of co-eluting components in the nanogram range [45, 88, 89].

Fluorescence detection provides selectivity for individual PAHs and the possibility of identification of specific compounds in complex mixtures. Using HPLC with fluorescence wavelength programming, May and Wise [44] quantified 13 PAHs in an air particulate material, now available as a Standard Reference Material (SRM 1649-Urban Dust/Organics). The same method was used for the determination of PAHs in diesel particulate samples. Nowadays, programmable fluorescence detectors capable of providing up to 16 changes of excitation and emission wavelength conditions are available, permitting selective determination of a great number of PAHs in one chromatographic run. Detection limits of the fluorescence detection of PAHs are in the picogram range [44, 89, 90]. An HPLC/fluorescence detection chromatogram of PAHs in airborne particulate matter is illustrated in Fig. 3 [44].

The potential of the combination of liquid chromatography with mass spectrometry (LC/MS) for the separation and identification of organic compounds has generated considerable research in the past few years. Although only few applications of LC/MS involving PAHs have been reported so far, this technique will find greater application in the determination of higher molecular weight PAHs that are not amenable to GC analysis.

Finally, HPLC provides a useful multidimensional liquid chromatographic technique for the isolation of specific PAH fractions for subsequent analysis by other chromatographic or spectroscopic methods. Multidimensional chromatographic techniques (i.e. the sequential use of different chromatographic separation modes) constitute an advantageous approach to the analysis of complex mixtures, like the air particulate samples [38, 39, 90].

Organochlorine Pesticides and Polychlorinated Biphenyls (PCBs)

The major analytical tool for chlorinated pesticides and PCBs is capillary gas chromatography with electron capture detection (GC/ECD) [7, 6, 43, 91, 92]. The ECD is a very sensitive detector, selective for halogens, thus minimizing the risk of interference from extraneous organic compounds. However, the ECD still responds to a wide variety of organic compounds that would be encountered as interferences during GC/ECD analysis (e.g. PCBs may interfere with the determination of pesticides). PCBs may be separated from other organochlorine pesticides by column chromatography on silicic acid, whereas organophosphorus and carbamate pesticides can be removed by column chromatography on alumina.

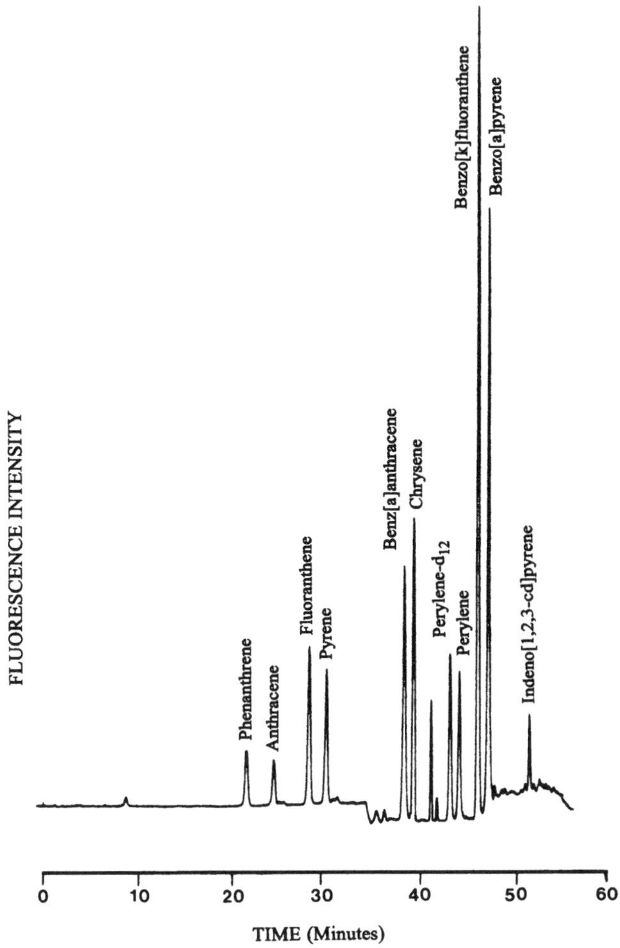

Fig. 3. Reversed-phase LC analysis with fluorescence detection of the total PAH fraction from St. Louis air particulate sample (SRM 1648) isolated by normal-phase LC clean-up [44]

Pesticide mixtures are quantified by comparison of the total heights or areas of GC peaks with the corresponding peaks in the best-matching standard. For confirmation, a second GC column is often used. A Hall electrolytic conductivity detector operated in the reductive mode may be substituted for the ECD for improved specificity. However, limits of detection will be reduced by at least an order of magnitude.

The fact that PCBs as well as certain chlorinated pesticides (e.g. toxaphene and chlordane) are complex mixtures of individual compounds can cause difficulty in accurately quantifying by GC/ECD a particular formulation in a multiple component mixture [43]. However, most isomeric PCBs, particularly those of environmental and toxicological relevance such as nrs 28 (2,4,4′-trichlorobiphenyl),

52 (2,5,2′,5′-tetrachlorobiphenyl), 138 (2,3,4,2′,4′,5′-hexachlorobiphenyl), 180 (2,3,4,5,2′,4′,5′-heptachlorobiphenyl), etc., and chlorinated pesticides (e.g. DDT, DDE, DDD) can be distinguished by HRGC/MS. Moreover, gas chromatography in combination with Fourier-transform infrared spectrometry (GC/FT-IR) offers great advantages in differentiation of those isomers that are indistinguishable by GC/ECD and GC/MS due to similar retention indices and mass spectra [93].

Polychlorinated Dibenzo-*p*-dioxins (PCDDs) and Polychlorinated Dibenzo-*p*-furans (PCDFs)

Analysis of PCDDs and PCDFs has now developed to high standards. Combination of high resolution gas chromatography (HRGC) with high- or low-resolution mass spectrometry (HRMS or LRMS) using either electron ionization (EI) or negative chemical ionization (NCI) are the generally accepted techniques [68, 94–101]. Detection limits have fallen over the past decade with the availability of increasingly sensitive instruments. At present, levels of the order of parts per trillion (ppt) and, in some instances, parts per quadrillion (ppq) of 2,3,7,8-TCDD can be detected.

Most frequently, encountered potential inteferences are co-extracted sample components, e.g. PCBs, chlorinated naphthalenes, methoxy-biphenyls, DDT, DDE, etc. Because very low levels of PCDDs/PCDFs must be measured the elimination of interferences is essential. Column chromatographic procedures are used to remove some co-extracted components, whereas previous fractionation by HPLC is often necessary when interfering substances are still present after the clean-up steps [68, 102–104].

Isotopically labelled PCDD and PCDF congeners, such as ^{13}C-2,3,7,8-TCDD, ^{13}C-2,3,7,8-TCDF, etc, are used as internal standards.

Hagenmaier et al. [103] observed that 2,3,7,8-TCDD demonstrates different adsorption behaviour (selective loss) on alumina in comparison to other PCDDs

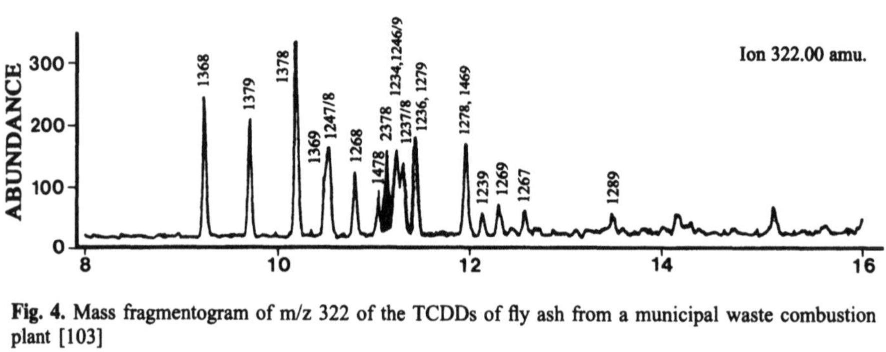

Fig. 4. Mass fragmentogram of m/z 322 of the TCDDs of fly ash from a municipal waste combustion plant [103]

and PCDFs. Since almost all clean-up procedures for PCDDs/PCDFs described in the literature use basic aluminium oxide, and as long as the determination of this congener is of specific interest, its analysis should not be performed without the use of ^{13}C-2,3,7,8-TCDD as internal standard.

A mass fragmentogram of the tetrachlorodibenzo-*p*-dioxins of the fly ash emitted from a municipal waste incinerator is shown in Fig. 4 [103].

It is generally acknowledged that time and costs involved in PCDDs/PCDFs analysis are considerable depending mainly on the sensitivity required and the number of individual congeners to be determined.

Analytical Method Validation

Reliable analytical measurements of environmental samples are an essential ingredient of sound decisions since they involve many facets of society including safeguarding the public health, improving the quality of the environment and facilitating advances in technology.

Analytical objectives for environmental samples are governed by the necessity of obtaining reliable measurements at very low concentration levels in complex matrices. At these levels, many factors are of critical importance in influencing the reliability of the results. Therefore, the analytical procedure and its individual steps should be subjected to detailed evaluation regarding its efficiency (method validation).

Unfortunately, only few procedures for the determination of organic compounds in airborne particulate matter have been described with adequate method validation, e.g. an experimental process to evaluate the procedure's performance characteristics, such as accuracy, precision, sensitivity, recovery, interferences, matrix effects, etc.

Different criteria are used to determine the efficiency of the extraction and clean-up steps. One method often used is the addition of a known amount of a compound as a surrogate standard (e.g. a chemically inert compound not expected to occur in the sample), the percentage recovery of which is considered representative of the recovery of all determinants. Surrogate recovery is also used to monitor for unusual matrix effects, gross sample processing errors, etc. It is evaluated for acceptance by determining whether the measured concentration falls within the acceptance limits.

Dibromobiphenyl and decafluorobiphenyl have been successfully utilized as surrogates for PAHs analysis by GD/FID, GC/MS and HPLC, while deuterated analogues of selective PAHs are suggested for GC/MS (deuterated PAHs cannot be used as surrogates for HPLC analysis due to co-elution problems). Isotopically labelled PCDD and PCDF congeners are also available and are used for recovery efficiency tests, while octachloronaphthalene is an appropriate standard to check recovery efficiency of organochlorine pesticides from airborne particular matter.

However, this method assumes that the surrogate standard behaves in an identical way to the compounds of interest. That is usually true of the clean-up step, but may not be true of the initial extraction.

Another method for recovery efficiency assessment is spiking of a blank or loaded filter with a standard solution of the compounds of interest at one or more concentration levels. Thus, the recovery efficiency of individual determinands at various spike levels can be established and appropriate correction of the amount found can be performed.

Recovery efficiency tests do not directly assess accuracy as is commonly believed, but rather procedure efficiency. They indicate analytical accuracy only when the analytical response for a determinand in an unspiked sample is due to the determinand alone (i.e. no interferences) and is not subject to any bias.

Analytical accuracy is normally measured directly by analysis of certified reference materials or by confirmatory testing together with circumstantial evidence. The latter consists in applying two inherently different analytical procedures to the same set of samples containing a range of determinand concentrations and comparing the resulting data statistically, in particular by regression methods.

However, standard reference materials of airborne particulate matter, although available, are not always affordable, while the application of a second inherently dissimilar procedure to measure analytical accuracy is not always feasible. Cross checks are therefore necessary to confirm any presence of bias. Many of the errors usually observed are ascribed to simple calibration bias. Use of standard methods is no guarantee of accuracy and precision if descriptions of calibration procedures are lacking. The contribution of calibration to procedure inaccuracy is among factors prompting legislation, and the directives of environmental agencies specify the precision and accuracy required for the measurement of many determinands including organic compounds in airborne particulate matter [43, 69].

The validity in terms of accuracy and precision of a daily calibration procedure can be assessed by examining the data of replicate determinations of a calibration standard on each consecutive day. If these data are of satisfactory accuracy and precision, then determinations employing the calibration procedure are judged to be of satisfactory accuracy and precision, too.

In addition to this, the bias arising from matrix effects and/or the measurement instrument used can be demonstrated and quantified by the method of standard additions performed on treated samples.

The detection limit is one of the most important performance characteristics of an analytical procedure. Detection limit can be determined statistically as the threefold standard deviation of the blank values, but it is more frequently estimated as twice or three times the signal-to-noise ratio. Since the method detection limit for an analytical procedure may vary as a function of the sample type (matrix), intersample variation within a given matrix type should be appreciated.

In general, a detailed evaluation of the analytical procedure employed regarding efficiency is necessary to obtain reliable results. Even if a standard method is employed, the operation of a quality assurance/quality control programme is required to evaluate and document quality data.

References

1. Schultzle D, Cronn DR, Crittenden AL, Charson RJ (1975) Envir Sci Technol 9:838
2. Pandis SN, Harley RA, Cass GR, Seinfeld JH (1992) Atmos Environ 26A:2269
3. Grosjean D (1992) Atmos Environ 26A:953
4. Hutzinger O, Friedler H (1989) Chemosphere 18:23
5. Laflamme RE, Hites RA (1978) Geochim Cosmochim Acta 42:289
6. Bidleman TF, Billings WN, Foreman WT (1986) Environ Sci Technol 20:1038
7. Foreman WT, Bidleman TF (1987) Envir Sci Technol 20:869
8. Storey JM, Pankow JF (1992) Atmos Environ 26A:435
9. Hites R (1990) Acc Chem Res 23:194
10. Pankow JF, Bidleman TF (1991) Atmos Environ 25A:2241
11. Masclet P, Pistikopoulos P, Beyne S, Mouvier G (1988) Atmos Envir 22:639
12. Yamasaki H, Kuwata K, Miyamoto H (1982) Environ Sci Technol 16:189
13. Van Vaeck L, Van Cauwenberghe K (1978) Atmos Environ 12:2229
14. Konig J, Funcke W, Balfanz E, Grosch B, Pott F (1980) Atmos Environ 14:609
15. Pupp C, Lao RC, Murray JJ, Pottie RF (1974) Atmos Environ 8:915
16. Lao RC, Thomas RS, Monkman JL (1975) J Chromatogr 112:681
17. Otson R, Leach JM, Chung LT (1987) Anal Chem 59:1701
18. Knecht V, Lammler C, Tobias S (1987) Fresenius Z Anal Chem 326:25
19. Lee MK, Novotny MV, Bartle KD (1981) Analytical Chemistry of Polycyclic Aromatic Compounds Academic Press, New York, London, Toronto, Sydney, San Fransisco
20. Knecht V, Woitowitz HJ (1988) Fresenius Z Anal Chem 331:8
21. Fox MA, Staley SW (1976) Anal Chem 48:992
22. Arey J, Zielinska B, Atkinson R, Winer A (1987) Atmos Environ 21:1437
23. Coutant RW, Brown L, Chuang JC, Riggin RM, Lewis RG (1988) Atmos Environ 22:403
24. Lewis RG, Jackson MD (1982) Anal Chem 54:592
25. Billings WN, Bidleman TF (1989) Environ Sci Technol 14:679
26. Billings WN, Bidleman TF (1983) Atmos Environ 17:383
27. Chuang JC, Kuhlman MR, Wilson NK (1990) Environ Sci Technol 24:661
28. Lans DA, Schroeder WH, Johnson ND (1992) Atmos Environ 26A:31
29. Davis CS, Fellin P, Otson R (1987) J Air Poll Control Ass 37:1397
30. Bidleman TF (1985) Trace Analysis. Lawrence JF (ed), Academic Press New York
31. Brockhaus A (1974) Atmos Environ 8:521
32. Konig J, Funcke W, Balfanz F (1980) Atmos Environ 14:609
33. John W, Reischl G (1978) Atmos Environ 12:2015
34. Lee FS-C, Schuetzle D (1983) Handbook of Polycyclic Aromatic Hydrocarbons, Bjorseth A (ed) Marcel Dekker Inc. New York, NY
35. Niehaus R, Scheulen B, Durbeck HW (1990) Sci Total Environ 99:163
36. Hanson RL, Clark CR, Carpenter RL, Hobbs CH (1981) Environ Sci Technol 15:701
37. Ligocki MP, Pankow JF (1985) Anal Chem 57:1138
38. Thrane KE, Mikalsen A, Stray H (1985) Int J Environ Anal Chem 23:111
39. Ciccioli P, Brancaleoni E, Cecinato A, Di Palo C, Buttini P, Liberti A (1986) J Chromat 351:451
40. De Raat WK, Schulting FL, Burghardt E, Meijere FA (1987) Sci Total Environ 63:175
41. Grimmer G, Naujack KW, Schneider D (1981) Int J Environ Anal Chem 10:265
42. Grimmer G, Naujack KW, Schneider D (1982) Fresenius Z Anal Chem 311:475
43. Winberry WT, Murphy Jr, Murphy NT, Riggan RM (1988) Compendium of Methods for the Determination of Toxic Organic Compounds. EPA/600/4-89/017, US EPA, NC
44. May WE, Wise SA (1984) Anal Chem 56:225
45. Krstulovic AM, Rosie DM, Brown PR (1983) Anal Chem 48:1383
46. Yang SYN, Connell DN, Hawker DW, Kayal SI (1991) Sci Total Environ 102:229
47. Breuer GM (1984) Anal Letters 17:1293
48. De Raat WK, Meijere FA (1991) Sci Total Environ 103:1
49. Dong M, Locke DC, Ferrand E (1976) Anal Chem 48:368
50. Stanley TW, Meeker JW, Morgan MJ (1967) Environ Sci Technol. 1:927
51. WHO – Working Group on Air Standardization of Sampling and Analytical Procedure for Estimation of Polynuclear Aromatic Hydrocarbons in the Environment, Geneva, December 1969
52. Bartle KD, Lee ML, Wise SA (1981) Chem Soc Rew. 10:113
53. Tan JL (1988) Anal Letters 21:553

54. Nielsen T (1983) Anal Chem 55:286
55. Lankmayr EP, Muller K (1979) J Chromat 170:139
56. Chalot C, Castegnaro M, Roche JL, Fontagnons R, Obatan P (1971) Anal Chim Acta 53:259
57. Golden C, Sawicki E (1975) Int J Environ Anal Chem 4:9
58. National Institute of Occupational Safety and Health (1977) "NIOSH Mannual of Analytical Methods, HEW Publication No. NIOSH 77–157
59. Rappe C, Kjeller LO, Bruckmann P, Hackle KH (1988) Chemosphere 17:3
60. Olie K, Vermeulen PL, Hutzinger O (1977) Chemosphere 6:455
61. Hengstmann R, Hamann R, Weber H, Kettrup A (1980) Organohalogen Compounds 4:407
62. Buser HR, Rappe C, Bergqvist TA (1985) Environ Health Persp 60:293
63. Kirschmer P, Muller W, Eynck P (1992) Chemosphere 24:575
64. Levy JM, Rosselli AC (1989) Chromatographia 28:613
65. Chuang JC, Kuhlman R (1990) Environ Sci Technol 24:661
66. Ligocki MP, Pankaw JF (1989) Environ Sci Technol 23:75
67. Pankow JF, Kristensen TJ (1983) Anal Chem 55:2187
68. Krost KJ, Pellizzari ED, Walburn SG, Hubbard SA (1982) Anal Chem 54:810
69. VDI Handbuch Reinhaltung der Luft – VDI 3498 (1990)
70. Novotny M, Lee ML, Bartle KD (1974) J Chromatogr Sci 12:606
71. Lee ML, Novotny M, Bartle KD (1976) Anal Chem 48:1566
72. Hoffmann D, Wynder E (1960) Anal Chem 32:295
73. Grimmer G, Bohnke H (1972) Fresenius Z Anal Chem 261:310
74. Bjorseth A (1977) Anal Chim Acta 94:21
75. Natusch DFS, Tomkins BA (1978) Anal Chem 50:1429
76. Snyder LR, Kirkland JJ (1972) Introduction to Modern Liquid Chromatography, John Wiley & Sons Inc, New York
77. Kodama Y, Ishinishi N, Yoshikawa M (1983) J Chromat 261:103
78. Hoffman D, Wynder EL (1971) Identification and Measurement of Environmental Pollutants Westley B (ed) Symposium, Ottawa, Canada
79. Keller CD, Bidleman TF (1984) Atmos Environ 18:837
80. Severson RF, Snook ME, Arrenade RF, Chortyk OT (1976) Anal Chem 48:1866
81. Niles R, Tan Y (1989) Anal Chim Acta 221:53
82. Moriske HJ, Ruden H (1988) Chemosphere 17:1167
83. Harrison RM, Perry R (1986) "Handbook of Air Pollution Analysis" Chapman and Hall, London, New York
84. Stenberg U, Alsberg T, Blomberg L, Wannman T (1979) Polynuclear Aromatic Hydrocarbons Jones PW and Leber P (eds) Ann Arbor, Michingan
85. Romanowski T, Funcke W, Konig J, Balfanz E (1982) Anal Chem 54:1285
86. Viras LG, Siskos PA, Samara C, Kouimtzis T, Athanasiou K, Vavatzanidis A (1991) Environ Toxicol Chem 10:999
87. Tong HY, Karasek FW (1984) Anal Chem 56:2129
88. Rossi DT, Desilets DJ, Pardue HL (1984) Anal Chim Acta 161:191
89. Dong MW, Hetheringtoon N (1990) Perkin Elmer Environmental Application, ENV-6
90. Wise SA, Chesler SN, Hilpert LR, May WE, Rebbert RE, Vogt CR (1984) Polynuclear Aromatic Hydrocarbons: Mechanisms, Methods and Metabolism Cooke M Dennis AS (eds) Battelle Press, Columbus, Ohio, 1984, p. 1413
91. Annual Book of ASTM Standards (1983) Atmospheric Analysis. American Society for Testing and Materials, Philadelphia, PA
92. Riggin RM (1983) Technical Assistance Document for Sampling and Analysis of Toxic Organic Compounds in Ambient Air, EPA 600/1 83 027, US EPA, NC
93. Sherma J (1989) Analytical Methods for Pesticides and Plant Growth Regulators, Acad Press Inc California USA
94. Rappe C, Buser HR (1989) Chemical and physical properties, analytical methods, sources and environmental levels of halogenated dibenzodioxins and dibenzofurans, in: Halogenated biphenyls, terphenyls, naphthalenes, dibenzo-dioxins and related products, Kimbrough RD and Jensen AA (eds), Elsevier, Amsterdam
95. Smith AG, Stalling DL, Johnson JL (1984) Anal Chem 56:1830
96. Eltzer BD, Hites RA (1989) Environ Sci Technol 23:1396
97. Eltzer BD, Hites RA (1989) Environ Sci Technol 23:1389
98. Tierman TO, Wage DJ Vanness GF, Garrett JH, Solch JG, Harden LA (1989) Chemosphere 19:541

99. Lenoir D, Hutzinger O, Mutzenich G, Horch K (1989) UWSF-Z. Umweltchem Okotox 1:3
100. Ballschmiter K, Buchert H, Niemczyk R, Munder A, Swerev M (1986) Chemosphere 15:901
101. Maisel B (1990) Organohalogen Compounds 4:413
102. Oehme M, Kirschmer P (1984) Anal Chem 56:2754
103. Hagenmaier H, Brunner H, Haag R, Kraft M (1986) Fresenius Z Anal Chem 323:24
104. Kaupp H, Towara J, McLachlan MS (1994) Atmos Environ 28:585

Particle Counting and Particle Size Analysis

Werner Holländer

Fraunhofer-Institut of Toxicology and Aerosol Research, Nikolai-Fuchs-Str-1, D-30625 Hannover, Germany

1 Introduction . . . 254
2 Counting and Number Concentration Measurement . . . 254
2.1 Counting Statistics . . . 255
2.2 Coincidence . . . 256
2.3 Other Effects . . . 257
3 Particle Size Statistics . . . 257
3.1 Distribution Functions . . . 257
3.2 Moments . . . 259
4 Particle Sizing . . . 261
4.1 Imaging Methods . . . 262
4.1.1 Optical Microscopy . . . 262
4.1.2 X-ray Microscopy . . . 263
4.1.3 Electron Microscopy . . . 263
4.1.4 Atomic Force Microscopy . . . 263
4.2 Other Techniques . . . 264
4.2.1 Mobility Sizing . . . 265
4.2.1.1 Diffusional Sizing . . . 265
4.2.1.2 Electrical Mobility Sizing . . . 265
4.2.2 Relaxation Time Sizing . . . 266
4.2.2.1 Gravitational Sizing . . . 266
4.2.2.2 Impactors . . . 266
4.2.2.3 Aerodynamic Sizing . . . 267
4.2.3 Light Scattering . . . 268
4.2.4 Kelvin Sizing . . . 270
4.2.5 Epiphaniometer . . . 271
4.3 Equivalent Diameters . . . 272
4.4 Multiparameter Sizing . . . 272
5 Inversion Methods . . . 273
5.1 Least Square Fits and Singular Value Decomposition . . . 274
5.2 Modified Fit Techniques . . . 274
5.3 Constrained Regularization . . . 274
5.4 Parametric Inversion . . . 275
5.5 Merits and Drawbacks of the Various Methods . . . 275
6 Conclusions . . . 276
7 References . . . 276

1 Introduction

It cannot possibly be the objective of a short contribution like the present one to present a complete survey of such a large field of knowledge. Instead, this paper is intended to serve the purposes of

- presenting introductory information on the most prominent particle counting and size analysis techniques,
- giving references for the most frequent measuring tasks, and
- providing enough physical background for successfully approaching nonstandard problems, and problems most likely to be met in future.

Several books provide a wealth of information on aerosol measuring techniques as well as the physical background to a variable depth [1–9]. As far as publications on certain topics are concerned, reference is usually made to the latest paper only, from which earlier papers can be retrieved. Also, we have tried to collect material which is usually not covered in standard textbooks on aerosols.

The companion paper 'Sampling of Airborne Particulate Matter' chapter 6 in this volume deals essentially with the fate of particles between the environment and the place where actual measurement is made. The most important question in this context concerned the bias introduced by the sampling process. The physical behavior of individual particles has been addressed to a limited extent in the section 'Elements of Single Particle Mechanics', and possible changes of the particles in the subsection 'Particle Population Dynamics'. Where necessary, reference will be made to this companion text.

2 Counting and Number Concentration Measurement

In the present paper, we assume that the particles have been transported to the measuring place, where we want to size and count them, which can be done individually or on a particle collective. This is important since it constitutes the first step towards characterizing airborne particles, but it is not always a straightforward procedure, since

- the presence of another particle nearby may interfere with the measurement of the first particle with respect to the counting and/or sizing process or
- the particle to be measured may be near to the detection limit of the measuring device.

All general criteria pertaining to measuring time, calibration, outliers, detection limit, reproducibility, comparability, selectivity etc. have to be observed [10].

While it is intuitively clear what it means to count a particle, in reality difficulties may arise if particles are close to the size detection limit of the instrument, or if particle concentrations are rather high or low. Usually the particle size detection limit is defined as the size of a reference aerosol for which 50% of the particles are counted. Limitations with respect to concentration are discussed below.

2.1 Counting Statistics

The binomial distribution is the fundamental frequency distribution governing random events, and the other frequency distributions can be derived from it. The Poisson distribution $P(x)$ for instance is the limiting case of the binomial distribution for those processes in which the probability of occurrence is very small ($p \ll 1$) and the average value $\bar{N}$ is fixed. The probability of finding x events (e.g. particles in a volume) where $\bar{N}$ are expected is

$$P(x) = \frac{\bar{N}^x}{x!} \exp(-\bar{N}) . \tag{1}$$

The Poisson distribution is skew for small $\bar{N}$ values but becomes symmetric for larger values (see Fig. 1). If a certain x is observed, one can also calculate the confidence interval for finding different expectation values $\bar{N}$. To give an example, if $x = 2$, with 95% confidence $0.4 \leq \bar{N} \leq 6$ [11].

If a volumetric sample of size V is taken from a population with an average true concentration $\bar{c}_{\text{true}}$, one would expect an average particle count $\bar{N} = \bar{c}_{\text{true}} \cdot V$ for a whole measuring series. An individual measurement N_i will normally be $\neq \bar{N}$ due to statistical fluctuations. The best estimate for the standard deviation

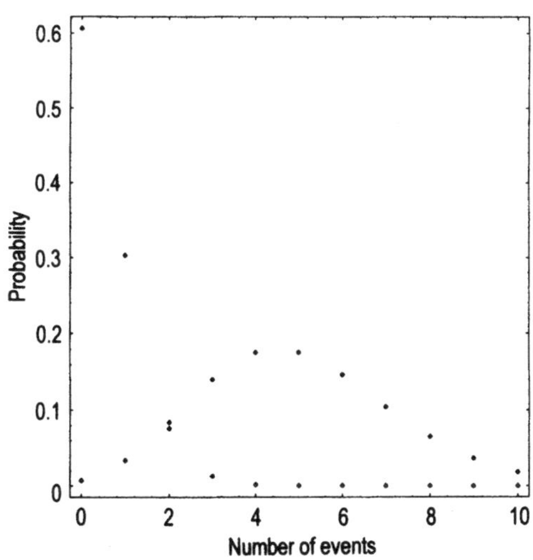

Fig. 1. Poisson distributions for expectation values $\bar{N} = 0.5$ and 5. For larger $\bar{N}$, the distribution becomes more symmetric

of the particle count σ_N obtained from m measurements is

$$\sigma_N^2 = \frac{1}{m-1}\sum_1^m (N_i - \bar{N})^2 . \tag{2}$$

(The same formula applies to the standard deviation of any other observable.) This means that the statistical uncertainty with respect to the number is $\propto \sqrt{N}$, i.e. the relative error is $\propto 1/\sqrt{N}$. For large N, the statistical error is negligible when compared to other measuring errors, but in clean environments statistics often become the limiting factor, and instruments have to be specially designed to cope with low concentrations. In addition to real particles, false counts may arise due to electronic detector noise or electromagnetic interference, ionizing radiation or power line disturbances. Depending on the instrument, false count rates range between several counts per minute and counts per hour. If the instrument detection probability for a particle is η then a measured concentration

$$c_{\text{true}} = \frac{c_{\text{meas}}}{\eta} = \frac{N_{\text{meas}}}{\eta V} \tag{3}$$

is obtained as an approximation to c_{true}. An important problem concerns the precision of measurements resulting from several independent random processes, where error propagation has to be properly accounted for [12]. For instance, the sum S, the difference D, the quotient Q, and the product P of two random variables a, b are given by

$$S \pm \sigma_S = a + b \pm \sqrt{\sigma_a^2 + \sigma_b^2} \tag{4}$$

$$D \pm \sigma_D = a - b \pm \sqrt{\sigma_a^2 + \sigma_b^2} \tag{5}$$

$$P \pm \sigma_P = a \cdot b \pm ab\sqrt{\left(\frac{\sigma_a}{a}\right)^2 + \left(\frac{\sigma_b}{b}\right)^2} \tag{6}$$

$$Q \pm \sigma_Q = \frac{a}{b} \pm \frac{a}{b}\sqrt{\left(\frac{\sigma_a}{a}\right)^2 + \left(\frac{\sigma_b}{b}\right)^2} . \tag{7}$$

Since there are numerous books on statistics (e.g. [11]), and some of them are specially devoted to particles (e.g. [13, 14]), we will not go into details of statistical methods but will rather outline some of the most important applications.

2.2 Coincidence

Every particle measuring method receives its signal (which extends for a certain time) from a certain measuring volume. Therefore, it may happen that more than one particle is present in the measuring volume during the time measurement is being taken. As a result, the measurement may result in an erroneous particle size if, for instance, the scattered light intensity is used for sizing the particle. Similarly, the decay of the measuring signal may take longer than the time for which the particle is present in the measuring volume, causing a certain 'dead' time k which may 'paralyze' the detector (i.e. an event during the dead time renews the dead cycle) and reduce the particle count rate $\dot{N}$ in a way that

$$\dot{N}_{\text{meas}} = \dot{N}_{\text{true}} \cdot \exp(-k \cdot \dot{N}_{\text{true}}) . \tag{8}$$

For 'nonparalyzable' detectors (i.e. an event during the dead time is simply ignored) the result is [12]

$$\dot{N}_{\text{true}} = \frac{\dot{N}_{\text{meas}}}{1 - k \cdot \dot{N}_{\text{meas}}} . \tag{9}$$

Corrections can be applied up to certain dead time fractions, but beyond the corresponding concentration, the measuring signal is caused by a particle assembly rather than individual particles, and analog evaluation and calibration techniques should be used.

2.3 Other Effects

The situation becomes more involved if measuring signals produced by collectives of particles of different sizes are interpreted as counts of individual particles. Depending on the physical measuring principle, the 'size' attributed to this 'phantom particle' may be entirely meaningless and even seriously disturb the whole measurement even for relatively infrequent events of such kind (see Sect. 4.2.2).

3 Particle Size Statistics

Aerosol particles differ widely with respect to size, chemical composition, morphology, electrical charge etc., and there is a certain probability of finding a particle in a property range between x and $x + dx$ for a continuous variable x like size, or with a well defined property n like number of elementary charges in case of a discrete variable. Depending on the actual property and the aerosol under consideration, different distribution functions are used to describe reality as well as is possible with as few as possible parameters.

3.1 Distribution Functions

Depending on which property the distribution describes we may have a number concentration distribution $f_{\text{N}}(x)$ or a mass concentration distribution $f_{\text{m}}(x)$ which is related to the volume and the number size distributions by

$$f_{\mathrm{m}}(x) = \rho(x) f_{\mathrm{V}}(x) = \frac{\pi}{6} x^3 \rho(x) f_{\mathrm{N}}(x) \,. \tag{10}$$

These distributions can be assumed to have continuous variables whereas the electric charge, for instance, is a discrete variable. The diameter x where the distribution function has its maximum is called the modal diameter, and the diameter which divides the distribution into the larger 50% and the smaller 50% is called the median diameter.

The most popular size distribution is without doubt the so-called Gaussian normal distribution defined by (here normalized, i.e., $(\int_{-\infty}^{\infty} f(x)\,dx = 1)$

$$f_{\mathrm{G}}(x) = \frac{1}{\sqrt{2\pi} \cdot \sigma} \exp\left[\frac{u^2}{2}\right] \tag{11}$$

with the standardized variable u

$$u = \frac{x - \bar{x}_{\mathrm{a}}}{\sigma} \tag{12}$$

where σ is the standard deviation and $\bar{x}_{\mathrm{a}}$ the average arithmetic diameter.

Often, for interpreting data the Gaussian sum-curve $F_{\mathrm{G}}(x)$ is used, defined by

$$F_{\mathrm{G}}(x) = \int_{-\infty}^{x} f_{\mathrm{G}}(\xi)\, d\xi \,. \tag{13}$$

It has the special values $F_{\mathrm{G}}(\bar{x} - \sigma) \approx 0.16$, $F_{\mathrm{G}}(\bar{x}) = 0.5$, and $F_{\mathrm{G}}(\bar{x} + \sigma) \approx 0.84$. Although nowadays even on hand-held calculators x can be calculated from F_{G}-values, the so-called Logit-transform $L(x)$ is still popular due to the fact that it can be easily inverted [15] and integrated:

$$L(x) = \left[1 + \exp\left(-\frac{\pi u}{\sqrt{3}}\right)\right]^{-1} \tag{14}$$

and

$$\frac{\pi u}{\sqrt{3}} = ln\left(\frac{L(x)}{1 - L(x)}\right) \,. \tag{15}$$

The numerical approximation to $F_{\mathrm{G}}(x)$ is good enough for many applications since $|L(x) - F_{\mathrm{G}}(x)| \leq 0.023$.

Unfortunately, the Gaussian distribution extends from $-\infty$ to ∞ which is unphysical since there are no negative particle sizes. In order to remedy this, the so-called lognormal distribution is introduced which is defined by (here as number concentration)

$$f_{\mathrm{N}}(x) = \frac{N}{\sqrt{2\pi} \cdot x \cdot ln\sigma_{\mathrm{g}}} \exp\left[\frac{u_{\mathrm{LN}}^2}{2}\right] \tag{16}$$

with the standardized variable u_{LN}

$$u_{LN} = \frac{ln x - ln \bar{x}_g}{ln \sigma_g} \tag{17}$$

where σ_g is the geometric standard deviation and $\bar{x}_g$ the average geometric diameter. Note that σ_g is a pure number, and $F_{LN}(\bar{x}_g/\sigma_g) \approx 0.16$, $F_{LN}(\bar{x}_g) = 0.5$, and $F_{LN}(\bar{x}_g \cdot \sigma_g) \approx 0.84$. For strictly monodisperse particles $\sigma_g = 1$, while normally [16] distributions with $\sigma_g \leq 1.15$ are called technically monodisperse, those with $1.15 \leq \sigma_g \leq 1.5$ are called quasimonodisperse, and for $\sigma_g > 1.5$ they are called polydisperse. Other distributions frequently used over limited size ranges are the Junge distribution

$$f_N(x) = A \cdot x^{-v} \tag{18}$$

with the parameter v typically in the range $2 < v < 4$ and the normalizing constant A. It describes, usually reasonably well, the total number size distribution including the range of condensation nuclei.

For better approximations to actual skew atmospheric size distributions Twomey [17] introduced the Gamma distribution $n_\gamma(x)$ with the three positive parameters α, β, γ

$$n_\gamma(x) = A \cdot x^\alpha \exp(-\beta x^\gamma)\,. \tag{19}$$

Very often, one wants to find out whether a certain observed distribution is significantly different from an expected (normal or Poisson) distribution. An appropriate method to test this hypothesis is Pearson's Chi-square test, where large values of

$$\chi^2 = \sum^{i} \frac{[(\text{observed value})_i - (\text{expected value})_i]^2}{(\text{expected value})_i} \tag{20}$$

indicate difference. Extended versions of this test allow application of the method to arbitrary distributions.

3.2 Moments

For any size distribution $f(x)$ the k-th moment M_k is defined as

$$M_k = \int_0^\infty x^k f(x)\, dx \tag{21}$$

and any function can be completely reconstructed from the complete set of moments M_k. Some of the lower moments are intuitively interpretable like the zeroth moment of the number density distribution $M_0^{(N)} = \int_0^\infty f_N(x)\, dx = N$ which is the total number concentration. Similarly, the first (second, third) moment of the number density distribution is proportional to the mean arithmetic particle diameter (the surface, the volume or mass if the density is independent of particle size). Similarly the volume concentration distribution is the zeroth moment of the volume size distributions $f_V(x)$ or $f_V(V)$

$$V = M_0^{(V)} = \int_0^\infty f_V(V)\,dV = \int_0^\infty f_V(x)\,dx\,. \tag{22}$$

When higher moments are calculated with respect to mean values of the distribution and are normalized with the standard deviation, the values of these so-called 'central moments' M_k^c can be used to characterize the shape: unsymmetric ones have a 'skewness' $= M_3^c \neq 0$, and curves 'blunter' or 'sharper' than normal ones have a 'kurtosis' $= M_4^c - 3 \neq 0$ [11].

An example of a mass size distribution of diesel soot is shown in Fig. 2. It can be seen that the aerosol mass can be characterized by an approximate lognormal size distribution with a geometric mass median diameter of 0.19 μm and a geometric standard deviation $\sigma_g = 1.62$.

The mass sedimentation flux density j is proportional to the mass of the particles and their sedimentation velocity (see Eq. (*31)[1]), i.e. according to

$$j = \int_0^\infty v_{sed}(x) \cdot m_p(x) \cdot f_N(x)\,dx \;\propto\; \int_0^\infty x^2 \cdot x^3 \cdot f_N(x)\,dx \tag{23}$$

proportional to the fifth moment of the number size distribution. At the same time, it is proportional to the 2/3rd moment of the volume distribution function $f_V(V)$ (provided the particle density ρ_p does not depend on the size):

$$j = \int_0^\infty v_{sed}(V) \cdot \rho_p(V) \cdot f(V)\,dV \propto \int_0^\infty V^{2/3} \cdot f_V(V)\,dV\,. \tag{24}$$

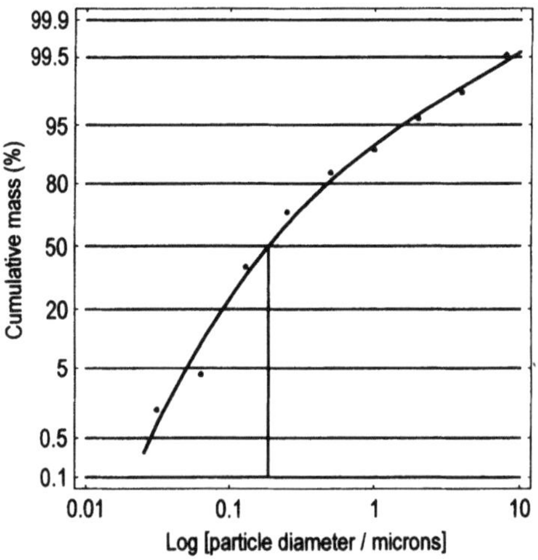

Fig. 2. Mass size distribution of a typical diesel soot from a light duty engine with $\bar{x}_g = 0.19$ μm and $\sigma_g = 1.62$ obtained from the diameter ratio of d_{84}/d_{16}. The scaling of the vertical axis is such that lognormal distributions give a straight line. The curve is a polynomial fit to the experimental data points scaled in the above way.

[1]References to equations and tables in the companion paper are denoted by * preceding the equation and table number

Simple distributions are completely characterized by a few moments: for the lognormal distribution the k-th moment is given by [18]

$$M_k = M_0 \cdot \bar{V}^k \cdot \exp(4.5k^2 ln^2 \sigma_g) \tag{25}$$

$$\bar{V} = \frac{M_1^2}{\sqrt{M_0^3 M_2}} \tag{26}$$

$$ln^2 \sigma_g = \frac{1}{9} ln \frac{M_0 M_2}{M_1^2} \; . \tag{27}$$

The above approach is very useful since

- many aerosol processes can be written in terms of moment equations [19, 20] which can be solved as set of coupled linear differential equations instead of the original nonlinear integro-differential general dynamic aerosol equation and
- most aerosol measuring instruments actually measure special moments of the aerosol size distribution, and combining several measurements can provide good information about the distribution.

4 Particle Sizing

Intuitively, sizing means first of all the application of an imaging method; the result obtained is a geometric size in this case. On the other hand, every size-dependent physical particle property can in principle also be employed for sizing. Often this is easier and cheaper than using labor- and cost-intensive imaging methods, but the measurement will usually result in an 'equivalent' diameter, which has to be converted somehow into the geometric diameter (or any other diameter of interest). A general problem encountered is the limited size range and the limited size resolution of any measuring device. Therefore any interpretation of a continuous particle size distribution on the basis of a measurement with limited resolution has a certain degree of arbitrariness, and it is important to know how reliable the conclusions drawn under such circumstances are.

One important characteristic of a sizing method is the sizing accuracy which is the difference between the measured size and the true size. It may depend on particle density, shape, refractive index etc.

The size resolution Δd_p is another relevant instrument parameter; it describes the size difference between particles necessary for distinguishing them; it depends to a great extent on the physical sizing principle and the construction details of the instrument. As is intuitively clear, it is proportional to the slope of the measuring signal J when the particle size changes, i.e.

$$\Delta d_p = \frac{\Delta J}{\partial J(d_p)/\partial d_p} \tag{28}$$

where ΔJ is the instrumental resolution of the measuring signal. An example is discussed in the section on aerodynamic sizing.

In some cases, it is not the absolute measuring signal resolution ΔJ which determines instrument performance rather than the relative resolution $\Delta J/J$, like in electrical measurements, and often one is interested in large particle size intervals so that a dimensionless resolution

$$R = \frac{\partial \ln J}{\partial \ln d_{\mathrm{p}}} \tag{29}$$

may also be useful.

4.1 Imaging Methods

The application of this principle has a long tradition, and as early as in the 17th century, light waves were focused by lenses or mirrors to work in microscopes and telescopes. Here the interaction (scattering, absorption) between the light waves and the object changes the light intensity at the object periphery in the projection, giving essentially two-dimensional information about the object. In the 20th century, finer details could be resolved by using the shorter wavelength of electron beams. Recently, the strategy of imaging the whole object in one step was complemented by a 'pixel' (image element) oriented approach: with the advent of cheap, high speed control electronics and digital storing devices, at one time information from one object point only was collected at the expense of scanning the whole object. This approach, on the other hand allows large scale digital image processing. In principle, resolution far below the wavelength can be achieved by using light-sources of sub-wavelength dimensions but implementation of the techniques has proved difficult [21]. Also, efforts to obtain information on the object's third dimension were increasingly successful. There is an abundant literature on all imaging methods, which cannot be covered in this short survey [22].

Recently, efforts have been successful to use tomographic methods for obtaining spatial distributions, but resolution still leaves much to be desired [23].

4.1.1 Optical Microscopy

Standard microscopy uses absorbing structures in the object in order to produce the imaging information. For this purpose, great ingenuity was applied to stain cells and other structures of interest to make them visible. Nonabsorbing structures could be observed using the technique of phase-contrast microscopy which uses the fact that coherent light waves with different phases interfere and transform the phase shift into an intensity change. In the last few years, confocal

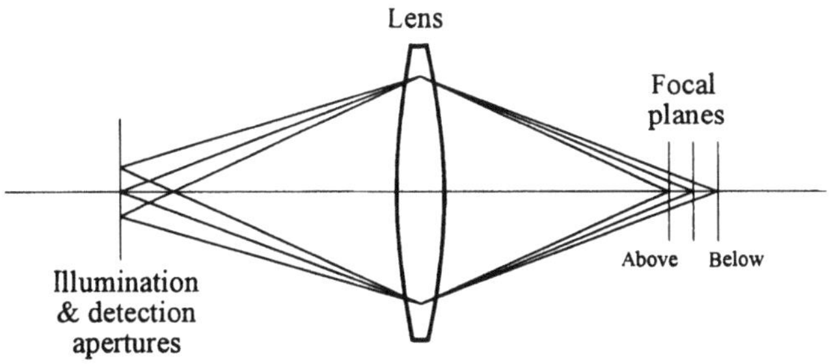

Fig. 3. Scheme of confocal microscopy. Only light passing the illumination and detection apertures are used for image formation, so that 3D information will be obtained by distance scanning

microscopy using the scanning principle has been used to provide three-dimensional information. Its principle is shown in Fig. 3 (adapted from [22]).

Another way of obtaining 3D information is by holography which was invented by Gabor [24].

4.1.2 X-ray Microscopy

Soft X-rays have a wavelength ten times shorter than light, and their potential resolution is better by the same amount. Since X-rays are not refracted well, it is difficult to make lenses. The technical problems of manufacturing Fresnel zone plates have prevented widespread application so far. On the other hand, samples need no special preparation, and the information derived from the image (specific chosen elements like carbon or nitrogen) should complement electron microscopy nicely, where light element detection is a problem.

4.1.3 Electron Microscopy

Unlike to electromagnetic waves which travel in all nonabsorbing media, electrons need a vacuum in which the object must be exposed for the imaging process, and not all objects stand this treatment without changes. Transmission electron microscopy (TEM) has a very high resolution, but the disadvantage of needing very thin probes. This means that replicas of thick samples have to be made which is not only a time-consuming procedure, but may give rise to artefacts as well. Biological specimens have to be specially stained in order to reveal the structures one is interested in.

4.1.4 Atomic Force Microscopy

Since the first report on a scanning tunneling microscope (STM) by Binnig and Rohrer [25], this technique has been modified and extended to what is called atomic force microscopy (AFM) and can handle nonconducting samples and samples in liquids. Not only can these instruments image surfaces with unprecedented vertical resolution (fractions of a nanometer) but they also allow one to manipulate individual atoms on surfaces. Because of this wide range of applications the inventors of the STM have been awarded the Nobel prize.

4.2 Other Techniques

Nonimaging sizing techniques use a size-dependent property of the particle which can be used to

- provide a characteristic measuring signal which is directly used for the sizing process or to
- select particles with a certain property out of a population. Here, a yes/no response of the detection instrument following the selector is sufficient since the sizing is done by the property selector.

In principle, virtually every size dependent physical process can be used but in practice only a few are actually used. Presently, the most popular methods follow closely the particle effects one is interested in, and employ determination of

- mobility as a measure of hydrodynamic interaction (responsible for the deposition of small particles)
- relaxation time as a combined measure of particle mass times mobility (which determines the deposition behavior of large particles)
- particle light scattering (which determines to a large extent atmospheric optics) and
- condensation behavior (which plays a role in cloud physics, and can be used for counting extremely small particles).

Some of the physical parameters mentioned above have already been discussed in the companion paper, and the rest will be dealt with here. Most of the measuring methods available at the time were discussed in [26]. Another new method uses the attachment of radioactive atoms from radon decay to particles; this process is proportional to the 'effective' particle surface. The instrument based on this technique is the 'epiphaniometer'; it will in our opinion most likely find more users in the future [27]. In general, the measuring techniques should determine quantities of actual relevance (like light scattering, condensation behavior etc) since any conversion from a diameter derived from one physical process to another one is problematic and prone to errors. Therefore, in the following, we selected those techniques which fulfil the above requirements.

4.2.1 Mobility Sizing

The mobility B as defined in Eq. (*13) is a measure of the hydrodynamic particle resistance, and as such does not depend on the particle material properties, but on its size and shape exclusively.

4.2.1.1 Diffusional Sizing

This technique uses the diffusional deposition of small particles in a long narrow tube or a wire screen. From the measured particle penetration $P = n/n_0$ (i.e. the fraction of particles crossing the channel) with Eqs. (*52), (*51) and (*10) the average mobility of the particle collective, and hence the apparent diffusional size of the whole population, can be calculated. This method is very simple and can be used for small particles for which $D_p \propto d_p^{-2}$, while for larger particles $D_p \propto d_p^{-1}$. The determination of size distributions is not very easy for diffusion battery measurements, since the size resolution is rather poor because of the low power dependence of D_p on particle size (see Table 1 in the companion paper) and the fact that the particles fill the whole cross section so that the distances before particles hit the walls vary widely.

4.2.1.2 Electrical Mobility Sizing

The history of this sizing technique goes back very long (see collection of articles in [6, 7]). It has become very popular since the widespread introduction of the electrical aerosol analyzer (EAA), the differential mobility analyzer (DMA) and the scanning electrical mobility sizer (SEMS) [28]. The performance of the different devices is reviewed in several papers in [3]. The principle is very simple: an electrical force due to a field E acting on particles with an electric charge q leads to a migration velocity

$$v_{mig} = E \cdot Z = E \cdot qB = E \cdot neB \tag{30}$$

which is proportional to the number of elementary charges n on the particle, and to their mobility B. Normally, the system geometry is such that v_{mig} is perpendicular to the flow direction so that particles will be separated according to the electrical mobility Z. If the charge n is known, the mobility can be determined. This will normally be the case when particles are rather small so that most of them carry no charge at all ($n = 0$) and the rest carry only one charge ($n = 1$), irrespective of whether charging was done by exposure to a unipolar or a bipolar ion atmosphere. For larger particles, by a reduction of ion strength, single charging can be achieved even for relatively large particles up to 1 µm [29]. A general problem stems from the fact that one is normally interested in the size distribution of *all* particles independent of their charge. So, if there are uncertainties with respect to the ratio of uncharged and (multiply) charged particles under the specific charging conditions, the procedure is questionable, even if multiply charged particles are taken into account and inversion procedures are employed to uncover size distributions as explained in the following sections. On the other hand, size resolution of electrical mobility sizing techniques is very good, and operation is easy which explains their success. As detection techniques, electrical measurement (for comparatively large concentrations only) and particle counting by condensation nuclei counters (CNCs) are standard.

4.2.2 Relaxation Time Sizing

For the physical background of this method, see the corresponding Sect. 3.3 in the companion paper.

4.2.2.1 Gravitational Sizing

For particles of sufficient size, gravity has the largest impact (see e.g. Table*1), and their sedimentation velocity v_{sed} (which is proportional to the relaxation time) can be used for measuring their aerodynamic size in elutriators. For small particles, gravity may be enhanced by using centrifuges (see e.g. [30, 31]), where particle sedimentation time t_{sed} is much larger than particle relaxation time τ, which means stationary particle motion, while in all following methods particle motion is not stationary.

4.2.2.2 Impactors

The impactor principle consists of impinging an aerosol laden gas jet on a surface. The jet forms a stagnation flow but the particles, depending on their size, cannot follow the sharp bend and impact onto the stagnation plate as shown in Fig. 4.

The important parameter for impaction is the Stokes number – see Eq. (*37). Properly designed impactors have rather steep cutoff curves which makes them suitable for the determination of particle size distributions when they are cascaded. There are many publications on properly performing measurements with impactors (e.g. in [6, 7, 32]), and limited advices are given here: impaction stages need to

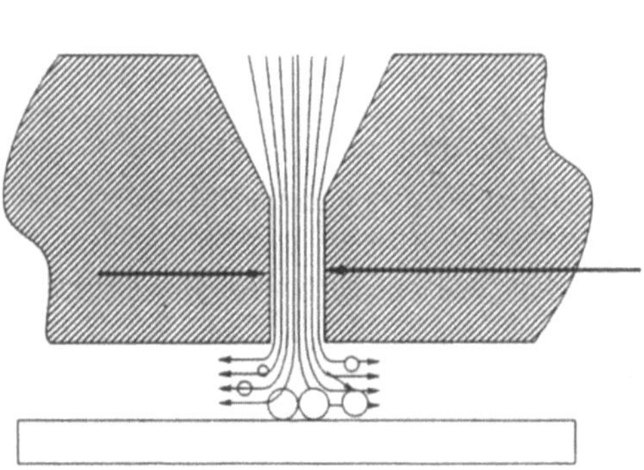

Fig. 4. Schematic of an inertial impactor. Particles beyond a certain inertia cannot follow the gas streamlines and are separated from the flow which results in their collision with the impaction plate

be coated with sticky material in order to prevent particle bounce [33]; impactors should not be used for measuring total airborne concentrations, since the wall losses are not negligible in cascade impactors; impactors used for stack sampling must be heated before use in order to avoid dew point problems.

Recently, impactors with sheath air operated at low, variable pressure and very high resolution have been described in the literature [34], and this approach seems very promising but needs more development before becoming available commercially.

4.2.2.3 Aerodynamic Sizing

The term 'aerodynamic sizing' has been extensively used in the literature, but not always in the strict meaning of the definition, which refers to the settling velocity *at normal gravity conditions.* The term 'relaxation time spectrometry' seems more appropriate, since this is the physical quantity measured.

The instrument described by Mazumder and Kirsch [35] consists of a laser Doppler anemometer which measures the motion of particles exposed to a stationary sound wave. The phase shift $\phi = \arctan(\omega\tau_p)$ between particle and sound wave allows particle size determination. The disadvantage of the proposed method is the complex equipment, and that only single particles can be measured.

During the last few years, so-called 'aerodynamic particle sizers' have become very popular. In this type of instrument the particles are accelerated in a very short nozzle for the time Δt. According to Eqs. (*27) and (*28) the velocity of the particles starting with v_0 develops as

$$v = v_\infty + (v_0 - v_\infty)\exp\left(-\frac{\Delta t}{\tau_p}\right) \tag{31}$$

where the particle relaxation time is again

$$\tau_p = m_p B \tag{32}$$

(v_∞ is the gas velocity at the nozzle exit). If $\Delta t/\tau \ll 1$, the velocity of the particles has not reached the final value and can be employed to size the particles. This can be done by measuring the time of flight ($TOF = \Delta x/v$; v is given by Eq. (31)) they need to cross the distance Δx between a pair of laser beams. Its use for sizing particles is illustrated in Fig. 5.

Particle coincidences in the TOF measuring volume may result in apparently large particles which can even distort the volume size distributions which are calculated from number size distributions.

Unfortunately, the situation is even slightly more complicated: due to the high velocity difference between particle and gas in the accelerating nozzle of the order of 100 m/s, the hydrodynamic force needs a correction term

$$F_{\text{highspeed}} = -\frac{3\pi\mu d_p}{C(d_p)} \cdot K_{\text{in}} \tag{33}$$

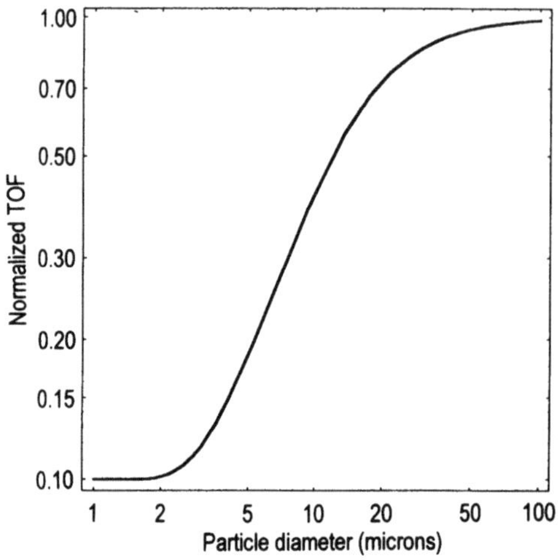

Fig. 5. Time of flight (TOF) required for a particle of diameter d_p lagging behind gas velocity to cross a distance Δx (schematic). Particles can be resolved if their TOF-difference is larger than the time resolution Δt of the built-in electronics, which is typically a few nanoseconds. Size resolution is best where the slope is maximum. Typically, TOF is resolved in a thousand channels, so that optimum resolution may be ≈ 0.01 μm, if all particles are accelerated along the same streamline

where the inertial correction factor K_{in} is

$$K_{\text{in}} = 1 + 0.158\ Re_{\text{p}}^{2/3} \tag{34}$$

with the particle Reynolds number

$$Re_{\text{p}} = \frac{\rho_{\text{g}} d_{\text{p}} |v_{\text{g}} - v_{\text{p}}|}{\mu} \tag{35}$$

(valid for $Re_{\text{p}} \leq 1000$) [4]. This means that the particle velocity will no longer be a pure function of the relaxation time alone but also depend on the hydrodynamic diameter. Therefore, particles of different density with the same aerodynamic diameter will behave differently [36, 37]. (It should be mentioned that the same effect acts in impactors, but has not been observed there due to the poor resolution of this instrument.)

Presently, two commercial instruments based on this principle are widely used [36, 38], and a VDI guideline on this instrument type is under preparation [39].

4.2.3 Light Scattering

The atmospheric visibility under ideal circumstances is several hundred kilometers and limited by Rayleigh scattering of the air molecules. The fact that usually atmospheric visibility falls far short of this limit indicates that airborne particles scatter much light, although their concentration is many orders of magnitude smaller than that of the molecules. This is due to the fact that, incidentally, many aerosol particle have a size comparable to the wavelength of the sun's light which has a maximum spectral emission at ≈550 nm.

Light interacts with matter by absorption and scattering, the latter being the dominant factor except for combustion aerosols (soot). The sum of absorption and scattering is called extinction, and the extinction cross section of a particle is proportional to the geometric cross section of the particle and given by

$$\sigma_e = \frac{\pi}{4} d_p^2 \cdot Q_e \tag{36}$$

where Q_e is the extinction efficiency factor, which is a function of the product of the difference of refractive indexes between particle (m) and its environment (normally close to unity) and the so-called Mie parameter α which is defined as the ratio of particle perimeter to light wavelength

$$\alpha = \frac{\pi d_p}{\lambda}. \tag{37}$$

For absorbing particles, m is a complex number, with the imaginary part $\mathrm{Im} > 0$. For small nonabsorbing particles, in the Rayleigh regime $Q_e \propto d_p^4$. At larger particle sizes there are many resonant structures which eventually settle at a value of 2 for large particles, which is called the extinction paradox. Absorbing particles do not exhibit the rich resonance structures, and their extinction for small particle sizes is due solely to absorption. It is interesting to note that the general structure of Q_e of nonabsorbing particles does not much depend on the refractive index if the dimensionless particle size $\alpha(m - 1)$ is used for the abscissa.

Two examples of a nonabsorbing ($\mathrm{Im} = 0$) and an absorbing ($\mathrm{Im} = 0.1$) particle with an index of refraction of $m = 1.44$ are shown in Fig. 6; they were calculated with the Mie-program given in [41].

As a consequence, $\sigma_e \propto d_p^6$ for small particles, $\sigma_e \propto d_p^4$ for intermediate 'Mie' particles, and $\sigma_e \propto d_p^2$ at the geometrical optical limit. More details on light scattering theory can be found in e.g. [40, 41].

Between the Mie and the geometrical optical regions there is a range where one σ_e corresponds to up to three particle sizes. Using white light (which consists of many wavelengths) and/or specially selected scattering angles, the ambiguities can eventually be removed. The disadvantage of white light instruments is the limited illumination intensity and the difficulty of avoiding stray light in the measuring cell.

There is a great variety of different optical particle counters (OPCs); their size detection limit can be pushed down with decreasing scattering volume which reduces the background from the gas molecules. Since laser light can be focused down to the diffraction limit, laser counters have become very popular in the last decade. A further reduction in particle size detection limit is possible by increasing the intensity using parts of the laser cavity as scattering volume, and by improving the signal to noise ratio using time-correlated detector arrays. From theory and present technical limitations a detection limit for OPCs of around 20 nm is expected, although at the expense of considerably reducing the flow rate. There is an enormous number of publications on OPCs and we recommend several papers on the various aspects [3, 6, 7].

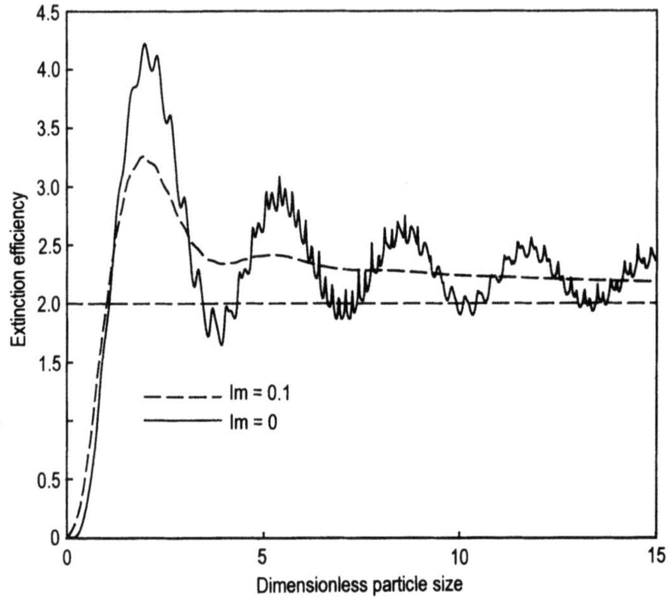

Fig. 6. Extinction coefficient Q_e for a nonabsorbing ($Im = 0$) and an absorbing ($Im > 0$) particle as a function of dimensionless particle size $\alpha(m-1)$

Particles smaller than the OPC detection limit must be magnified before they can be detected individually. This is usually done in condensation nucleus counters (CNCs, see below) with relatively volatile liquids (e.g. water or butanol [42]), or in particle magnifiers with nonvolatile liquids [43, 44].

4.2.4 Kelvin Sizing

In very small droplets, the molecules do not have as many neighbors as in bulk liquid with the consequence that they can escape more easily which leads to an equilibrium vapor pressure $p_{eq}(T, d_p)$ higher than the bulk liquid saturation vapor pressure $p_s(T)$ (see Sect. 6.2.1 in the companion paper):

$$S_{eq} = \frac{p_{eq}(T, d_p)}{p_s(T)} = \exp\left(\frac{4\sigma M_v}{\rho_{liq} R T d_p}\right) . \tag{38}$$

If the actual saturation ratio $S_{act} = p_{v,act}/p_s(T)$ exceeds S_{eq}, a particle will be covered with a thin liquid film which makes it behave like a droplet of the pure liquid, and it will start to grow by condensation. If enough time is available for growth (see Eqs. (*57)–(*64)), the droplets become big enough to be detected by light scattering or other methods: this is the principle of the so-called condensation nucleus counter (CNC). Normally, they are designed so as to allow detection of the smallest possible particle, limited only by condensation on ions (the 'Wilson' chamber operating conditions) or by spontaneous droplet formation without the presence of particles. CNCs are unique in that they allow detection of single extremely small particles due to the built-in magnification. There are different procedures to achieve supersaturations, $S_{act} > 1$: humid air may be adiabatically expanded leading to a temperature reduction with a consequent decrease in $p_s(T)$ according to Eq. (*56). Another method uses isobaric cooling of the vapor-laden flow where (transient) supersaturation is also obtained if the diffusivities of the vapor and of heat are different. Design procedures for the latter were published by Liu and Ahn [45, 46].

Kelvin sizing on the other hand uses different S_{act} for growing different fractions of the particle population (see Fig. 7). Most Kelvin sizers use variable expansion ratios [47], but locally variable supersaturations have also been proposed [48]. Depending on the affinity of the particle material to the vapor, the Kelvin size may be quite different from the mobility size [49].

4.2.5 Epiphaniometer

The epiphaniometer [27] measures radon progeny (of subnanometer size) attaching to particles. Since this is basically a coagulation process, the attachment rate is $\propto d_p^2$ for small particles and $\propto d_p$ for large particles, so that, depending on the exact size distribution, a moment M_N^x is measured with $1 \leq x \leq 2$. In essence, an epiphaniometer gives the effective surface for mass exchange, and is therefore of application to atmospheric chemistry.

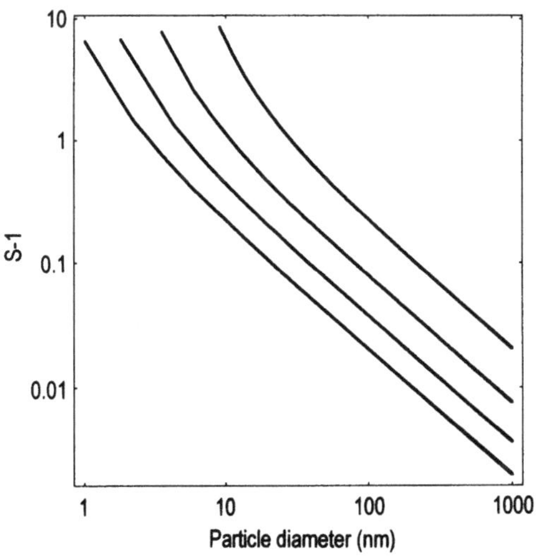

Fig. 7. Kelvin diameter of particles from different vapors from Table *2. In ascending order water, butanol, glycerol and DEHS. A water droplet of around 3 nm diameter for example is in equilibrium with the vapor phase at a supersaturation $S - 1 = 1$; smaller droplets will evaporate, larger ones will grow

4.3 Equivalent Diameters

All size information obtained from a certain measuring technique bears the signature of the underlying physical process: it is an equivalent diameter with respect to this process, and is reliable only in the range where this process works well. For example, a Kelvin sizer relates supersaturations necessary for particle growth to sizes according to Eq. (38), or an OPC calibrated with Latex particles of a certain refractive index will give 'Latex particle optical equivalent' sizes; the actual particle size may be different according to the refractive index, the angles under which the scattered light is collected etc. Diffusion battery measurements will give a diffusion size which should be equivalent to the electrical mobility size according to Einstein's relation. Slightly more complicated is the situation with differential mobility analyzers, since one has to be sure that the particles have an exactly known electric charge (see above), and that there is no preferred orientation (of nonspherical particles) induced by the electric field in the analyzer [50].

Another problem refers to the conversion of number size distributions into volume size distributions. A comparison between low pressure impactor (LPI), and differential mobility particle sizer/aerodynamic particle sizer (DMPS/APS) combination showed that channel-by-channel conversion from number to mass distribution results in good agreement with LPI data [51].

4.4 Multiparameter Sizing

When comparing the equivalent diameters obtained from different techniques one can infer other physical particle properties relevant to the particular measurement. Taking a monomobility sample of small particles[2] with a differential mobility analyzer, one can, for instance, determine the refractive indexes of particles by sizing them subsequently in an OPC [52] or the density by applying inertial classification [53]. Using an impactor or an aerodynamic particle sizer, one gets complementary information on density or fractal particle dimension [54, 58] as well as optical properties [56]. For larger particles, the mass can be determined with Millikan's method, and by applying an electric field the fractal dimension can be found again [55]. Exposing a monomobility fraction of particles to elevated humidity and measuring size changes in an OPC gives information about the hygroscopic properties [57]. There are many more interesting combinations which have not yet been explored.

[2] This is necessary in order to have negligible fractions of multiply charged particles (see Eq. (*22))

5 Inversion Methods

In ideal separators, the cut-off curve is very steep so that a perfect discrimination of particles can be achieved. Also, in sizers with ideal characteristics, all particles are correctly sized. In reality, however, there is a certain probability of particles entering the wrong size category. For recovering the true size distribution from erroneous data, a procedure called inversion should be employed. Since the use of inversion procedures is not very easy to do mathematically and requires a well characterized instrument, many users simply assume a sharp cut-off function at the size d_{50} where the cut-off curve has a 50% probability for both rejection and acceptance. It is the intention of this section to give a short introduction into this difficult subject and to provide enough references for the reader to chose the inversion technique most suitable for his purposes.

If one tries to characterize an aerosol population with size distribution $f(x)$ by a series of m measurements, the measuring values y_i with $1 \leq i \leq m$ will reflect both the contribution of the size distribution and the (size dependent) instrument weighting characteristics $W_i(x)$ according to

$$y_i = \int_0^\infty W_i(x) f(x) dx + \delta y_i ; \ (i = 1, 2, \ldots m) \,. \tag{39}$$

Here, $W_i(x)$ may be a particle penetration curve through diffusion battery number i or a size specific particle response like the scattering cross section $\sigma_i(x)$ contributing to the signal of detector i. Equation (39) is a Fredholm equation of first kind which has the unpleasant property of being ill-conditioned, notably with measuring errors δy_i present.

Normally, one would not search for analytical solutions rather than numerical ones which can be obtained by discretizing Eq. (39) at n sizes x_k where $1 \leq k \leq n$:

$$y_i = \sum_{k=1}^{n} W_i(x_k) f(x_k) + \delta y_i ; \ (i = 1, 2, \ldots m) \, k = 1, 2, \ldots n \tag{40}$$

or shortly in matrix notation

$$\vec{\mathbf{y}} = \mathbf{W}\vec{\mathbf{f}} + \delta\vec{\mathbf{y}} \,. \tag{41}$$

$\vec{\mathbf{y}}$ is the measuring vector and $\vec{\mathbf{f}}$ is the true size distribution. The basic idea is now to look for a solution vector $\vec{\mathbf{f}}'$ which minimizes the above equation i.e.

$$|\vec{\mathbf{y}} - \mathbf{W}\vec{\mathbf{f}}'| = \mathbf{Min} \,. \tag{42}$$

The least squares solution of an under- or an overdetermined system is given by

$$\mathbf{W}^{\mathrm{T}}\vec{\mathbf{y}} = \mathbf{W}^{\mathrm{T}}\mathbf{W}\vec{\mathbf{f}}' \,. \tag{43}$$

$\mathbf{W}^{\mathrm{T}}$ is the matrix transpose of $\mathbf{W}$. Unfortunately, this simple approach usually leads to highly oscillatory solutions with even negative concentrations!

5.1 Least Square Fits and Singular Value Decomposition

For singular matrices under certain conditions, a solution may be found by means of the so-called pseudoinverse which is essentially a generalisation of the inverse for singular matrices. This can be done by a method called singular value decomposition. The ratio of the largest to the smallest of these singular values is the so-called condition number which determines the accuracy of inverses. Programs for the calculation of the pseudoinverse, which is the key to the inversion problem, are available (e.g. in [59]).

5.2 Modified Fit Techniques

Paatero et al. [60, 61] proposed a method (which is available for PC use from the author) giving a solution range instead of a single solution. First, in a fit procedure an optimal solution is determined, the quality of which is assessed by its χ^2-value. All solutions in a certain (not explicitly plausible) χ^2-neighborhood of the optimum solution are then admitted as solution range.

5.3 Constrained Regularization

Constraints concerning the smoothness of the admitted solution were introduced by Tikhonov [62]. Instead of Eq. (42), the minimum of the expression

$$|\vec{y} - \mathbf{W}\vec{\mathbf{f}}'| + \lambda\boldsymbol{\Omega}(\mathbf{f}') = \mathbf{Min} \tag{44}$$

has to be found. A very similar approach was used by Twomey [63] and Phillips [64].

A simple choice for the regularization functional $\boldsymbol{\Omega}(\mathbf{f}')$ is, e.g.,

$$\boldsymbol{\Omega}(\mathbf{f}'(\mathbf{x})) = \int_0^\infty \left(\frac{\partial^2 \mathbf{f}'}{\partial \mathbf{x}^2}\right)^2 \mathbf{dx} \tag{45}$$

which describes the overall smoothness of the distribution curve. The larger the (arbitrary) λ-value the smoother the solution, but still physically meaningless negative concentrations are not a priori excluded. Crump and Seinfeld [66, 67] and then Wolfenbarger and Seinfeld [68] additionally introduced nonnegativity of the solution.

Helsper et al. [65] proposed a linear programming method with the constraint of nonnegative concentrations $f(x_k)$ which assures physically meaningful solutions but does not control the smoothness of the result, and is easier to apply than the above techniques.

5.4 Parametric Inversion

This is certainly the simplest and most stable approach: led by experience one choses a size distribution (Lognormal, Junge, Gamma etc.) and fits the distribution parameters so that the sum of the squared errors is minimized:

$$\sum_{i=1}^{n} \left(y_i - \int_0^{\infty} W_i(x) f(x, \text{parameters}) dx \right)^2 = \text{Min} . \tag{46}$$

The weighting factors $W_i(x)$ have to be determined by calibration. The problem with this approach is that one has no way of estimating how good the fit is, since the choice of the distribution may be inappropriate. Nevertheless, parametric inversion is widely used due to its simplicity and availability [69].

5.5 Merits and Drawbacks of the Various Methods

For instrument kernels very close to delta or step functions, no inversion is necessary at all. This approach is chosen for cascade impactors and differential mobility analyzers by most aerosol scientists. However, one should make sure from time to time that the errors are not unacceptably large by comparing the direct results with those of inversion techniques as discussed above.

If one is absolutely sure about the shape of the size distribution, a parametric inversion may be justified; it also provides the fastest solution. For most cases linear programming techniques with constraints or least square fits with singular value decomposition seem the most effective and reliable approaches. For the most demanding problems with poor instrument kernels, constrained regularization should be used, and the MICRON code was found to work very reliably [70].

It has to be made clear, however, that even the best inversion procedure, because of the inherent instability of the ill-conditioned Fredholm integral equation, cannot recover more precise results than are contained in the quality of the data. This means that one should make the greatest effort to improve the instrument kernel $\boldsymbol{W}$.

6 Conclusions

The above survey showed that adequate techniques are available for most single-parameter measurements presently of interest. High resolution number distributions allow even reasonable reconstruction of volume size distributions [51] (or of mass size distributions if the density is known).

The situation is different if multi-parameter characterization of individual particles with respect to size, material, morphology etc. is required, and there is a clear need for the development of new instrumentation with these capabilities. In principle, size distributions can be characterized by and reconstructed from the whole set of their moments (which are obtained from measurements based on different physical principles) to any degree of precision desired. In many cases, a few moments from a suitable combination of instruments like CNC, epiphaniometer, gravimetry, dust fall gauge etc. would provide sufficient information, but this approach has not been used so far although, in our opinion, it seems promising.

7 References

1. Fuchs NA (1989) The mechanics of aerosols. Wiley, New York
2. Hinds W (1982) Aerosol technology. Wiley, New York
3. Willeke K, Baron PA (1993) Aerosol measurement: principles, techniques and applications. Van Nostrand, New York
4. Friedlander SK (1977) Smoke dust and haze. Wiley, New York
5. Marple VA, Liu BYH (eds) (1983) Aerosols in the mining and industrial work environments. Ann Arbor Science, Ann Arbor
6. Liu BYH (ed) (1976) Fine particles. Academic, New York
7. Lundgren DA, Harris FS, Marlow WH, Lippmann M, Clark WE, Durham MD (eds) (1979) Aerosol measurement, University Presses of Florida, Gainesville
8. Shaw DT (ed) (1978) Recent developments in aerosol science. Wiley, New York
9. Shaw DT (ed) (1978) Fundamentals of aerosol science. Wiley, New York
10. VDI guideline 2449/part 1 (1991) Ermittlung von Verfahrenskenngrößen für die Messung gasförmiger Schadstoffe (Immission)
11. Stange K (1970) Angewandte Statistik. Springer, Berlin
12. Evans RD (1955) The atomic nucleus. McGraw-Hill, New York
13. Irani RR, Callis CF (1963) Particle size: measurement, interpretation, and application. Wiley, New York
14. Herdan G (1953) Small particle statistics. Elsevier, Amsterdam
15. Stevens DL, Moss OR (1983). In: [5]
16. VDI guideline 3491/part 1 (1990) Particulate matter measurement; methods for characterization and monitoring test aerosols (survey)
17. Twomey S (1977) Atmospheric aerosols. Elsevier, Amsterdam
18. Lee KW, Chen H, Gieseke JA (1984) AST 3: 53
19. Pratsinis S (1988) J. Colloid Interf. Sci. 124: 416
20. Frenklach M, Harris SJ (1987) J. Colloid Interf. Sci. 118: 252
21. Lewis A, Lieberman K (1991) Nature 354: 214
22. (eds) Duke PJ, Michette AG (1990) Modern microscopies. Plenum
23. Herman GT (1980) Image reconstructions from projections. Academic, New York
24. Gabor D (1948) Nature 161: 777
25. Binnig G, Rohrer H, Gerber C, Weibel E (1982) Appl. Phys. Lett. 40: 178
26. Holländer W (1982) VDI-Berichte 429, p. 35
27. Gäggeler HW, Baltensperger U, Emmenegger M, Jost DT, Schmidt-Ott A, Haller P, Hofmann M (1989) J. Aerosol Sci. 20: 557

28. Wang SC, Flagan RC (1990) AST 13: 230
29. Gupta A, McMurry PH (1989) AST 10: 451
30. Stöber W, Flachsbart H (1971) J. Aerosol Sci. 3: 1280
31. Stöber W, Martonen TB, Osborne S (1978). In: [8], p. 263
32. Lodge JP, Chan TL (1988) Cascade impactors: sampling and data analysis. Am. Ind. Hyg. Assoc. publication L87028, Akron
33. Turner JR, Hering SV (1987) J. Aerosol Sci. 18: 215
34. Rao NP, Fernandez De La Mora J, McMurry PH (1992) J. Aerosol Sci. 23: 11
35. Mazumder MK, Kirsch KJ (1977) Rev. Sci. Instr. 48: 622
36. Baron PA (1986) AST 5: 55
37. Wang HC, John W (1989) AST 10: 501
38. Cheng YS, Barr EB, Marshall IA, Mitchell JP (1993) J. Aerosol Sci. 24: 501
39. VDI guideline 3489/part 8/draft (1994) Relaxationszeitspektrometer
40. Kerker M (1969) The scattering of light and other electromagnetic radiation. Academic, Orlando
41. Bohren CF, Huffman DR (1989) Absorption and scattering of light by small particles, Wiley, New York
42. Agarwal JK, Sem GJ (1980) J. Aerosol Sci. 11: 343
43. Kogan YI, Burnasheva AG (1960) Phys. Chem. Moscow 34: 2630
44. Okuyama K, Kousaka Y, Motouchi T (1984) AST 3: 353 (+J. Aerosol Sci. 13: 232)
45. Ahn KH, Liu BYH (1990) J. Aerosol Sci. 21: 249
46. Ahn KH, Liu BYH (1990) J. Aerosol Sci. 21: 263
47. Niessner R, Helsper C (1985) J. Aerosol Sci. 16: 201
48. Holländer W (1980) J. Aerosol Sci. 11: 403
49. Porstendörfer J, Scheibel HG, Pohl FG, Preining O, Reischl G, Wagner PE (1985) AST 4: 65
50. Wen HY, Kasper G (1984) AST 3: 397
51. Peters TM, Chein HM, Lundgren DA, Keady PB (1993) AST 19: 396
52. Emets EP, Kascheev VA, Poluektov PP (1991) J. Aerosol Sci. 22: 389
53. Kelly WP, McMurry PH (1992) AST 17: 199
54. Emets EP, Kascheev VA, Poluektov PP (1992) J. Aerosol Sci. 23: 27
55. Colbeck I, Nyeki S, Wu Z (1992) J. Aerosol Sci. 23: S365
56. Covert DS, Heintzenberg J, Hansson HC (1990) AST 12: 446
57. McMurry PH, Stolzenburg MR (1989) Atm. Env. 23: 497
58. Cheng YS, Chen BT, Yeh HC, Marshall IA, Mitchell JP, Griffiths WD (1993) AST 19: 255
59. Wolfram S (1991) Mathematica. Addison Wesley, Reading
60. Paatero P, Raunemaa T, Dod RL (1988) J. Aerosol Sci. 19: 1223
61. Paatero P (1990) The extreme value estimation deconvolution method with applications in aerosol research. Univ. Helsinki Report Series in Physics
62. Tikhonov AN (1963) Soviet Math Dokl 4: 1035
63. Twomey S (1963) J. Assoc. Comput. Mach. 10: 97
64. Phillips DL (1962) J. Assoc. Comput. Mach. 9: 84
65. Helsper C, Fißan H, Kapadia A, Liu BYH (1982) AST 1: 135
66. Crump JG, Seinfeld JH (1982) AST 1: 15
67. Crump JG, Seinfeld JH (1982) AST 1: 363
68. Wolfenbarger JK, Seinfeld JH (1990) J. Aerosol Sci. 21: 227
69. Whitby E TSI model 390045
70. Pohlmann G (1993) Ein einfaches Verfahren zur Bewertung von Inversionmethoden in der Aerosolmeßtechnik. Dissertation Universität Hannover

Particulate Emission Control

Th. Kouimtzis[1] *and A.I. Zouboulis*[2]

[1] Environmental Pollution Control Laboratory, Department of Chemistry, Aristotle University, GR-540 06 Thessaloniki, Greece.
[2] Laboratory of General and Inorganic Chemical Technology, Department of Chemistry, Aristotle University, GR-540 06 Thessaloniki, Greece

List of Symbols and Abbreviations 280
Problem Definition 281
Introduction 281
Particulates Emission Sources 281
Particulate Emission Standards 282
Selection of a Control System 283
Properties and Characteristics of the Carrier Gas 283
Particle Properties and Characteristics 286
Particle Removal Mechanisms 287
Collection Efficiency 289
Mechanical Collectors 290
Description 290
Applicability 291
Design-Performance 291
Reliability 292
Cyclones (Centrifugal Separators) 293
Description 293
Applicability 296
Design-Performance 297
Reliability-Maintenance 299
Filtration Processes 301
Description 301
Applicability 305
Design-Performance 307
Reliability 309
Electrostatic Precipitation Process 309
Description 309
Applicability 314
Design-Performance 316
Reliability-Maintenance 321
Scrubbing Processes 323
Description 323
Applicability 331
Design-Performance 332
Reliability 333
Comparison of Treatment Methods: Overview and Discussion 333
Further Reading 335

Summary

In recent times, a strong technology base has developed rapidly for air quality control; particulate emission control dominates a major part of the whole problem. This chapter focuses on particulate emission controls for stationary sources of air contaminants. The main methods and techniques, i.e. mechanical collectors, cyclones, filtration, electrostatic precipitation and scrubbers, are briefly described, examining also their applicability, design, performance and reliability. A comparison/general discussion between these methods/techniques is also given.

List of Symbols and Abbreviations

A : collecting surface area (m^2)
C : particle concentration in the gas effluent
C' : Cunningham slip correction factor (dimensionless)
C_D : drag coefficient
C_t : total concentration of particles in the gas entering the treatment equipment
d_c : characteristic dimension of the collector, e.g. diameter (cm)
d_p : particle diameter (μm)
E_o : overall collection efficiency
g : acceleration of gravity (9.806 m/s^2)
K_a : empirical constant of Anderson equation calculated from measured efficiencies, depending on the specific process and set of operating conditions
K_p : inertia parameter (dimensionless)
n : collection efficiency (%, weight fraction collected)
ΔP : pressure drop (cm water gauge, or kPa)
Q : gas volume flow rate (m^3/s)
Q_L/Q_G : L/G ratio (L/m^3)
r_c : cylinder radius (m)
Re : Reynolds number
S : separation factor
t : time (s)
u_i : gas inlet velocity (m/s)
u_o : gas velocity (m/s)
U_G : throat velocity of the gas in scrubbers (m/s)
w : migration velocity (m/s)
μ : gas viscosity (poise or kg/ms)
ρ_ρ : particle density (g/cm^3)

Problem Definition

Introduction

Particulate matter is generated by a variety of physical and chemical mechanisms from numerous sources and is composed of finely dispersed liquids and solids. Small dust particles are usually kept suspended in air for long periods of time by the viscous force or drag of the air, which resists any forces tending to precipitate them. Sometimes forces such as electrostatic charge also act to keep the particles suspended. Dusts suspended in air are not in a stable state and will tend to be separated from the air in time. Particle separation processes and equipment are designed to make this time as short as possible.

A large amount of the particles, mainly in the atmosphere of urban areas, is emitted by various man-dependent sources such as central heating, industries, traffic, incineration plants, etc. These particles, together with the secondary particles formed in the atmosphere, contribute to the atmospheric pollution. The effect of this aerosol on human health is sometimes dramatic (smoke and photochemical smog episodes in many cities).

Solid particles or liquid droplets naturally present in the atmosphere exist in low concentrations such as to be harmless. They may, however, play a vital part in various natural cycles of growth and decay together with other air constituents. Of course, increased concentrations of particles in nature do occur, as for example during explosion of volcanos or big forest fires, but these are rather exceptional circumstances. It is well documented that primary high concentrations of particles can adversely affect human health, plant material, visibility and climate. These detrimental effects are mainly determined by particle size, concentration and chemical composition.

In order to minimise the contribution of the emission from man-dependent sources, various control systems have been developed. In addition, emission standards have been introduced in an attempt to guide the development of the most appropriate control systems.

Particulate Emission Sources

The main particulate emission sources (mobile or stationary) can be generally classified into the following groups.

— **Process-point sources**: these sources are individually defined and include furnaces, reactors, etc., which discharge emissions through a vent-pipe of a stack. They can be controlled through the use of add-on control equipment.

— **Process-fugitive sources**: these sources are also individually defined and produce particles that escape from pumps, access ports and feed or discharge openings to process. They also include ventifans from rooms or enclosures containing emission sources. These sources can be controlled by add-on control devices once the emissions are captured and then transferred to the appropriate control equipment.

— **Area-fugitive sources**: these sources are characterized by large surface areas from which emissions occur. They include loading and unloading areas, raw material storage, roads, uncovered trucks, etc.

Particulate Emission Standards

Historically, the first measures to protect the atmosphere from various emissions referred to particulate emissions. They go back to the 14th century when King Edward I of England prohibited the burning of coal in London during sessions of Parliament because of the smoke produced. The first emission standards again referred to particulate emissions. They prohibited the emission of black or dark smoke. Definition of terms was first devised by Maximillian Ringelmann in 1898.

Today, particulate emission standards are included in all air pollution legislation and regulations. These standards are classified into many categories, depending on the existing national or local regulation and the type of sources, such as particulate emission standards from any process (Fig. 1), fugitive dust standards, particulate emission stack height standards, etc. The composition and the size distribution of the emitted particulate matter is also considered in setting emission standards. Thus, the presence of toxic compounds in the particulate matter leads to setting even more stringent standards.

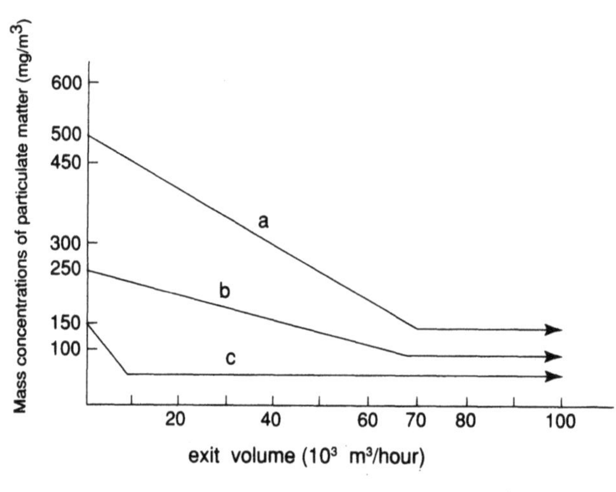

Fig. 1. Emission standards for particulate matter in effluent air or gases (Federal Republic of Germany): (a) total particulate matter; (b) particulates < 10 μm; (c) total particulates

Selection of a Control System

The treatment methods used for air pollution control in general can be classified according to the applied technology: (1) by changing the distribution pattern of the pollutant to be removed, (2) by using alternative technology, hence preventing the generation of a pollutant (strategies for preventing pollution were found to include regulatory (restrictive and prohibitive), economic (incentives and sanctions) and educational measurements), and (3) by removing/collecting/separating the pollutants.

The selection of a control system for collecting/separating particulate matter is based upon the relationship of the process emission to the required control equipment performance and emission limitations imposed by the existing regulations. Therefore careful characterization of contaminants in the effluent streams and establishment of requirements for the removal efficiency must precede selection of the appropriate equipment. (Fig. 2). After identifying the requirements which must be met, it is very important to characterize both the effluent stream to be controlled and the particulate matter to be collected for all their characteristics and properties which affect the operation of collectors. In addition, particle removal mechanisms have to be especially considered among all the other parameters.

Disposal facilities for collected dust are also of importance for the selection of control systems. More specifically, considering the disposal of the dust, a selection between dry or wet collection equipment should be made. In a wet system there is a need for a liquid storage pond, in order to facilitate sluicing of the collected dust. Thus, wet collection and disposal would be favoured if wet processing is involved. On the other hand, in dry processing, because of existing technology and expertise at the plant, dry collection and disposal would be preferred.

Finally, after the definition of the problem in terms of what particulate matter is emitted and what must be removed, it is necessary to choose among the several collector alternatives. Materials of construction should be adequate to handle the contaminants, as many are highly corrosive. Equipment should be as maintenance-free as possible. Equipment weight may also be an important consideration, particularly if it is to be placed on a roof. An engineering approach to the selection of control equipment is presented in Table 1.

Properties and Characteristics of the Carrier Gas

The main properties/characteristics of the carrier gas are the following.

1. **Gas flow rate**: the flow rate of the gas stream essentially controls the size of the equipment which will be selected for the particle removal. Optimization

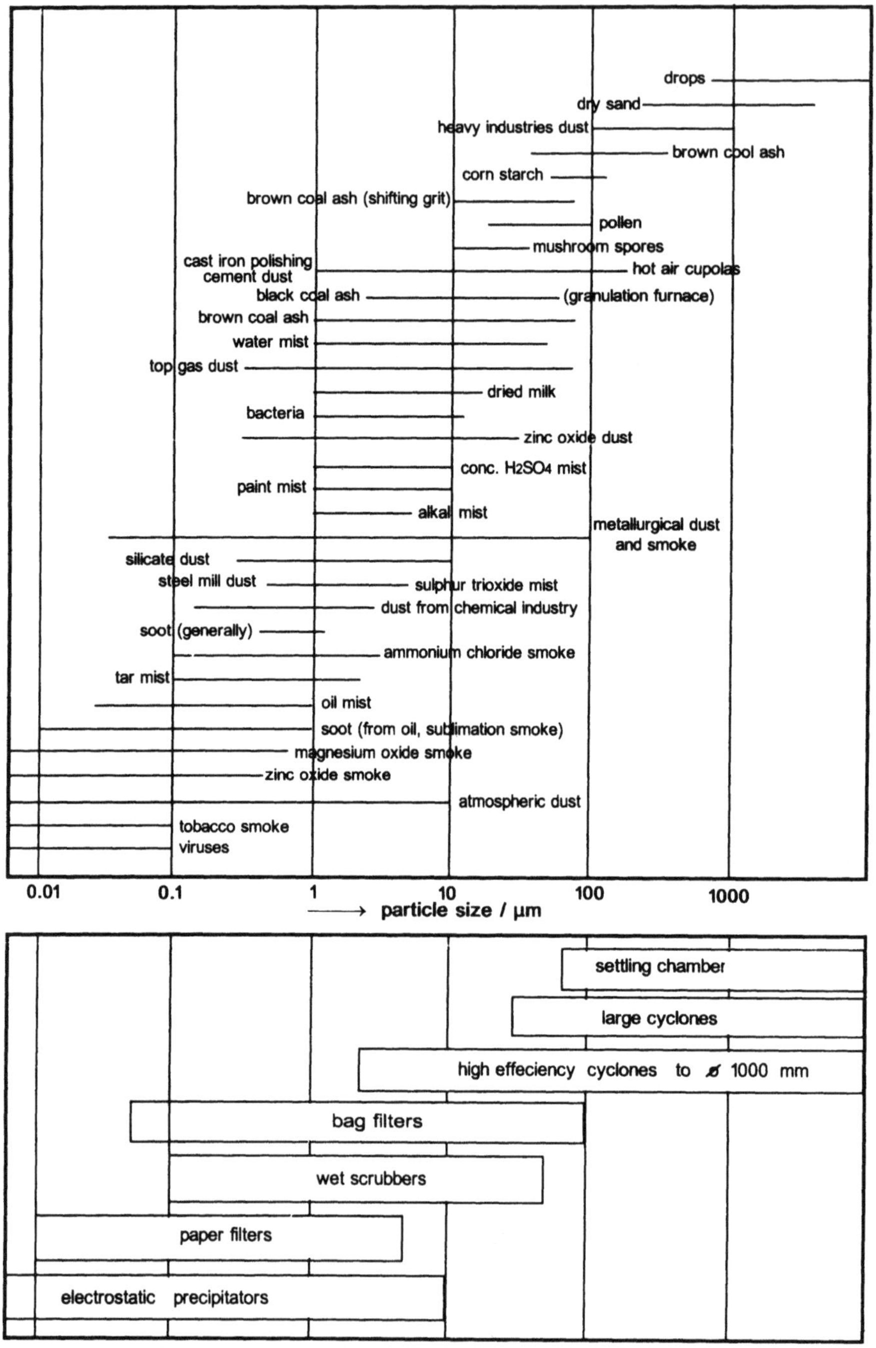

Fig. 2. Origin/size range of various particle and natural aerosols and particulate emission control systems

Table 1. Engineering requirements for design and selection of control equipment

1. Known properties of the pollutant: chemical species, physical state, particle size, concentration, quantity of conveying gas.
2. Known effects of the pollutant on the surrounding environment.
3. Design for likely future collection requirements.
4. Determination of the advantages and disadvantages of alternative collection techniques: collection efficiency; ease of reuse or disposal of recovered material; ability of the collector to handle variations in gas flow and loads at required collection efficiencies; equipment reliability and freedom from operational and maintenance attention; initial investment and operating cost.
5. Recover or convert contaminant into a saleable product.
6. Application of known engineering principles even in areas of extremely dilute concentrations.

of the various parameters involved in the relationship between size of the equipment and gas velocity leads to the following.

a) The effect of the velocity on the removal mechanisms involved.
b) Any size reduction of the equipment leads to an increased pressure loss and, as a result, to increased power requirements.

For example, high velocities favour removal mechanisms by inertial forces, up to the point of turbulence. Then further increased velocities result in decreased efficiency. Regarding the venturi scrubbers, removal efficiency is directly proportional to velocity throughout the system, while the performance of an electrostatic precipitator decreases with increasing flow rate. Consideration should also be given to the effect of flow rate variation on equipment efficiency and pressure drop. Various control techniques are used to adjust flow changes. In the cases where flow rate variations are inescapable, it is necessary to choose devices which are inherent positive in their operation, or are designed for extreme conditions.

2. **Temperature**: the temperature of the gas stream is an important parameter which determines the selection of the appropriate collector. The mechanisms involved in the particle removal are also temperature dependent. On the other hand, the temperature influences the volume of the gas stream and the choice of the construction material of the collector. For example, if the gas stream temperature is higher than the affordable fabric fibres, the latter are not recommended, while wet processes cannot be used where liquid would boil or evaporate rapidly.
3. **Pressure**: in general, gas stream pressure is not of prime importance in the particle collection systems. However, it may be of importance in some certain cases, such as in high efficiency scrubbers.
4. **Density**: particle density is much greater than gas density, and thus any usual changes in gas density have a negligible effect.
5. **Humidity**: humidity of the gas stream is of importance to the performance of control equipment, as high humidity may lead to blocking because of mud and cake formation. In cases where humidity is a serious problem, scrubbers are the most appropriate control systems.

6. **Composition reactivity**: the reactivity of the carrier gas presents special problems. The magnitude of the problems may be greater when they combine with high temperatures. For example, the presence of gaseous fluorides in the gas stream eliminates the possibility of using glass fiber fabrics for filtration, due to corrosion problems. In some cases, additives have been used to change carrier gas characteristics prior to passing through the control equipment. For example, dolomite is used for preliminary neutralization, in order to protect fabric filters from gas streams containing acid components.

Particle Properties and Characteristics

The main properties and characteristics of the particles that may influence the selection of the control system are the following.

1. **Particle size**: this is one of the most important considerations in the selection of the control system. In all types of control equipment the efficiency is essentially dependent on particle size distribution, as for example cyclones can collect particles at reasonable efficiencies only when their sizes are greater than 10 μm. On the other hand, medium energy wet scrubbers can control gas streams containing particles in the size range of 1 to 5 μm. Particle size is also an important parameter for the design of electrostatic precipitators, because the power input causing migration of the particles to the collection plates is inversely dependent on the diameter of the particles. Generally, the smaller the size of the particles to be removed, the greater the expenditure that will be required for equipment or power, or both. Particle shape and density are also important factors that determine the selection and design of the control equipment.
2. **Hydroscopicity**: hydroscopic particles will cause difficulties in some control equipment, especially in dry collectors. Hydroscopicity may influence how readily particulate will collect or accumulate in equipment if moisture is present. Such accumulation may completely blind the control equipment and prevent normal gas flow.
3. **Electrical phenomena**: electrical properties of the particles are considered to be a contributing factor, influencing the performance of any control system. In filtration, electrostatic properties influence to a different extent the various mechanisms involved in a particle collection. They also have direct and observable influence upon the process of cake formation and its subsequent removal. In electrostatic precipitators the electrical properties of the particles are very important in determining collection efficiency and greatly influence the cleaning of the collection plates.
4. **Solubility**: solubility of the particles will provide a secondary removal mechanism to aid the basic ones. The most important influence of the solubility is relevant to wet scrubbing control systems.

Particle Removal Mechanisms

Various mechanisms may contribute to the removal of the particles from a gas stream (Fig. 3). Initially, particles are removed from the flowing gas and transferred to the collection space, to the walls of the separator or to the collecting and precipitating surfaces inside the equipment. Then the particles are fed into a dust hopper and the collected dust is removed, either continuously or at intervals, with either manual or automatic control.

The most important principles and forces used in particle removal mechanisms are inertia, gravitation, diffusion, interception, electrostatic, wettability and coagulation. In addition, various interfering phenomena, such as turbulent flow, properties of the boundary layer, adhesiveness of the particles etc., influence the process efficiency. Whatever the type of control system, particles are separated by means of one or more of the following basic mechanisms.

1. **Gravitational sedimentation**: particles in a gas stream are subject to two main forces: the carrying force of the flowing gas and the gravitational one. The influence of the gravity force is strongly dependent on the mass of the particles. This mechanism is usually of little consequence for any small diameter particle (below 50 μm).
2. **Centrifugal Deposition**: particles may be separated from a gas stream by centrifugal force induced by a change in gas flow direction. Large-scale changes, as would be encountered in a cyclone separator, are not very effective on particles smaller than about 5 μm in diameter.
3. **Inertia**: the principle of Newton's law of inertia can be used for the removal of particles from gas streams. According to this, a body tends to remain in its original dynamic state, provided that an external force does not act upon it. A common method is to use a movable wall in order to introduce curvature into the pathway of the particles. Different curvatures in the pathways are used for particle removal in various dry mechanical separators such as in cyclones, dust collectors and louvre separators. Another method is to introduce a body, such as a filter fiber, into the pathway of the particles. The larger particles collide with the body because of their inertia, whereas the smaller particles are able to follow the gas streamlines around the body. This mechanism of particle collection by inertia is also called **impaction**. Many examples of impaction can be found in nature as well as in air pollution control, such as particle collection on the stages of a cascade impactor.
4. **Interception**: the direct capture of particles on a body, e.g. a filter fibre, placed in the gas stream, without crossing any lines, is called stream interception. This occurs when the line, the particle is travelling on, passes within a distance of one particle radius from the surface of the body.
5. **Diffusion**: small solid particles are rarely collected by inertial impaction or interception because they follow the streamlines surrounding the collecting body. On the other hand, these particles diffuse as a result of the turbulence

of the flowing gas (turbulence diffusion) and of the thermal motion of the gas molecules (molecular diffusion). Some of these particles hit the collecting body and are removed from the gas. The diffusion principle is especially employed in filtration through fibrous layers.

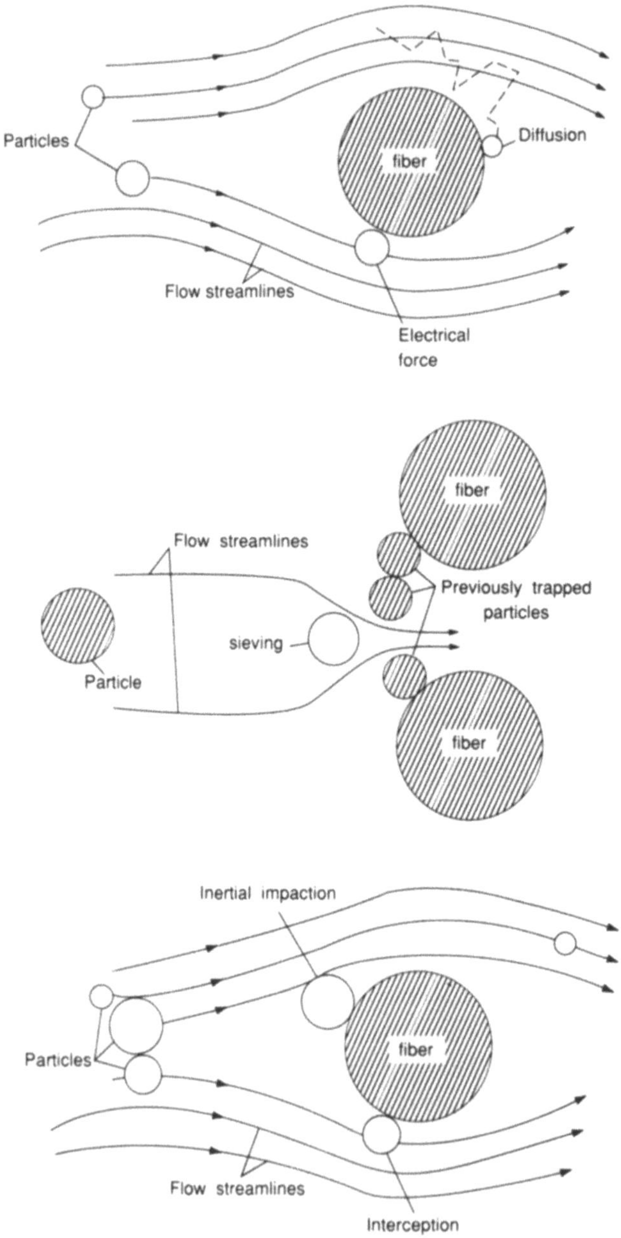

Fig. 3. Main particulate removal mechanisms

6. **Thermophoresis**: if there is a heat transfer from the gas to the liquid, there will be a corresponding temperature gradient, and fine particles will be driven towards the cold surface by differential molecular bombardment due to the gradient.
7. **Electrostatic deposition**: in a non-homogeneous electric field a force in the direction of increasing field intensity act on particles against the resistance of the medium. This effect is relatively weak for uncharged particles, but increases if the particles are electrically charged. This principle is used for particle removal from a gas stream. The removal process may occur in two phases. In the first phase the particles are being electrically charged and in the second phase these charged particles are removed from the gas stream on to the collecting walls by the action of a non-homogeneous electric field.
8. **Coagulation of particles**: particle coagulation enhances the removal of particles from gas streams. Coagulation processes result from thermal, turbulent, sedimentary, electrostatic or acoustic forces. As a result of these forces, two or more particles arrive at relative positions and distances where adhesive forces between the particles can result in formation of aggregates. Coagulation of small particles with larger ones, which are introduced into the gas stream, is also used for particle control system. The collection of small ones on the larger particles occurs either through impact (collision) or through diffusion. The removal of the larger particles produced is much easier than the removal of the original small particles.
9. **Sieving**: the sieve effect is another phenomenon employed in particle collection, especially in various filters used to purify air.

Most types of particle collection equipment use more than one of these mechanisms and it is not uncommon for the controlling mechanism to change when the collector is operated over a wide range of conditions. It is therefore more convenient to classify equipment according to type rather than according to the underlying mechanisms.

Collection Efficiency

The collection efficiency (E) for a given control equipment in a particular application is a measure of that equipment's performance. It is defined as the ratio of the amount of particulate matter collected to the amount fed into the separation equipment, expressed as percentage.

The most suitable parameter for the practical evaluation of collection efficiency is the amount of dust per unit volume of gas. Thus, the total concentration (C_t) of dust in the gas entering the equipment and the concentration (C) in the gas effluent are determined. Then the **overall collection efficiency (Eo)** is given by the equation

$$E_o = \frac{C_t - C}{C_t} \times 100 \ . \quad (1)$$

Mechanical Collectors

Description

The term "mechanical collectors" generally refers to devices in which particles are collected by the action of gravity, or by inertial effects upon the particles, or perhaps by a combination of both.

In **gravity collectors**, particles simply settle slowly out of the gas stream due to their weight under the influence of gravity. Settling chambers utilizing this mechanism are practical only for relatively coarse material and have high space requirements. In **impingement** or **inertial collectors**, otherwise called **momentum collectors**, the flowing stream of particles is subjected to a sudden change of direction, deflecting around the impingement body. The resulting inertia of the particles, as well as impaction on a target and possible centrifugal forces created, cause the particles to tend to be thrown out of the gas stream and to be collected on the surface of the body. The separated particles may be retained on the impaction target, separated into a dust bin, or separated in a dust-enriched side stream which is conducted to a secondary high efficiency dust collector. Collectors employing mainly centrifugal force (cyclones) are an important special case of the inertial effect and will be examined separately.

Collection by gravity is usually carried out simply by providing a large chamber through which the gas stream moves slowly, thus giving particles an opportunity to settle into hoppers in the bottom while the cleaned gas passes on. Obviously, such a simple system works only on the most simple of problems – those in which the dust consists of relatively large, heavy particles. The distance of settling required may be reduced by placing a number of parallel horizontal trays in the chamber, effectively creating a set of multiple chambers in parallel (Fig. 4). The entire gas flow is divided up and passes through the parallel spaces between the plates. The spacing between the plates may be of the order of one-tenth of the chamber height. The concept is sound, but the practical difficulty

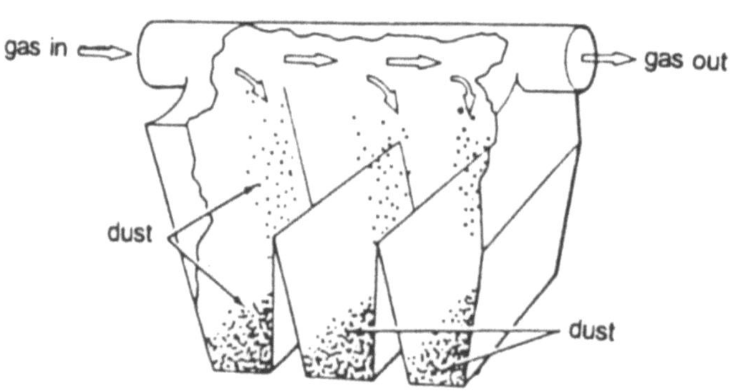

Fig. 4. Settling chamber with flow baffles

is that of removing the particles that settle on the plates. Baffles may also be placed in gravity chambers in order to change the direction of gas flow and bring about an inertial effect to enhance that of gravity. Other designs involve the use of louvers, shutters, or impingement surfaces in order to create an inertial effect.

Applicability

Mechanical collectors are useful primarily for removing coarse particles from a gas stream, i.e. as an aid or adjunct to more efficient subsequent gas cleaning equipment. The particles, which can be removed by the settling chamber itself, are seldom of air pollution significance. In most air pollution control problems, there is a need to collect much finer (in the neighbourhood of 1 μm) particles, so other types of collectors are usually required. However, the mechanical collector may be used as a primary collector in series with another type in order to reduce the load on the latter. This is especially true for heavily loaded gas streams. Mechanical collectors can work effectively for long periods of maintenance-free operation at low energy costs. Settling chambers are occasionally used on natural draft exhausts from kilns, but they are gradually being replaced by low pressure drop cyclones or other more compact dust collectors.

Design-Performance

The design of a settling chamber is based on a calculation of the forces acting upon a particle and the velocity of downward motion the net force produces. According to Newton's law the net downward acceleration of a particle is determined by the resultant of the weight of the particle, as offset by buoyancy, and of the drag force opposing downward motion. In gases the buoyancy effect may be neglected. The drag force is estimated by using the **drag coefficient** (C_D) as related to the **Reynolds number** (Re) for particle motion. As a particle settles it will accelerate vertically until the drag force becomes equal to the weight, after which further motion is at constant speed. This speed is called the **terminal settling velocity** of a particle. The time required for it to be reached in gases is very short.

Three factors mainly determine the inertial collection efficiency: (1) the velocity distribution of the gas flowing into the collector, which varies with the Reynolds number of the gas, (2) the trajectory of the particle, which depends on the mass of the particle, its air resistance, the size and shape of the collector and the flow rate of the gas stream, and (3) the adhesion of the particles to the collector – usually assumed to be 100%.

At high Reynolds numbers the parting of the gas streamlines occurs close to the collector. Except near the collector surface, the flow pattern corresponds to that of an ideal gas. The sudden spreading of the streamlines at high Reynolds number enhances the influence of particle inertia and therefore causes a higher collection efficiency. At low Reynolds numbers, flow is governed by viscosity (viscous flow) and the effect of disturbance created by the collector is noticed at relatively large distances upstream.

The dimensionless inertia parameter K_p is given by the following equation:

$$K_p = \frac{C' - Pd_p{}^2 u_o}{9\,\mu d_c}\,. \tag{2}$$

The gas velocity (u_o), which is the undisturbed upstream gas velocity relative to collector, also characterizes the motion of the particle and has a physical meaning; it is the ratio of the particle stopping distance to the radius of the collector. The particle **stopping distance** is the distance it would travel before coming to rest, if injected into a still gas at velocity u_o, when all forces on the particle except the drag force are zero.

The nature of the grade **efficiency relationship** obtained for a simple settling chamber will depend on the assumption made to describe the flow pattern of the gas. Four specific cases may be considered: (a) plug flow, no mixing; (b) laminar flow with a parabolic velocity profile, no mixing; (c) turbulent plug flow with vertical back mixing of uncollected particles; (d) turbulent plug flow with complete mixing of all uncollected particles. The performance of a real chamber will involve turbulence, some degree of mixing, and some distortion of plug flow.

The actual performance of settling chambers is hardly ever subjected to experimental measurement or testing. At best such devices provide only a preliminary cleaning of gas to remove the coarsest and heaviest particles. The particles escaping a settling chamber must subsequently be collected by other devices. It is the performance of these which is of critical importance in the system.

The range of performance of **inertial separators** is similar to that of high efficiency cyclones. They are designed for a pressure drop in the range of 0.25 to 3.8 cm water, depending on the type and application and are limited to removing dusts that are predominantly larger than 20 μm in diameter. Particular designs may be more suited for particular applications because they may occupy less space, or may be specially designed to compensate for particular problems such as erosion, necessity to minimize pressure drop, etc.

Reliability

All types of mechanical collectors are characterized by simplicity of construction and operation. They are relatively inexpensive in comparison with other kinds of collectors. The energy required for operation is also relatively low. It is due only to a rather small pressure drop, required to move the gas through the equipment and consisting primarily of entrance and exit losses. However, the space required for such chambers is large, and they are seldom used to remove particles smaller than 40 to 50 μm in diameter. In general, those without moving parts may be constructed of any reasonable material of construction, while those with moving parts must include considerations of stress and accelerated corrosion and erosion of the moving parts. Most inertial collectors can be operated wet, if desired, and under such conditions they exhibit a higher collection efficiency due to the same factors that operate in the case of cyclone dust collectors.

Careful design of the settling chamber is necessary to provide good distribution of gas entering and leaving the settling chamber. The usual types of design include gradual transitions, splitters, or perforated distributing plates. For a given volumetric air-flow rate, the collection efficiency of a settling chamber depends on the total plan cross section of the chamber and is independent of the height. The height need to be made only large enough so that the gas velocity in the chamber is not so high as to cause re-entrainment of separated dust. The gas velocity in the chamber should generally be restricted to 3 m/s or less in order to prevent excessive re-entrainment. Settling chambers with horizontal plates offer greater efficiency in a smaller space, but present difficult cleaning problems; probably the most effective method of cleaning is to flush the plates with water sprays.

Cyclones (Centrifugal Separators)

Description

The most widely used general purpose type of dust-collection equipment is the cyclone, which is an inertial separator without moving parts, where the gravitational acceleration is replaced by centrifugal acceleration. A particle of mass (m) moving in a circular path of radius (r) with a tangential velocity (u_T) is acted upon by a centrifugal force of mu_T^2/r. For typical values of $u_T = 15$ m/s and $r = 0.6$ m, this force is 38.8 times greater than that of gravity. It can therefore be used as a greatly enhanced collecting force in comparison with a simple settling (gravitational) chamber. It is used to remove medium-sized entrained particulate matter (usually of the order of about 15–40 μm) from gases. It usually consists of a cylindrical shell fitted with: (1) a tangential inlet through which dusty gas enters, which produces the vortex, (2) an axial exit pipe for discharging the cleaned gas, and (3) a conical base dust discharge facility (Fig. 5).

In cyclone collectors the gas is introduced into a cylindrical vessel at a velocity of about 10–30 m/s and is given a spinning or **vortex** motion to impart a centrifugal force to the particles. The centrifugal separating force or acceleration may range from 5 times the gravity force in very large diameter, low resistance cyclones, up to 2,500 times the gravity force in very small, high resistance units.

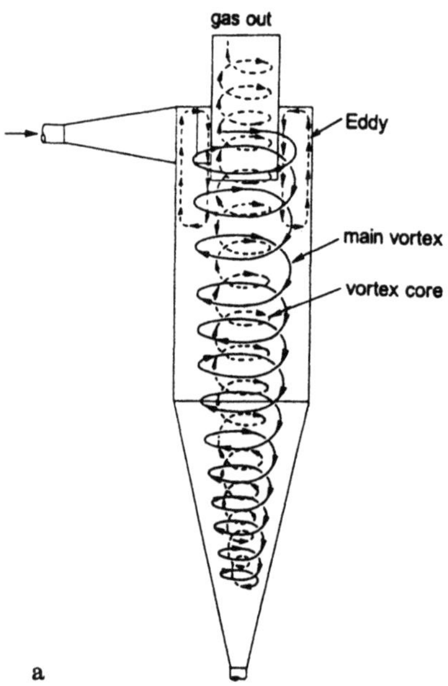

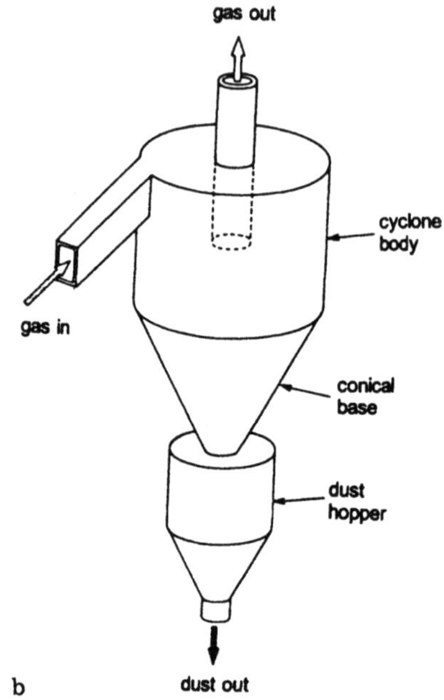

Fig. 5. **a** Vortex and eddy flows in a cyclone. **b** Typical cyclone design

This is done either by introducing the flow tangentially into a circular chamber, or by passing the gas through radial vanes along an axial flow.

It has been found that the separating force increases with the cube force of the particle diameter, but the resistance to particle flow toward the cyclone wall increases only linearly with the particle diameter. In any practical cyclone, the particulate matter is spread over the width of the dirty gas inlet, so that the radius of rotation, tangential velocity and distance from the cyclone wall would vary from the layer entering next to the outer wall of the cyclone to that nearest the wall of the gas outlet. Therefore, the cyclone separator is considered as a poor classifier and does not make a sharp size cut between particles separated and passed.

The **flow pattern** is complex in even the simplest cyclone: when the dust-laden gas enters tangentially, it creates a descending helical current in the body and cone, called the "main vortex". The entrained particulate matter, because of inertial forces, tends to move toward the outside wall, from which it is led to a receiver, i.e. it is the vortex that does the work in separating the particulate matter from the gas. Near the apex of the cone, the spinning gas flows radially inward and then forms an ascending helical current, a "secondary or inner vortex", ultimately being discharged from the top of the cyclone. Particles thrown against the wall are collected by sliding down into a hopper. This device is usually mounted vertically.

In the case of an **axial flow** device, the gas stream enters at the centre of one end of a cylinder. It flows through vanes which impart the spinning motion. Collected particles are carried out by a peripheral stream, while the clean gas exits through a central stream at the opposite end from the entry. This device may be mounted at any position. There are a number of proprietary designs which are variations of these basic designs.

In addition, there is an upward gas flow along the body wall surface near the top of the cylinder. This upward flow, known as an **eddy**, carries gas and dust particles up along the body wall, inward across the top, and downward along the

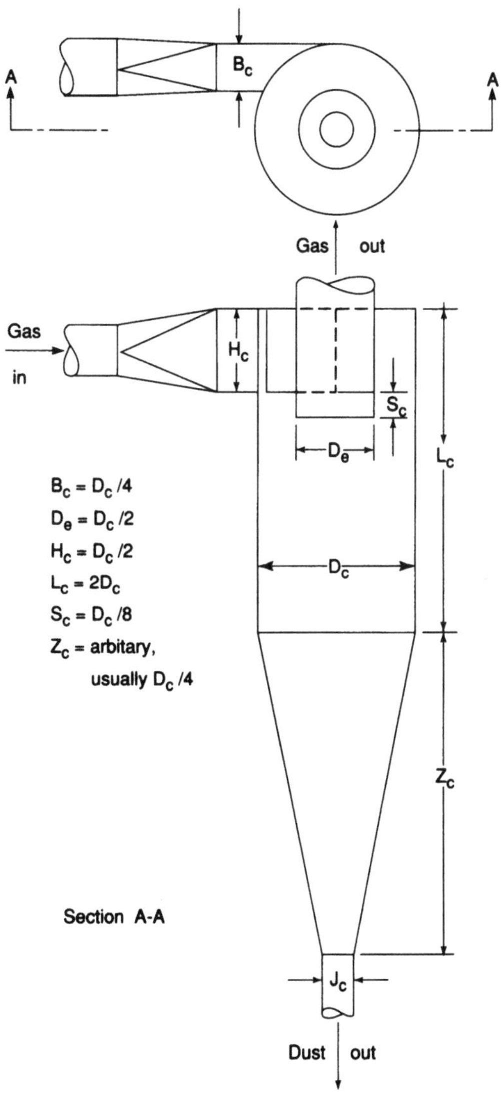

Fig. 6. Cyclone typical dimension ratios

gas outlet wall. From this point, dust particles are lost into the gas outlet. The longer the gas outlet projection into the body, the more pronounced the eddy; elimination of the gas outlet protrusion, however, does not completely eliminate the eddy.

The tangential velocity of gas in the vortex increases as the radius decreases from the radius of the cylindrical body to a maximum at some intermediate radius; and from this intermediate point inward to the axis of the cylinder, the tangential velocity decreases. The vortex core is generally smaller in diameter than the gas outlet. The radius of the core is between 0.2 and 0.4 times the radius of the gas outlet, and the radius of the maximum tangential velocity is 0.4–0.8 times the radius of the gas outlet (Fig. 6).

The application of centrifugal force to drive particles out of the gas stream towards the walls of a cyclone collector, results in a concentrated dust layer swirling slowly down the walls of the cyclone body. The dust is yet to be finally separated from the gas stream. The purpose of the discharge is to retain the dust or liquid in a container and to prevent its re-entrainment into the gas stream at the base of the vortex. The length and dimension ratios of the cyclone body and cone, of course, affect such re-entrainment. Smoothness of the inner walls of the cyclone is essential to prevent small eddy currents, which would bounce the dust layer out into the active zone of the vortex. Recirculation of dust or in-leakage of gas into the dust outlet will be harmful to attempts to discharge the dust without re-entrainment, and conversely, a small purge flow of gas outward from the dust outlet will be helpful.

Applicability

Cyclones for removing solids from gases are generally applicable when particles of over 5 μm in diameter are involved, although some of the multiple-tube parallel units attain 80–85% efficiencies on particles of 3 μm in diameter. In collecting particles of over 100–200 μm in diameter, cyclones may be used, but gravity chambers are usually satisfactory and less subject to abrasion. In special cases where the dust shows a high degree of agglomeration, or where high dust concentrations (over 3700 g/m^3) are involved, cyclones will remove dust having a much smaller particle size. In certain cases efficiencies as high as 98% have been realized on dusts having an ultimate particle size of 0.1 to 2.0 μm because of the predominant effect of agglomeration.

Various modifications can be made to improve the operation of the cyclone separator in special cases. If there is a large proportion of fine material present, a bag filter may be attached to the clean gas outlet. Alternatively, the smaller particles may be removed by means of a spray of water which is injected into the separator. In some cases, the removal of the solid material is facilitated by running a stream of water down the walls and this also reduces the risk of the particles becoming re-entrained in the gas stream. The chief difficulty lies in wetting the particles with the liquid.

Design-Performance

Cyclones in common use can be **classified** into four general types: (1) the common cyclone, having tangential inlet with axial dust collector, (2) tangential inlet with peripheral dust discharge, (3) axial inlet through swirl vanes, with axial dust discharge, and (4) axial inlet through swirl vanes, with peripheral dust discharge. Cyclones can also be classified as **conventional** and **"high efficiency"**. The latter have a smaller body diameter (up to about 23 cm) and a larger body length, hence increasing the inlet velocity, the detention time and the centrifugal force on the particles, in order to achieve greater separating forces (the smaller the diameter, the sharper the changes of direction that the dust particles will have to make in order to remain in the gas stream), but there is no sharp dividing line between the two groups.

Cyclones as a class of equipment provide the lowest collection efficiency, as well as the lowest initial cost for devices in general commercial use to control particulate air pollution sources. The performance of cyclone collectors is primarily dependent on particle size, and system design considers collection efficiency, pressure drop and unit size. Ranges of efficiency to be expected from cyclone collector installations are shown in Table 2. Optimizing a cyclone generally means achieving a required efficiency on a known dust dispersion at minimum pressure loss, cost factors of design excluded.

In general, cyclone **performance increases** with an increase in the following: (1) dust particle size/diameter (one of the primary parameters affecting the performance of a cyclone), or density of the particulate matter, (2) pressure drop, i.e. gas inlet velocity into the cyclone, (3) cyclone body or cone length, (4) number of gas revolutions, (5) ratio of cyclone body diameter to gas outlet diameter, (6) amount of gas entrained in carrier gas, (7) smoothness of inner cyclone wall, and finally (8) reduction of the gas outlet duct diameter will increase both collection efficiency and pressure drop.

Table 2. Efficiency range of cyclones

Particle size range, μm	Efficiency range (wt% collected) Conventional	"High efficiency"
Less than 5	Less than 50	up to 80
5–20	50–80	up to 95
15–40	80–95	up to 99
Greater than 40	95–99	95–99

The performance of the cyclone **decreases** with increases in carrier gas viscosity or density, cyclone diameter, gas outlet diameter, gas inlet duct width and inlet area. Pressure drop across a cyclone usually ranges between 25 to 80 mm water gauge. Pressure drop increases with the square of the inlet velocity. Removal efficiency also increases, but not so rapidly as pressure drop. All devices intended to minimize pressure drop result in decreased dust removal efficiency.

Cyclones are sized using a **separation factor** which is based on the premise that the centrifugal force applied to particles varies as the square of the inlet velocity and inversely as the radius of the cyclone. This relationship is expressed as:

$$S = u^2/r_c g. \tag{3}$$

Though not expressed by a specific equation, collection efficiency generally varies directly with the separation factor.

Cyclone **design** consists of selecting a configuration, then determining the size, grade efficiency, pressure drop and power requirement of each cyclone to be used. These determinations are based on a given gas flow rate, composition, temperature, pressure, and grain loading, together with data on the particle size distribution in the feed. The design will also give a predicted overall efficiency of collection, emission rate, outlet grain loading and particle size distribution. These latter items will provide the basis for design of a secondary collector following the cyclone, if one is to be used. The design method must also be based on a knowledge of the spiral path followed by a particle in the collector. Cyclone dimensions are determined by a selection of the configuration ratios coupled with a selection of the desired gas velocity at the inlet. For high collection efficiency the inlet velocity should be as large as possible, without causing excessive rebounding or re-entrainment of particles. The immediate entrance to a cyclone, is usually rectangular.

The main **design factors affecting efficiency and pressure drop** are the following: (1) body diameter and dimension ratios, (2) cone design, (3) inlet design, (4) dust discharge, (5) gas outlet design, and (6) effect of internal roughness.

The main **operating variables** and its effect on cyclone performance are the following: (1) flow rate, (2) physical properties of the gas, (3) properties of the dust, and (4) dust loading.

The overall collection efficiency (E_0) of an operating cyclone may be determined rather easily by **measuring**: (1) the total cyclone catch collected and weighed over a period of time (kg/min), (2) the total particulate concentration in the inlet stream (kg/m^3), and (3) the gas flow rate entering (m^3/min). Then **E_0 = (total catch/particulate concentration) × flow rate.**

Two or more cyclones of the same configuration and size may be **used in parallel**, the flow being divided equally between them. In this case, each cyclone can be of smaller diameter than if only one was used. The collection grade efficiency for each cyclone is the same, and is greater than if only a larger one was used. A large number of very small cyclones in parallel (called a "**multicyclone**") may be used to increase efficiency for very large flows of gas. However, it may be difficult to maintain an equal flow of gas in each cyclone of the parallel array, especially if they all discharge dust into a common hopper. Two or three cyclones may also be used in series, the emission from the first becoming the feed to the second, and so on. In this case, the size of the cyclones is made successively smaller downstream, in order to improve the collection efficiency of the finer particles which have penetrated upstream. The collected product from each cyclone has a different size distribution. This arrangement may also be used to separate a feed particulate into two or three size fractions.

Cyclones in series may be justified in cases where the dust has the following properties: (1) it has a broad size distribution including a sizable portion of particles finer than 15 μm, in which case a single large diameter cyclone is used to collect the coarse fraction and reduce the loading to a small diameter multiple-tube unit, or (2) it has a high tendency to flocculate in the equipment preceding the cyclones, as well as in the cyclones themselves.

Cyclones operated wet for the collection of dry dust improve efficiency, prevent wall build-up and fouling, and reduce erosion. The efficiency is considerably higher because the dust particles are trapped in a liquid film and are not easily re-entrained. Care must be exercised to direct spray nozzle patterns and flushing streams so that no portion of the cyclone or duct is merely moist, since this condition will lead to caking and plugging. If operated wet, a conventional cyclone will yield efficiencies comparable to that of the high efficiency cyclone.

The operation of wet cyclones presents a number of **advantages** provided that the droplet size is sufficiently large and that re-entrainment of liquid from the cyclone is prevented. The liquid droplets collect on the cyclone wall and form a continuous film, which is less subject to re-entrainment in the gas stream than is a dispersion of dry dust. In addition, liquid can be drained from any point in the cyclone bottom, thus permitting a location away from the axis of the vortex core. Under such conditions wet cyclones operate with a higher efficiency and less erosion and plugging difficulties than do dry cyclones. The major **disadvantage** of wet cyclone operation is caused by the corrosion problem. If corrosive dusts or gases are handled, the presence of water usually makes that problem much worse. Other fluids may be advantageous but are more expensive. The other disadvantage is the additional cost of water, and the cost of recirculating or disposing of the containment water.

Reliability-Maintenance

Cyclones are usually used as low-cost collector devices in applications where only a small portion of the dust to be collected is below 5 μm. In groups, or as multi-cyclones, these devices are often used for pre-collection purposes, in order to lower the dust load for the subsequently located more efficient and expensive devices like cloth filters or electrostatic precipitators, thus enabling smaller units to be specified. A specific area of application for cyclones is the cleaning of very hot gases that have high dust loads.

Cyclone collectors offer one of the least expensive means of dust collection from both an operating and an investment viewpoint. Cyclones have been employed to remove solids and liquids from gases and solids from liquids; they can be successfully used in **applications** where: (1) dust must be collected in dry form, (2) temperatures are high (up to 1000 °C), (3) dust concentrations are high, (4) gas is under high pressure (up to 50 atm), and (5) dust or gas becomes corrosive when wet.

Inversely, cyclones should **not be specified** for conditions where: (1) dust will adhere to cyclone and dust hopper walls because of its surface properties, or because temperatures drop below the gas dew point; the dust then wets, cakes and plugs the outlet (this will not be a problem if there is good control of moisture in the feed to the cyclone), (2) dust is very fine (below 1–5 μm, depending on the dust density and the gas flowrate), and (3) small collectors can also plug up if the dust loading in the gas is very high and the cyclones remove particles faster than they can fall out of the bottom.

Since cyclone collectors generally operate dry, there are seldom any corrosion problems. Breakdowns are generally due to high gas temperatures or abrasiveness of the dust particles. The most critical factor is the avoidance of moisture condensation on the internal surfaces of the cyclone, which invariably start corrosion and initiate dust agglomeration and build-up. In general, cyclones are very reliable and have negligible maintenance costs.

Cyclones are usually **built** of ordinary carbon steel, but any type of metal or even ceramic material may be used, if necessary, in order to withstand high temperatures, abrasive particles, or corrosive atmospheres. It is important that the interior surface must be smooth. There are no moving parts, so operation is usually simple and relatively free of maintenance. A cyclone will operate equally well on the suction or pressure side of a fan if the dust receiver is air-tight.

Problems may be encountered in cyclone application: (1) when using multiple cyclones, (2) with fouling and caking, and (3) with erosion.

Multiple cyclones are designed with the assumption that each cyclone will handle its prorated share of gas and dust and that the overall efficiency will be the same as in an individual cyclone. Since cyclone efficiency increases with flow and dust loading, it is possible with poor distribution to obtain much poorer efficiency in a cyclone group from those units which are lightly loaded. Wall caking, unequal flow, and partial plugging of the cone or cyclone inlets can all cause unequal apex pressures, which can cause gas flow from cyclones with higher pressures into the dust hopper and up through the apex outlet of lower pressure cyclones. Such short-circuiting can result in heavy dust re-entrainment and decreased efficiency.

Erosion in cyclones is caused by the impingement and rubbing of dust particles on the cyclone wall. Erosion is worst with high dust loadings, high inlet velocities, and large or hard dust particles. Any defect in cyclone design or operation which tends to concentrate dust moving at high velocity will accelerate erosion. It is possible to design a cyclone to reduce erosion by increasing the diameter of the cyclone body without increasing the diameter of the gas outlet. This results in reduced velocity at the body wall without reducing maximum velocities and separating force of the vortex. It also results in increased pressure drop. Consequently, at high loadings of abrasive dust, large-diameter cyclones are required to control erosion. Erosion-resistant linings consisting of a trowelled or cast refractory may be used.

Fouling of cyclones results in decreased efficiency, increased erosion, and increased pressure drop. Fouling is generally found to occur either by plugging of the dust outlet or by the build-up of materials on the cyclone wall. In general, the finer and softer the dust, the greater the tendency to cake on the wall. Chemical and physical properties will also affect this behaviour. Condensation of moisture on the walls of the cyclone will also contribute to the accumulation of material. In many cases, build-up of sticky materials on the walls can be minimized by keeping the inlet velocity above 15 m/s. Smoothness of the cyclone walls is also important.

Filtration Processes

Description

Filtration is one of the oldest and most widely accepted methods used for particle removal from dusty gas streams and is considered among the most reliable, efficient and economical methods by which particulate matter may be removed from gases. Fabric filters are capable of maintaining mass collection efficiencies of greater than 99%, down to a particle size approaching 0.3 μm in most applications. This efficiency is largely insensitive to the physical characteristics of the gas and dust, and, depending on fabric cleaning method, to the inlet dust loading. However, it is not suitable for applications requiring extremely high efficiency capture of very small particles at moderate cost.

Since the pores in the filter cloth are generally much larger than the size of the dusts being collected, the cloth itself cannot trap fine particles. Therefore, when the dust-laden gases first pass through the filter, the efficiency of separation will be low, until enough particles have been removed to build-up what corresponds to a "precoat" in the fabric pores. As this dust layer or "cake" builds up, particle penetration drops to a very low level, i.e. it is this layer of dust that actually removes fine particulates from the gas stream (Fig. 7). Build-up of this layer may take place in as little as a few seconds. This cake builds up on the filter fabric as a consequence of individual particles being trapped on yarns or fibres, thus enabling bridging across pores bounded by the filters. As pores are bridged, cake builds up as a permeable mass on the fabric. Because the pore spectrum for the cake is usually of a smaller average diameter than that for the underlying fabric,

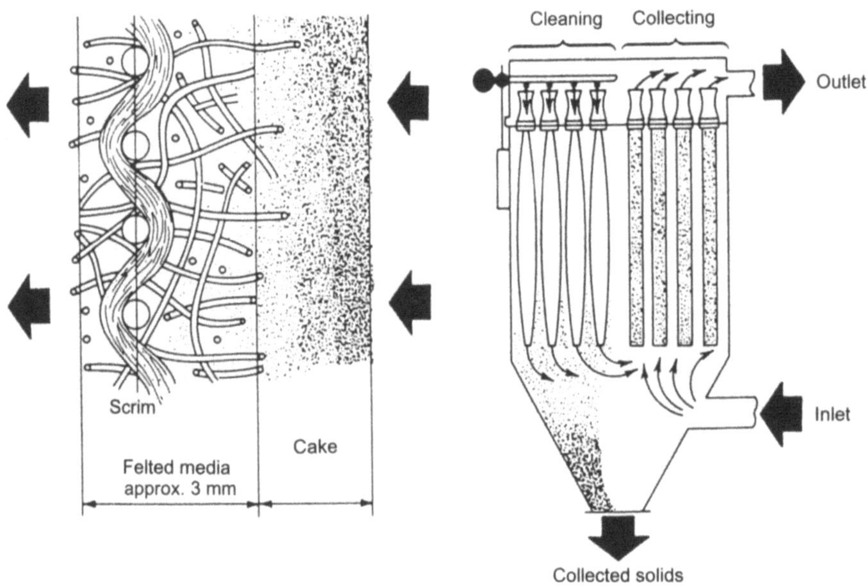

Fig. 7. a Filter media action on dust removal. **b** Filter type collector with pulse jet – on stream collecting and cleaning dust

penetration through the cake is much less than that through the fabric before the cake is formed.

When fine, uncharged particulate matter suspended in a gas stream approaches a fabric, there are several **mechanisms** by which it may be captured, like interception, inertial impaction, diffusion (Brownian motion), electrical precipitation, thermal precipitation, gravitational settling and sieving (Fig. 3). The most important mechanisms which occur in a single particle/single fibre collection process are probably inertial impaction, diffusion, and direct interception. Inertia is not considered significant in collecting particles below 1 μm in diameter, but it becomes increasingly important as size increases. Collection by direct interception is paramount for 1 μm down to about 0.2 μm particle diameters. Gravity influences the collection of relatively large particles only. Very small particles – less than about 0.2 μm in diameter – have a random movement and they are subject to enhanced chances for contact with a collecting surface. Thermal forces are usually minor, electrical forces can range from minor to major and sieving does not apply.

The **electrostatic properties** of both the dust and the filtration media influence filtering and cleaning performance. Electrostatic attraction will draw particles from the gas stream to the fibres if the two are oppositely charged. Even if only one of the materials, i.e. the particles or the fibres, is charged, an induced charge will be created on the other, producing a polarization force that results in attraction and particle deposition. Electrostatic charge also influences particle agglomeration which increases the likehood of particle entrapment. Charges may be developed through frictional effects during the filtering operation, even if not initially present. The charge intensity developed on either the fabric or the dust, or both, depends upon the processing conditions, as well as on the nature of the materials themselves.

The major differences in various fabric filtration **equipment** lie mainly in the areas of (1) **filtration media** and (2) **cleaning methods**. These differences normally result in differences in the operational mode.

(1) **Filtration media**. Historically, natural fibres, wool and cotton, were the first employed in fabric filter bag-houses. Today, although the use of synthetics now dominates because of their greater range of temperature and pH adaptability, both natural and synthetic fibres are employed in fabric filters, such as graphitized fibre, polyethylene, cotton and nylon. The latter two, although the least expensive, have poor resistance to acid attack. Orlon, Dacron and polyethylene show higher resistance when attacked by acids and may be suitable at low temperatures. Fibre-glass and Teflon offer higher temperature acid resistance, although Teflon is more costly. With the exception of wool and fibre-glass, most of the aforementioned materials provide good resistance when attacked by alkaline substances. Gas stream characteristics, such as temperature, acidity, alkalinity, and particulate matter properties (e.g. abrasiveness and hydroscopicity), also determine the fabric type to be used. Each filter material must be selected for a particular use from the viewpoint of both economics and collection characteristics. A non-woven cloth is often the more favourable filter medium because the pressure loss after cleaning is lower and remains essentially constant if a string cleaning method, for example, a pulsating reverse air jet, is employed. In many instances, several fabric types will be appropriate, and a final selection will be chosen only when the cleaning method and the desired air-to-cloth ratio are considered.

When the temperature of the gas stream is low, cotton (the least expensive fibre) or polypropylene is usually applied; when the temperatures are higher, teflon and silicone- or graphite-coated glass fibre cloth is commonly employed (in applications up to 250 °C). Metal screens have been used up to 550 °C, but only in small installations. Acceptable cloth life is generally considered to be one year or more. Both the average operating temperature and the excursion limits must be considered in fibre selection. Exposure of the filter medium to either temperatures or pH conditions outside its operable range can result in the loss of bag function in a matter of hours or less. Upset conditions must also be accommodated in the system design to keep the bag-house operating. For example, if the gas stream to be cleaned is above 330 °C (today's upper limit of commercially available bags), then cooling of the gas stream must be employed before applying the bag-house. Once the filter has been selected, the next step is to determine what fabric design and weight (density and thickness) should be employed; for example, should a felt or woven filter medium be employed. The performance of any fabric is greatly influenced by thread density, fibre composition, and nap, i.e. the hairy or downy texture of the cloth surface.

(2) **Cleaning methods.** As dust accumulates on the filtering elements, the pressure drop across the bag compartment increases and the gas velocity decreases concomitantly to a prescribed lower limit, until the filtering operation must be stopped and the filter cleaned by a dust-dislodging operation and the cycle is then repeated. Usually a timer is used to control the cleaning cycle, or the pressure drop is monitored, so that cleaning occurs when some maximum desirable values is reached.

The cleaning must be performed so as to accomplish the following: (a) to keep the pressure drop, and thus operating costs, within reasonable limits, (b) to clean bags as gently and/or infrequently as possible in order to minimize bag wear and to maximize efficiency, and (c) to leave a sufficient dust layer on the bags to maintain filter efficiency and to keep the instantaneous Air-to-Cloth (A/C) ratio immediately after cleaning from reaching excessive levels, if woven fabric with no backing is used.

There are three common cleaning methods of which the oldest is the shaking of the filter media to dislodge the dust. The other two consist of blowing the dust away from the filter media, i.e. back-flushing with low pressure air, or by impacting the dust with a jolt of compressed air (Fig. 7). In addition to these methods, others have been suggested which, either for economic or technical reasons, are not widely used today. Combining two cleaning methods is sometimes used, e.g., the combination of reverse air and shake in difficult cleaning applications. Generally, large-scale bag filter installations are cleaned by shaking or by countercurrent air scavenging. A cleaning cycle usually requires only a very few minutes.

Fabric filters can be broadly **classified** into one of two kinds – fabric (or cloth) filters and in depth (or bed) filters. The former is represented by various fabric bag arrangements, while the latter is most frequently encountered as a fibrous array, a paper-like mat, and occasionally, as a deep packed bed.

Gas filtration can also be accomplished by granular bed filters, passing the gas through one or more layers of granules that are large in comparison to the particles to be separated from the gas stream; these are periodically cleaned of accumulated particles. They are prepared using dry aggregates (crushed stone or brick, wire screens, or fibres of many types), arranged individually or in combination and can operate statically (cleaned by replacing, vibrating, agitating with reverse flow, or washing the filter media) or dynamically (fluidized or continuously moved downward for cleaning). Applications of this type are restricted to special conditions, such as high temperatures (above the 250–300 °C range) or corrosive gas treatment. They tend to be bulky and heavy, as well as requiring careful design and operation in order to maintain high efficiencies for small particles.

The separation of dust in a bed of granular solids may be due to: (1) gravity settling, (2) diffusional deposition, (3) inertial deposition, (4) flow-line interception, and (5) electrostatic precipitation. The last factor is important when the bed, made up of particles with high electrical resistivity, is agitated to develop a high surface charge. With fine packed-beds operated at low gas velocities, gravity settling and diffusional deposition will predominate, and collection efficiency would be expected to decrease, as the gas velocity increased. With coarse-packed beds operating at higher velocities, but still below fluidizing velocities, separation by inertial deposition and interception are the controlling mechanisms and increased velocities would be expected to increase collection efficiency, provided that the gas velocity is not so high as to re-entrain collected material.

A wetted packing provides an impingement surface that prevent reentrainment. The liquid also provides a means for washing off dust and conveying it away as a slurry or solution. The packing may be fixed or it may be a fluidized bed of low density spheres. The advantages are low cost, simplicity, corrosion resistance, and no moving parts.

Applicability

Bag filters are normally used where high efficiency is required and where process conditions enable the filter to operate without harm to itself or to the process. Fabric filters are extensively employed to control harmful emissions in operations involving abrasives, irritating chemical dusts and the exhausts from electric furnaces, oil-fired boilers, and oxygen-fed converters for steel making. Although liquids in the system have often been cited as detrimental to filter operation, they can be tolerated with proper design. Many field problems can be circumvented by altering operating procedures, e.g., cooling hot gases to a temperature suitable for the installed filter medium.

Applicability mainly **depends** upon the emission limitations, temperatures, moisture content and the bag replacement schedule of the fabric filter. Temperature of the gas imposes limitations upon usable materials of construction for bags, as well as affecting system size with regard to volumetric flow. Temperatures greater than 250–300 °C can also destroy the fabric or shorten its life to an uneconomical degree. Gas or particle characteristics that prevent proper cleaning of the fabric, e.g., sticky particles not dislodged, influences negatively the operation of the bag-filters. The chemical and physical properties of the dust can also be crucial to the successful operation of a bag-house. For example, if the dust is sticky or hydroscopic, it can lead to inadequate cleaning that eventually causes blinding of the bags. Moisture or other condensibles can affect performance of the fabric filter and may render it inoperative. Significant moisture content will definitely eliminate fabric filters from consideration. The air stream flow rate can also have considerable effect on the performance of the fabric filter, as wide fluctuations in the flow rate will cause a shift in the collection efficiency. However, if applicable, fabric filters provide high efficiencies with the retention of very fine particulates, which are collected in the dry form at relatively low pressure drops. Small particles can be collected, even in the Angstrom range, because of high diffusibilities.

There are three main types of bag filters. The simplest consists of a number of elements assembled together in a "bag-house". This is the cheapest type of unit and operates with a velocity of about 0.01 m/s across the bag surface. A more sophisticated and robust version incorporates some form of automatic bag-shaking mechanism, which may be operated by mechanical, vibratory or airpulsed methods. A heavier fabric allows higher face velocities (up to 0.02 m/s) to be used and permits operation under more difficult conditions than the simpler bag-house type can handle. The third type of bag filter is the reverse-jet filter, with face velocities of about 0.05 m/s and with the capability of dealing with high dust concentrations at high efficiencies; this type of filter can deal with difficult mixtures in an economic and compact unit. Use of a blow ring enables the cake to be dislodged in a cleaning cycle which takes only a few seconds.

Fibre mat filters are generally not cleaned; they are ordinarily used where particle concentrations are low, so that reasonable service life can be attained before changing the filter. Fabric filters are generally utilized with gas or airstreams having a dust loading of the order of 1 g/m^3; fibrous packings, paper filters and packed beds are applied when the particulate concentration is several orders of magnitude less, perhaps 1 mg/m^3. The typical range for woven cloth of the velocity at which the gases pass through the filter must be kept low, typically 0.005 to 0.04 m/s, in order to avoid compaction of the filtered solids on the filter medium and consequently high pressure drops, or to avoid local breakdown of the filter bed, which would allow large particles to pass the filter. Filters may also be **categorized** as to the type of service and frequency of bag cleaning (intermittent or continuous). Bag filters are also classified by the type of cleaning method and the shape of the filter bag.

Custom bag-houses are designed for continuous operation and normally use reverse air cleaning. Bag life is estimated (average) at 1.5 years. Power usage will depend on dust loading and cleaning cycle. Power requirements for shaker and blower motors are given as approximately 4 kW/1000 m^2 of cloth area. Fabric filter systems typically **consist** of cloth bags or envelopes, suspended or mounted in such a way that the collected particles fall into a hopper/bin for disposal, when dislodged from the fabric. The tubular bags range from 120 to 400 mm in diameter and may be up to 12 m long. The number of bags in a single installation of collectors may range from 12 to 70000. It is generally more economical to be conservative in specifying cloth area. Since pressure drop for a given service varies as the square of the velocity through the cloth, greater cloth area results in considerable reduction in shaking frequency and in increased bag life; incremental cloth costs are also relatively small, particularly in small installations.

Electrostatic precipitators are competitive with fabric filters in those cases where particle resistivity ranges from about 10^{11} to 10^{12} ohm-cm, the required collection efficiency is less than 99.5%, and the micrometer size particle population is relatively small. Scrubbers are competitive when small equipment size for large gas volumes is required, a liquid and sludge effluent can be tolerated, and where there is not a requirement for high capture efficiency for small particles.

Design-Performance

In order to design and/or select a bag-house, a thorough definition of the emission problem is required. Absolutely essential information, without which it is impossible to design or select a bag-house, includes volume flow (constant or variable and if variable, minimum and maximum values), temperature (average or normal operating values, but also peaks and durations) and specific application (large or sticky particulate matter, potentially flammable or explosive gases or particulate matter). The volume variations influence the gas-to-cloth ratio selection, while temperature variations impact on both filter media selection and system design parameters. With these data, the size of the unit, the selection of the filter medium and probably the cleaning method can be selected.

The **air-to-cloth (A/C) ratio, or filtration velocity** (expressed as $m^3/min/m^2$), is a traditional fabric design parameter defined as the actual volumetric flow rate divided by the total active, or net, fabric area. The A/C ratio is an important indicator of the amount of air that can be filtered in a given time, when considering the dust to be collected, cleaning method and fabric to be used, and the characteristics of the gas stream to be filtered for an individual situation. It has been demonstrated that there is a strong correlation between outlet concentration and face or filtration velocity for a given loading and type of fabric. The A/C ratio suggests the amount of air filtered over a period of time without specific regard to the cross-sectional area occupied by the fabric filter cloth. As a result, theoretical or empirical relationships are seldom used to size filtration units. Sizing is based on demonstrated experience of manufacturers for various applications and is recommended for a specific dust or fume and cleaning method. Although a relationship between pressure drop and air-to-cloth ratio does exist, as the pressure drop generally varies with the square of the filtration velocity (at constant dust loadings), it is difficult to relate directly pressure drop to the collection efficiency of the fabric filter. The filter pressure drop is also influenced by the frequency and completeness of bag cleaning.

Dust holding capacity is defined as the maximum dust load per unit area of the filter medium for a certain limit of pressure drop. Higher holding capacity is desirable in practice because of the longer life of the filter medium. Its value usually ranges between 0.3 and 2 kg of dust per square meter of filter surface. Ordinary mechanical filters may be shaken every 0.25 to 8 h, depending on the service. A manometer connected across the filter is useful in determining when the filter should be shaken. Fully automatic filters may be shaken every 2 min, but bag maintenance will be greatly reduced if the time between shakings can be increased to 15 or 20 min without developing excessive pressure drop. It is essential that the gas flow through the filter be stopped when shaking in order to permit the dust to fall off.

There are three major **performance criteria** for a filter, which are related to endurance and dust holding capacity.

1. The **lifetime of a filter** is very important from the economic standpoint, because the cost of the filtering medium is a major portion of the initial expense, as well as of long-term operating costs. Factors upon which fabric life depend are the operating temperature, the duration and manner of cleaning or removing the dust load from the cloth, the characteristics of the particulate matter, the nature of the gas, and the care with which the installation is designed for the particular application.
2. **Pressure loss** is usually expressed in terms of mm of water column and is directly proportional to the required fan or blower energy. The pressure loss is a major index of the operating cost of a filtration system. The permeability of the dust cake on the cloth varies with the operating conditions of the bag filter in a way that significantly affects the pressure. Common values of pressure drop for conventional bag filters under normal operations are between 100–200 mm water.
3. **Collection efficiency** is probably the single most important factor in the performance of a filter; it is usually evaluated in terms of the mass percentage of dust retained, but sometimes an optical density measure is employed. Once the particle bed has been formed on the filters, separation efficiency will rise to values near 99%. Efficiencies more than 99.9% are seldom encountered. Failure to achieve high collection efficiency is almost always due to excessive gas flow rate which produces pinholes within the deposited particle bed. Instantaneous efficiency values are usually higher than cumulative collection efficiencies for an operating cycle, because the collection efficiency generally increases with the dust deposit. Therefore, it is difficult to predict filter pressure loss and collection efficiency without making preliminary tests, but it has been shown that the less the volume fraction of fibres, the filter thickness, the superficial filtering velocity, and the distance between the two layers, the better the collection.

Pressure bag-houses are **designed** for 3 kPa gauge pressure and suction fabric filters are designed for a negative pressure of 5 kPa gauge. The fabric filter system may operate under negative or positive pressure, depending on the fan placement, which involves consideration of the emission's characteristics. The negative pressure system tends to have the most applications. When the emission gases are pushed through the system by a fan located upstream, the system unit is a positive pressure bag-house. This approach eliminates the need for ductwork, which reduces the need for space and other materials but increases difficulties for monitoring. The fan is subjected to the full dust load and the associated wear from abrasive actions. A negative pressure or suction bag-house is one where the fan is located on the downstream, clean side of the fabric filter.

Reliability

The production process must be taken into account during the problem definition stage. The fibrous filter is one of the more commonly used air filters, being particularly suitable for low dust concentrations. Maintenance is easy because of its simple construction. Many times, optimization of a fabric filter's collection efficiency occurs in the field after construction.

The reliability of a fabric filter system in meeting performance requirements depends mainly on: (1) the frequency of bag breakage, (2) the time taken to isolate the broken bags, which can be reduced if detectors are placed on the outlet of each compartment, indicating the opacity of the gas stream coming out from compartment, (3) the leakage of bypass dampers, (4) the high air-to-cloth ratio which can blind the system and reduce performance, while too low values can cause dust drop-out in the dusts, (5) the temperature – too high can destroy the bags, too low might cause dewpoint excursions, (6) the dust load – too high may exceed the unit's capacity to convey the dust from the bag-house, too low may cause extended emissions after each cleaning cycle, and (7) the particle size – too fine can cause blinding of the bags or excessive emissions at a given air-to-cloth value.

Troubleshooting a bag-house should include the following steps: (1) review of the original design specifications, (2) thorough inspection of the system and bag-house, (3) comparison of the original design inputs with those existing at the time difficulties arose, (4) identification of the problem symptoms, (5) listing of possible causes of each symptom, (6) final definition of the problem, (7) determination of alternative solutions and their associated cost, and (8) selection of the most suitable solution and means of implementation.

As with other collectors, bag filters should be set up on a routine maintenance cycle. At least once a month, bag tension should be checked and bags should be inspected visually for failures. Faulty bags should be replaced immediately because adjacent bags may be damaged by the bag that has failed. A leak is easy to find since it will often discolour adjacent bags or structures and cause build-up of dust at the bottom of the casing.

Electrostatic Precipitation Process

Description

Electrostatic precipitators use electrical rather than mechanical forces for the removal of suspended particulates from a gas stream and regardless of type are based on the same three underlying principles.

1. **A method of providing an electrical charge** on a particle by establishing an electric field and involving the addition of electrons to, or removal of electrons from, the material, or the attachment of ionized gas molecules. The electrical field is provided by the application of a high direct current voltage to a dual

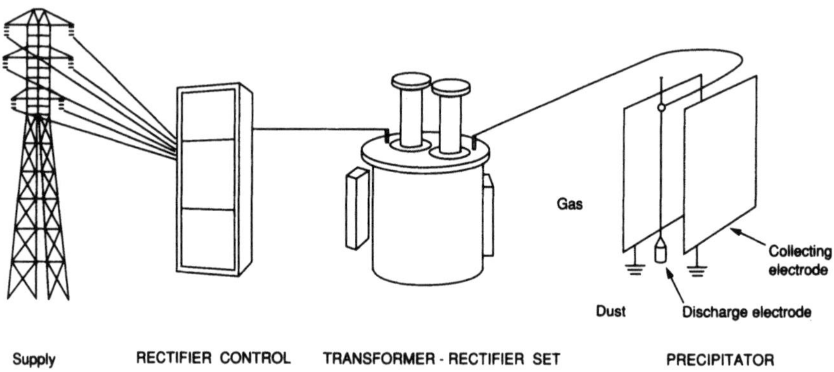

Fig. 8. Overall schematic of a wire and plate electrostatic precipitator

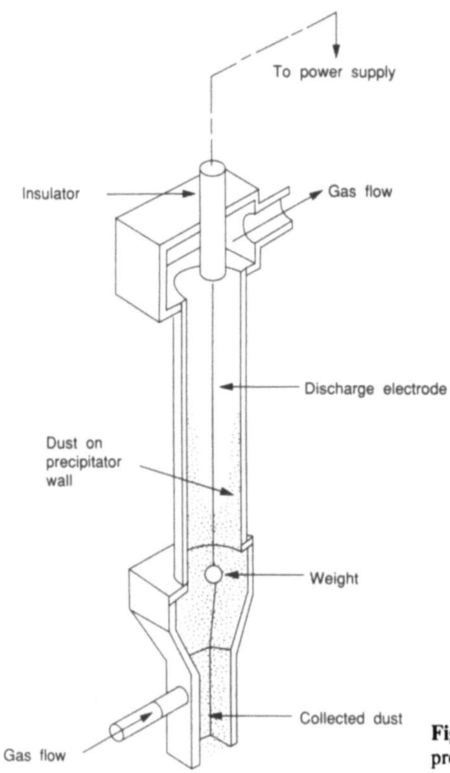

Fig. 9. Schematic of a wire and pipe electrostatic precipitator

electrode system: a discharge electrode, usually a small diameter wire, which is maintained at high voltage (30–75 kV, peak 70–100 kV, having 100–2000 mA current capacity), and a grounded electrode plate or pipe collecting the dust (Figs. 8 and 9).

Two **mechanisms of particle charging** are of importance in this area: charging mainly by **ion bombardment** (frequently referred to as **field charging**), and by **ion diffusion**, resulting from random thermal motion, which causes the molecules to diffuse through a gas and contact the particles. The rate of contact between a particle and a gas molecule is related to the density of ions in the vicinity of the particles and the root mean square velocity of the ions, the latter being related to the temperature and gas properties. As the particle becomes charged, it will repel additional gas ions and the charging rate decreases. The field charging process is predominant for particles larger than about 0.5 μm in diameter, the diffusion process for particles smaller than about 0.2 μm and both are important for particles in the intermediate range between 0.2 and 0.5 μm. The majority of industrial dusts contain size fractions in the range where both diffusion and field charging are significant.

2. **The collection of particles** on an electrode of opposite polarity.
3. **The removal of the collected particles** from the precipitator.

Most **industrial particles** are charged during the process of formation by such means as flame ionization, friction, and grinding, but usually only to a low or moderate degree. These natural charges are far too low for effective precipitation. It is therefore essential to provide a high degree of charging as a part of the collection process. When a precipitator is energized at low voltage with dirty gas entering, no collection will occur and no current will flow in the secondary circuit. No change occurs with increased voltage until gas ionization starts around the discharge electrode. This is known as **corona formation**. When a corona occurs, current starts to flow in the secondary circuit and some dust particles start to precipitate. The high-voltage DC corona is the most effective means for particle charging and is universally used for electrostatic precipitation.

An electrical **corona** is the discharge associated with gaseous breakdown near a highly electrically stressed electrode. A manifestation of the corona is a luminous glow which takes on a variety of shapes such as bright spots, brushes, streamers, or uniform glows and is accompanied by a rapid increase in current as voltage is increased beyond the point of corona onset. The appearance of the glow differs with the polarity of the corona electrode. The **corona discharge** as used in electrostatic precipitation occurs between small diameter wires and outer pipes/cylinders, concentric with the corona electrode, or between wires and plates located parallel to the plane of the corona wires, or other configuration, which gives the small radius of curvature required to produce a highly non-uniform electric field. The corona generated in the high field region provides the charges necessary for electrical collection. Electrons released from or near the wire surface move in the intense electric field of the inter-electrode region and produce enormous numbers of additional electrons by impact ionization with gas molecules.

The electrons thus produced are attracted to the outer positive polarity electrode and attach to gas molecules in the passive region between the glow and the pipe surface by electrical, mechanical and molecular forces. Electron attachment occurs because free electrons in the relatively low electric field in the passive region attach to the gaseous molecules instead of producing further ionization. This passive zone typically has an ion density of 10^7 to 10^8 ions/cm^3. Electron impact ionization in the glow region produces a positive ion for every electron generated. These positive ions move to the wire where they release new free electrons by impact with the wire surface or by photoelectric effect in the gas or at the wire surface. This regeneration action of the positive ions in releasing new electrons at the wire ensures that the corona discharge will be self-maintaining. The electric field in the inter-electrode region, acting in conjunction with the charged particle, produces a coulomb force which propels the dust particles to the grounded or collection electrode.

The factors influencing the potential at which the corona is initiated and the voltage-current relationship of the corona discharge are of most practical significance. The current-voltage relationships in a precipitator depend upon the geometry of the electrodes, the composition and conditions of the gas, the thickness and properties of the collected dust layer, and the concentration and particle size of the suspended dust. The voltage-current relationship can be altered by a number of factors. One of the principal sources of variation is the composition of the gas, which determines the molecular species of the charge carrier. The process of electron attachment to form negative ions is different for different gases.

As the potential is increased, current flow increases until, with increasing potential, a spark finally jumps from the discharge wire to the collecting surface. This is a **sparkover** which, if permitted to occur excessively, can result in the destruction of the internal parts of the precipitator. Precipitator efficiency improves with increasing potential and current flow. Maximum efficiency is achieved at a potential just short of heavy sparking. The **flow of corona current** through the collected dust layer causes a voltage drop which is proportional to the current density, dust resistivity and thickness. Electrical breakdown of the interstitial gases in the dust layer occurs when the electric field exceeds the breakdown strength of the gases. For most gases encountered in industrial precipitators, breakdown of the precipitated layer occurs when the electric field exceeds about 20 kV/cm. The exact value of the breakdown strength depends on the particle size and extent of packing of the dust and on the gas composition.

Corona discharge is accompanied by a relatively small flow of electric current, typically 0.1 to 0.5 mA/m^2 of collecting electrode area. Sparking usually involves a considerably larger flow of current which cannot be tolerated except for occasional periods of a fraction of a second duration and then only when suitable electrical controls are provided to limit the current. However, when suitable controls are provided, precipitators have been operated continuously with a small amount of sparking to ensure corona. Besides the disruptive effects on the electrical equipment and electrodes, sparking will result in low collection efficiency because of reduction in applied voltage, re-dispersion of collected dust and current channelling.

The magnitude of the **particle charge** depends primarily upon the particle size and the magnitude of the electric field modified by the dielectric constant of the material being collected. The time required for a particle to reach its saturation charge varies with the ion density in the region where charging takes place. For normal conditions where relatively high currents can be maintained, charging times are short, often a few milliseconds. However, if currents are limited because of high dust resistivity or other factors, particle charging times can be relatively long, such that the material to be collected can travel several metres through the precipitator before approaching its saturation charge. **Resistivity** considerations apply only to dry electrostatic precipitators. Resistivity is a measure of a particle's ease in conducting electricity, and higher values signify less ability of the particle to transfer the charge. Another factor influencing particle charging time is the variation in the electric field with time. Despite the implication of the term electrostatic, the input voltage applied to a precipitator is 460 V (three phase), having an alternating frequency of 50–60 Hertz and it is rectified. Consequently, the voltage waveform is a time-varying function. The maximum rated current output is usually in the range of 250 to 1500 mA.

Temperature and pressure of the gas modify both corona starting potential and the voltage current relationship. One effect of pressure and temperature is to change the gas density and hence the electron mean free path. This, in turn, alters the electric field required to accelerate an electron to the velocity required for ionization. The effect of this factor is that, at lower gas densities, electrons travel further before attachment; thus, the higher mobility of the electron appears as an increased velocity of the current carriers. A second effect is to alter the voltage-current curves by changing the effective mobility of the charge carriers. Effective ion mobilities increase with: (1) decreasing gas density at constant temperature and field strength, (2) increasing temperature at constant density and field strength, and (3) increasing field strength at constant temperature and gas density.

Functionally, electrostatic precipitators can be classified as single-stage or two-stage, depending on whether the collecting electric field is an extension of the corona field or is separated from it. The first type is known as a **single-stage or Cottrell precipitator**. The second type is called a **two-stage arrangement**, because the charging and collecting are carried out in separate stages. In this case, a charging section consisting of conventional wire and plate electrodes is followed by parallel plates alternately connected to a high voltage source and to the ground, to provide the field for particle collection.

In the **single-stage wire/plate** electrostatic precipitator the discharge electrodes may be round wire, square twisted rods, ribbons, etc. The most common collection electrode is a smooth plate with vertical interlocking baffles, producing near-zero velocity conditions as the gas flows in a direction normal to them. Other special plate electrode configurations are rod curtains, zigzag plates, and various hollow electrodes. The high-voltage discharge electrodes are suspended vertically between each pair of collection electrodes. They are carefully centred between the collection electrodes to ensure proper corona gaps (usually 0.033 to 3.3 mA/m of wire). The choice of construction material for the electrodes is wide, mainly depending upon the degree of corrosive service.

Applicability

The increasingly stringent air pollution control legislation of the past decade has led to expanded fields of application of electrostatic precipitators and required much higher performance levels. Economic recovery of valuable materials has also played an important role. In Table 3 the main advantages and disadvantages of electrostatic precipitators are presented.

The **principal uses** of precipitators today are in gas-cleaning applications in which high collection efficiencies of small particles are required for processes

Table 3. Advantages and disadvantages of electrostatic precipitation

Advantages

1. Highly effective collection with efficiencies exceeding 99% in some cases
2. Low power requirements and associated low operating power costs
3. Capable of removing very small particles, even those not removable by other treatment schemes
4. Dry dust collection often used, which may be useful for product and byproduct recovery
5. Pressure drops are small
6. Temperature changes are small with dry systems; most impacted in spray-wet systems
7. Low maintenance requirements, due to few or no moving parts
8. System tolerant to high temperatures, with no special provisions below 535 °C.
9. Collection efficiency and system capacity can easily be expanded with larger or additional units
10. Some pollutants, such as acid and tar mists, are effectively collected by other schemes
11. Particulates are recovered in agglomerated form, rendering them more easily collectable in case of re-entrainment

Disadvantages

1. Capital investment is high
2. Space requirements are relatively large
3. Very low or high resistivity particles are difficult to remove
4. Relatively constant operating conditions are needed for efficient collection
5. Gaseous wastes are not affected
6. Safety is a concern due to high voltage
7. Mechanical, electrical or process problems can cause poor on-stream time and reliability
8. Wet systems produce sludges that may require dewatering before disposal

that emit large gas volumes. Main applications of electrostatic precipitators are in the following industries: electric power (fly ash collection from pulverized coal-fired furnaces), nonferrous smelters of copper, lead and zinc (particle collection from furnace operations), primary aluminium (cleaning of pot-line gases and anode furnace emissions), iron and steel (from furnaces, cupolas and sintering plants), chemical (sulphuric or phosphoric acids, etc.), cement production (rotary kilns and clinker coolers), paper mills (black liquor recovery furnaces), municipal incinerators (cleaning stack emissions), glass manufacture (recovery of emissions from furnaces). The greatest application for electrostatic precipitators is in the electric power industry for the elimination of flue gas particulate emissions from coal combustion.

Plate precipitators are utilized for collecting fly ash and other comparatively coarse and dry particulates. They are preferred for high volume gas application because of their lower cost. Both sides of the plate serve as a precipitating surface. The pipe type (Fig. 9) is used mainly for removal of liquid or sludge particles and volatilized fumes. For maximum efficiency, the precipitator should be operated at the highest voltage possible, but without excessive sparkover or arcing. Precipitator voltage must be varied to match changing conditions and dust properties. The most common, the plate type, is primarily for removal of dry dust.

The **single-stage wet wire/pipe** precipitator design is better suited for wet collection applications. This type of precipitator is build in a cylindrical shell. The collection electrodes consist of nested pipes, while the discharge electrodes are suspended axially in the collection electrode pipes. Water flows over carefully levelled weirs at the tops of the pipes to form a water film on the inner walls of the pipes. The charged particles are collected in the water film, where their charge is neutralized, and drained off from the bottom of the precipitator with the water. Because wet collection is involved, materials of construction are usually corrosion resistant.

Precipitators are generally purchased on the basis of a specified collection efficiency, exit dust loading, stack plume opacity, or all three. Specifications generally include the gas volume to be treated, gas composition, temperature and information relative to the dust properties or type of fuel burned. To meet these conditions, precipitators must be sized to provide sufficient collecting surface area. Precipitators have historically been sized by analogy with installations of a similar type.

The combination of high-collection efficiency, moderate energy use, ability to treat large gas flows at high temperatures, and to handle corrosive atmospheres and particles, accounts for the wide use and varied applications of electrostatic precipitation. Despite the many fundamental and practical advances made in the process since its origin in the early part of the century, wide areas for improvement remain in dealing with high-resistivity particles, reducing the size and cost of precipitator installations, reducing energy consumption, and other similar aspects. Another basic **advantage** of the precipitator is that its force is applied only to the particles to be collected, permitting large volumes of gas to be handled with almost no pressure drop and relatively little power requirement.

Problems usually encountered during the operation of **wet precipitators** include the enhancement of corrosion problems due to the water, which may require the use of expensive alloys for construction. The water use also creates a subsequent water treatment and waste handling problem, which at times can equal the cost and complexity of the precipitator installation itself. Spray rate and distribution can be critical in the precipitator and when not properly applied can require reduce voltage operation. Recirculated spray water can become supersaturated with low solubility compounds that can build up inside the precipitator. Suspended solids in the water can erode or plug spray nozzles. Wet precipitators have found their greatest use to date for treating mixtures of gaseous and submicron particulates, such as fibre-glass fume control, coke oven and metallurgical fumes and phosphate fertilizer emissions.

Design-Performance

The basic design problem for precipitators is determination of the principal parameters for: (1) precipitator sizing, which includes specific collection area, precipitation rate parameters, gas velocity and **aspect ratio** (defined as the ratio of the effective duct length to duct height and related to rapping loss), (2) the electrode arrangement, and (3) the electrical energization needed to provide specified levels of performance. Ancillary factors such as rappers, gas flow control methods, dust removal systems, and performance monitoring must also be considered.

Almost all industrial precipitators are of the **single-stage design, wire and plate type**, where charging and collection take place in the same section. The collecting surface consists of flat parallel plates spaced 0.20–0.28 m apart, with wire or rod discharge electrodes located between the plates. Discharge electrodes (wires) must be tensioned to maintain spacing, such as by attaching a weight to each wire at the bottom, held in a weight-spacing frame to prevent sway. The plates usually range from 4 to 7.5 m in height and 2 to 7.5 m in length. Long plates present collection problems, since the particles collected at the top of the plates must fall to the hoppers to be removed from the precipitator. As they fall, the particles tend to be carried across the plates by the gas stream, and in a very long fall they may be carried out of the precipitator before they reach the bottom. Plate-type precipitators are typically used for dry particulate collection. In **tube-type** precipitators, the collecting surface consists of a cylinder with the discharge electrode (wire) centred along its longitudinal axis. The tubes may vary in diameter from 150–300 mm, by 4.5–5.0 m long. Dirty gas generally enters the hopper at the bottom and travels up through the tubes. They are generally used for wet gas cleaning.

Most commonly used **corona electrodes** are round, square or barbed wires of the order of 2.5 mm in diameter. Pipe precipitators use pipes ranging in size from about 15 cm in diameter by 3 m long for small units, up to 40 cm in diameter by

6 m long for larger units. The number of pipes per precipitator varies from a few up to 100 or more, depending on application and amount of gas treated. Duct precipitators use vertically mounted collecting plates, spaced apart in rows, with corona electrodes placed midway between adjacent plates. Duct sizes are typically 10–20 m long by 10–15 m high by 20–40 cm wide. Plates are commonly 1–3 m wide by 10–15 m high.

Corona current densities should be maintained at the highest possible levels in order to maximize particle collection rates. In practice, it will be in the range of 0.05 to 1.0 mA/m^2, but operating current densities are limited by the following factors: (1) gas composition, temperature and pressure, (2) electric space charge effects of the suspended particles, (3) resistivity of the collected particle layers, (4) electrode and accuracy of alignment, (5) collection area energized per electrical set, (6) type and design of the high-voltage sets and controls, and (7) effectiveness of rappers in keeping the electrodes clean.

In both theory and practice, the **operation** of precipitators is governed largely by the magnitude of the charge, the electric field, and the extent of re-entrainment of collected dust. Factors that determine precipitator performance are those which establish limits on each of these parameters. These theoretical considerations should be used only for order-of-magnitude estimates, since a number of extraneous factors may enter into the actual performance. In actual installations, rectified alternating current is employed. Hence the electric field is not fixed but varies continuously, depending on the wave form of the rectifier, although it has been suggested that this is not a critical factor. In addition, irregularities on the discharge electrode will result in local discharges. Very high dust loadings increase the potential difference required for corona and reduce the current due to the space charge of the particles. This tends to reduce the average particle charge and reduces collection efficiency. This can be compensated for by increasing the potential difference when high dust loadings are involved.

Important gas stream and particle properties which affect the effectiveness of electrostatic precipitation of particulate matter include particle size distribution, gas flow rate, resistivity and temperature. The design must also consider how these characteristics affect the corrosiveness of the particles and the removability of particles from the collection electrodes. Gas velocity in the electrostatic precipitators is extremely important since collection is highly sensitive to velocity variations. The gas flow velocity, at which maximum efficiency can be attained, depends on such factors as plate configuration, precipitator size, and the judicious use of flow distributors required to minimize velocity gradients. The design velocity limit for high efficiency fly ash precipitators is about 1.5 to 1.8 m/s. Two factors limit the **gas velocity** through the precipitator (and consequently its capacity). First, the particles must have time to build up a charge and migrate to the collecting electrode before the gas leaves the precipitator. Secondly, the gas velocity must be low enough so that it does not re-entrain the collected particle. Not only may falling particles be picked up by the gas stream, but high enough velocities may scour them right off the collecting electrodes. Particles may even be pulled out of full hoppers. The highest permissible gas velocity is dependent on the material being collected. Different materials have differing densities, adhere differently to the collecting electrodes, dislodge and fall differently.

Uniformity of **gas flow** has a profound influence on precipitator performance, especially when low-density dust is being collected. Non-uniformity of gas flow can cause severe re-entrainment of dust due to direct scouring of the collected dust, variable treatment times and increased rapping losses in the higher velocity regions, the combination of which can cause marked decreases in precipitator efficiency. Other effects of poor gas flow are fallout or build-up of dust around turning vanes, elbows and distribution plates. These conditions further alter the gas flow pattern, causing even less uniformity. The best operating conditions for electrostatic precipitators would be provided by uniform gas flow. Such a condition is, however, never achievable in practice.

In general, increased **pressure** increases precipitation efficiency, although a somewhat higher potential is required, because it reduces ion mobility and hence increases the potential required for corona and sparking. Increased **temperature** reduces collection efficiency because ion mobility is increased, lowering critical potentials, and because gas viscosity is increased, reducing migration velocities.

The fundamental equation for describing the collection efficiency of electrostatic precipitators is the empirical **Anderson equation**:

$$n = 1 - \exp(-K_a t) . \qquad (4)$$

This equation was derived on the basis of theoretical considerations, the principal assumptions being that: (a) the particle concentration is uniform through the cross section, (b) the particles are fully charged immediately on entering the precipitator, and (c) there is no loss or re-entrainment of the collected particles.

In general, the **performance** of a given electrostatic precipitator is a function of the size (plate area and depth), the particle resistivity and size distribution, the electrical parameters defining particle charge and field strength, and proper operation and maintenance of equipment. Typical values of design parameters include among others: gas velocity 0.61 to 2.4 m/s, treatment time 2 to 10 s, draft loss 25 to 125 Pa, field strength 276 to 590 kV/m, gas temperature up to 370 °C (standard) or 540 °C (high), or 700 °C (special), efficiency up to 99.9% (although usually 90–98%). Values of these parameters for a given design will depend on particle and flue gas properties, total gas flow, and required collection efficiency. It has also been shown that the efficiency of particle removal is an exponential relationship of the form:

$$n = 1 - \exp[-(A/Q)w] . \qquad (5)$$

This equation is commonly referred to as the **Deutsch equation** and is widely used in precipitator design and analysis. It is generally used to estimate collection efficiency when this relies upon flow rate, plate area, and average electrical migration velocity. In most cases, however, field data indicate lower efficiencies than predicted by the Deutsch relationship and equipment suppliers rely on experience in similar applications to size commercial precipitators. The area of the collection plate (electrode) is also an important design consideration and can be evaluated by using the previous equation, as it is a function of the gas stream flow rate, the particle drift velocity and the design system removal efficiency.

The **electrical energy requirement** for an electrostatic precipitator is that necessary to produce an effective corona. Power requirements vary with application depending upon the resistivity of the dust, gas temperature and velocity, particles loading, precipitator geometry, and extent of corona current suspension. Average values are around 0.012 kW/m^3 for high efficiency electrostatic precipitators and 0.009 kW/m^3 for others. Current densities for wet-wall precipitators or dry precipitators collecting material of low resistivity can be as high as 550–660 $\mu A/m^2$. However, if the collected dust resistivity is high, current densities of 55–110 $\mu A/m^2$ or lower may be found in practice. Under conditions of extreme corona current suspension due to a large number of small particles, current densities of about 10–20 $\mu A/m^2$ may also be found.

Particle **deposits** occur on both the corona and the collection electrodes in thickness from a few millimetres to several centimetres. Deposits on the corona electrodes may interfere with the magnitude and uniformity of the corona current emissions. The problem is simply to keep the corona electrodes sufficiently clean. The most effective means is rapping the corona electrode supporting structure, which must be sturdily constructed to effectively transmit the rapping blows to the corona electrodes. Rapping methods usually consist of mechanically actuated "hammers" which are driven electrically, or pneumatically/impulse, or vibration type against the freely suspended collection electrode, so that the dust is released in the direction of gravity. Rappers are provided for the discharge electrodes as well. Rapping systems must be highly reliable, provide high-intensity rapping forces when needed, and be adjustable in intensity or frequency. Rapping frequency or interval is generally adjustable in all types of rapping equipment. Cycling times range from a few minutes to several hours, depending upon the type of dust and the rapping technique. In the case of some viscid dusts the removal of collected material can be accomplished by washing down the electrode with water or other fluids in what is termed a "wet" precipitator.

Electrical resistivities of the dusts encountered in industrial gas cleaning applications can differ considerably. Some materials, such as carbon black, have very low resistivity, so that on contact with a grounded metal surface the particles lose their charge and are easily re-entrained into the gas stream. At the opposite extreme, dusts of insulating materials, such as alumina, can have a sufficiently high resistivity so that the charge leaks off very slowly. In such cases, the electrical force holding the dust to the collection plate can be very high and the voltage drop across the dust layer can be sufficient to cause breakdown of the interstitial gases within the dust layer. Liquid particles and certain types of solid particles are intrinsically conductive and therefore cannot cause difficulty because of high resistivity. Many dusts and fumes met in industrial precipitator applications, however, are composed of silicates, metallic oxides, and similar inorganic compounds, which in the pure dry state are good insulators and therefore might be expected to cause trouble in precipitators. However, moisture and chemical impurities present in the gases and particles always provide at least some of the trace conductivity required. In other cases the gas temperature may be sufficiently high to ensure adequate conductivity. Thus high resistivity is a problem in only a fraction of applications.

Two modes of **electrical conduction** are possible in industrial dusts, depending upon the temperature and composition of the dust and flue gases. At elevated temperatures (above 220 °C) conduction takes place primarily through the bulk of the material. The resistivity of the dust in the temperature region where bulk conduction predominates is referred to as **volume resistivity**. At lower temperatures, moisture or other substances present in the flue gases are adsorbed and conduction occurs principally along the surface of the dust particles. In the temperature region where conduction takes place along the surface of the dust particles, the resistivity is referred to as surface resistivity. In the high temperature region, the resistivity depends on the chemical composition of the material. In the low temperature region, resistivity is thought to be an ion-transport phenomenon related to the adsorption of water vapour or other conditioning agents present in the flue gas. Higher moisture content in the flue gases and lower temperature give lower resistivities.

Major methods used in practice for overcoming high-resistivity problems are as follows.

1. Keeping electrodes as clean (dust-free) as possible, by frequent cleaning. The dust thicknesses can be reduced by heavy rapping to the order of only 1 mm or less, which is sufficient to reduce substantially the adverse effects of high resistivity.
2. Providing better electrical energization, by using pulse voltages, greater high-tension sectionalization, fast-acting spark-quenching circuits and effective automatic control systems.
3. Using chemical and moisture conditioning, such as adding sodium compounds, by alteration of the moisture-adsorption properties of dust surfaces.
4. Using temperature conditioning. Many industrial particles of mineral origin, which have high resistivities at temperatures in the 150–200 °C range, become sufficiently conductive for precipitation purposes at both lower and higher temperatures. In practice, it has been found that sufficient conductivity for precipitation can often be achieved by lowering the gas temperature below about 130 °C, or by raising the gas temperature above about 350 °C. These lower and higher temperature levels can in some cases be achieved by changes in the plant process, or at the design stage, by suitable location of the precipitator in the plant system. Cooling by water sprays is especially effective, because both moisture and lower temperature effects are obtained simultaneously.
5. Designing a larger than normal precipitator (oversizing the unit) to accommodate the lower precipitation rate.

Low resistivity can sometimes also be a problem. If the resistivity is below 10^4 ohm-cm, the collected particles are so conductive that their charges leak to the ground faster than they are replenished by the corona. The particles are no longer electrostatically pinned to the plate, and they can be swept away and re-entrained in the exit gas. The particles may even pick up positive charges from the collecting plate and then be repelled. Low resistivity problems are common with dusts of high carbon content, and may also occur in fly ash precipitators, which handle the ash from high-sulphur coal and operate at low gas temperatures. Low temperature in this case results from excessive condensation of electrically conductive sulphuric acid. Special design features for collection electrodes or coatings are needed to reduce reentrainment.

High-temperature precipitators, operating at 300–400 °C, where dust resistivity tends to be below the critical level of 10^{10} ohm-cm, have long been used for the collection of cement kiln and certain other dusts. This arrangement became known as a "hot" or "hot-side" precipitator. The primary advantage of the hot-side precipitator is avoidance of high-resistivity ash which often occurs with low-sulphur coals at flue gas temperatures of the order of 150 °C, typical for flue gases after the air preheater. Also, less cooling is required and a greater plume buoyancy is achieved. There are also serious disadvantages. Gas flows are about 50% higher at the high gas temperatures. Precipitator operating voltages are substantially reduced owing to the lower densities of hot gases. Gas viscosities increase with gas temperatures, thus reducing precipitation rate. Structural and mechanical problems have occurred and high costs associated with the necessarily long interconnecting flues needed between the precipitator and the boiler. But the major problem is the deterioration of precipitator performance which has been found to occur in a significant percentage of cases in the form of degraded electrical conditions and lowered precipitator efficiency.

Reliability-Maintenance

Electrical precipitators are probably the most versatile of all types of dust collectors. Very high collection efficiencies can be obtained regardless of the fineness of the dust, provided that they are properly maintained. The chief disadvantages are the high initial cost and, in some cases, high maintenance costs. Furthermore, caution must be exercised with dusts that are combustible in the carrier gas.

It is very important to stress that the performance actually attained in practice is also highly dependent on the mechanical and electrical quality of the precipitator. Properly designed and installed precipitators should require a minimum of maintenance and supply. One of the goals in the design of precipitators is to ensure maintenance-free operation between major plant maintenance outages, which occur at 1- to 2-year intervals. Experience shows that deficiencies in quality such as poor electrode alignment, warped electrodes, air in-leakage, poor gas flow distribution and mismatched or unstable rectifier sets are not infrequent and must be guarded against. In some applications allowance needs to be made for outage of one or two electrical sections of a precipitator by addition of redundant sections. In practice, precipitators are usually operated at the highest voltage practicable without sparking, since this increases both the particle charge and the electrical precipitating field.

The **main problems** encountered in a significant percentage of precipitator installations, can be classified into three major areas: fundamental, mechanical and operational, as summarized in Table 4.

There are also some cases in which a precipitator is not expected to work well. These include separating dusts with high resistivity that prevents proper electrical operation, and dusts that tend to build up and form hard substances impossible to remove from the collecting electrodes. These dusts can usually be handled by changing the gas temperature or moisture content to a more favourable level that lowers resistivity or adhesiveness. This is expensive unless it can be justified because of some other favourable circumstances in the process. There are other cases where a mist will pass through a precipitator at high temperature and condense later upon contact with the ambient air, forming an objectionable vapour plume. The temperature would have to be reduced to approximately the ambient temperature so that the precipitator could collect all the material in mist form. Safety precautions are always required since the operating voltages could be as high as 100000 V. The overall size of electrostatic precipitators is comparable to fabric filters (bag-houses).

Table 4. Commonly encountered electrostatic precipitator problems

Fundamental Problems

1. High-resistivity particles
2. Re-entrainment of collected particles
3. Poor gas flow
4. Insufficient or unstable rectifier equipment
5. Insufficient number of corona sections
6. Improper or incomplete rapping
7. Gas velocity too high
8. Aspect ratio too small
9. Precipitator size too small

Mechanical Problems

1. Poor electrode alignment
2. Distorted collecting plates
3. Vibrating or swinging corona wires
4. Excessive dust deposits on corona electrodes and/or collecting plates
5. Formation of dust mountains in precipitator inlet and outlet ducts
6. Gas turning vanes and/or gas distribution screens plugged with dust
7. Air in-leakage into hoppers, shells or gas ducts
8. Gas sneakage around precipitation zones and/or through hoppers

Operational Problems

1. Full or overflowing hoppers
2. Shorted corona sections (broken wires, etc.)
3. Rectifier sets or controls poorly adjusted
4. Precipitator overloaded by excessive gas flow
5. Precipitator overloaded by excessive dust concentration
6. Process upsets (poor combustion, steam leaks, etc.)

Scrubbing Processes

Description

A wet scrubber is a device that uses a liquid in the separation of particulate or gaseous contaminants from a gas. These devices are serving for applications requiring very high collection efficiencies or particles generally between 0.5 to 5.0 μm in diameter. The transfer of a contaminant requires the contacting of the dusty gas and liquid and their subsequent separation into cleaned gas and contaminated liquid streams. Particles will not be removed efficiently if the scrubbing-liquid droplets are much more than 200 times the particle diameter. In view of this very general definition, there are as many types of scrubbers as there are ways of contacting a liquid and a gas; this may be done by using baffles, packing, bubble caps, and the like (Fig. 10). The purpose of the liquid is to provide the collecting surface and/or to wash a solid collecting surface. In the latter case the primary means for particle collection is some type of dry collector, such as filter or cyclone separator.

The specific application of scrubbers examined here is restricted to the collection of particles. The chief difficulty in operating a gas washer is wetting the particles, because it is often necessary for the liquid to displace an adsorbed layer of gas. In some cases the gas is sprayed with a mixture of water and oil to facilitate wetting and it may then be bubbled through foam. This method of separation cannot be used, of course, in cases where a dry gas is required.

Typically, a liquid (usually water) is introduced upstream of the throat and flows down the converging sides into the throat where it is atomized by the gaseous stream; this method is called the "wetted approach". Alternatively, the liquid can be injected into the throat itself by use of nozzles directed at the throat; this approach is called the "non-wetted approach". The non-wetted approach works well when a gas is already close to saturation; however, this method requires that the liquid is free of particles that could clog the nozzles. Where inlet gases are hot and a significant amount of liquid needs to be evaporated, the wetted approach is preferred.

Once the liquid is atomized, it begins to collect particles from the gas impacting into the liquid as a result of the difference in velocities of the gas stream and the atomized droplets. As the mixture decelerates in the expanding section, further impaction occurs, causing the droplets to agglomerate. Once the particles have been trapped by the liquid, a separator (e.g. cyclone, demisters, swirl vanes) can readily remove the scrubbing liquid from the cleaned gas stream.

There are basically two **categories** of wet scrubbers: low energy and high energy types. The low energy type makes use of restricted passages through a liquid, a plate or packing on which a head of liquid is maintained, or a flow of gas through a chamber equipped with sprays. The high energy types are more commonly known as venturi scrubbers. These use a venturi for the intermixing

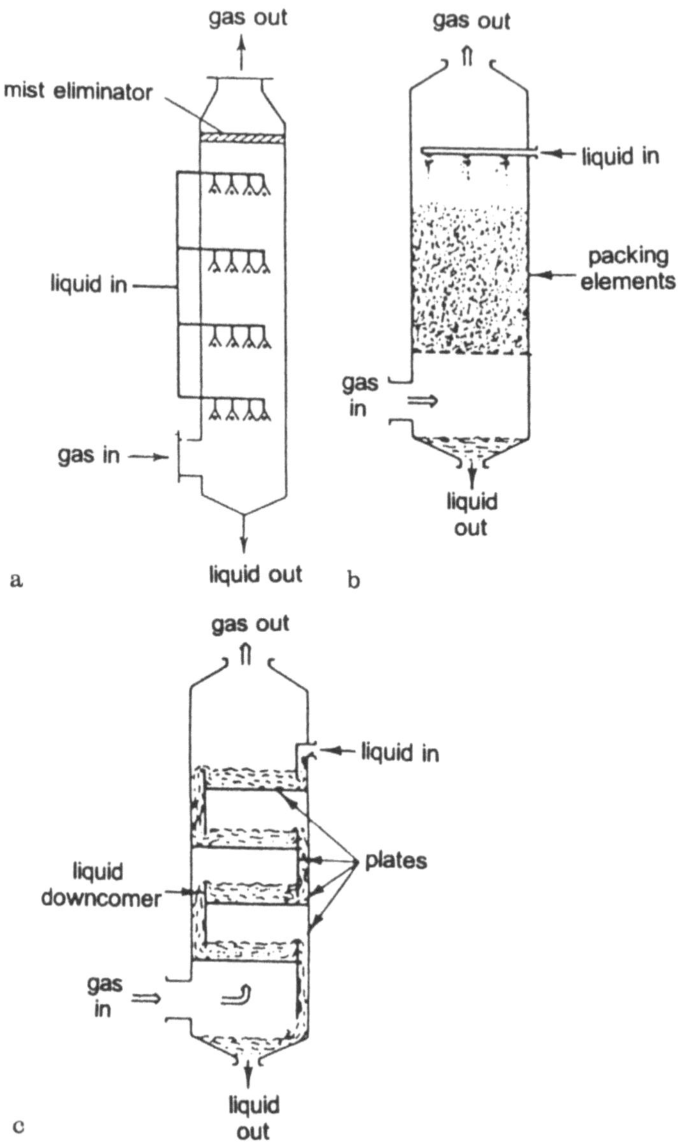

Fig. 10a–c. Wet scrubbers: **a** spray tower; **b** packed column; **c** plate column

of dust gas and scrubbing liquid by impacting high velocity gas on the injected liquid streams.

The unit mechanisms, i.e. the basic gas-liquid contacting situations in which particles are removed from the gas, are considered to be the following: (1) collection by drops (or liquid) moving through gas, (2) collection by cylinders (usually solids, lie wires), (3) collection by sheets of liquid (usually flowing over solid surfaces), (4) collection from bubbles of gas (usually flowing over solid surfaces), and (5) collection from gas jet impingement (either on liquid on solid surfaces). For each unit mechanism the particles are separated from the gas by one or more of the following particle collection mechanisms: gravitational sedimentation, centrifugal deposition, inertial impaction, interception, Brownian diffusion, thermophoresis, diffusion-phoresis and electrostatic precipitation.

Primary mechanisms are usually considered the inertia, diffusion and interception. In scrubbers operating at velocities in excess of 0.3 m/s, the predominant mechanism is inertial impaction. The other two are considered modifying parameters which can sometimes contribute to a significant extent. For example, diffusion is effective for particles smaller than 0.1 μm, when the relative velocity between the particle and collecting body is small. Interception can be significant when the particle path is obstructed by many collecting bodies of a diameter similar to that of the particle. The combined effects of diffusion and interception are, in most scrubber applications, minor when compared with the initial mechanism.

Scrubbers can be **classified** into the following generic groups (Table 5), according to their geometry and basic mechanism(s) for particle collection.

A brief description of these scrubber categories is given now.

1. **Plate scrubbers.** This type consists of a vertical tower with one or more horizontal plates (trays) inside. It is the most common type for gas absorption applications. Gas comes in at the bottom of the tower and must pass through perforations, valves, slots, or other openings in each plate before leaving through the top. Usually, liquid is introduced to the top plate and flows successively across each plate as it moves downward to the liquid exit at the bottom. Gas passing through the openings in each plate mixes with the liquid flowing over it. Gas-liquid contacting causes the mass transfer or particle removal for which the scrubber was designed. Particle collection takes place mainly by inertial impaction from gas jets impinging on the liquid or on solid members. It may possibly be aided by atomization of liquid in contact with the gas jets flowing through the plate. Collection efficiency increases as the perforation diameter decreases and

Table 5. Scrubber classification

Type	Mechanism(s) for particle collection
1. Plate	Jet impingement, bubbles
2. Massive packing	Sheets (curved or plane), jet impingement
3. Fibrous packing	Cylinders
4. Preformed spray	Drops
5. Gas atomized spray	Drops, cylinders, sheets
6. Centrifugal	Sheets
7. Baffle and secondary flow	Sheets
8. Impingement-entrainment	Sheets, drops, cylinders, jets
9. Mechanically aided	Drops, cylinders, sheets
10. Moving bed	Bubbles, sheets
11. Combinations	

can enable a **cut diameter** of about 1.0 μm for 3.2 mm diameter holes in a sieve plate (the cut diameter is that of a particle that is collected at 50% efficiency). A plate does not have the same efficiency for all particle sizes, but rather shows a sharp efficiency change around the cut diameter. Once particles larger than this size are removed from the gas, additional plates can do little good. This kind of behaviour is characteristic of most types of scrubbers and should be kept in mind whenever one is tempted to try two scrubbers in series. Care must be taken in selecting plates for systems that have a tendency toward scaling or adherence of solids to the plates, which can result in plugging of the perforations.

2. **Massive packing.** Packed-bed scrubbers are familiar as gas absorbers or fractionators (e.g. used for distillation) and can also be used, but rarely, as particle removal scrubbers. They may be packed with a range of manufactured elements, such as various ring- and saddle-shaped packings. The gas-liquid contacting may be con-current, counter-current or cross-flow. Collection in packing works mainly by centrifugal deposition due to curved gas flow through the pore spaces and around packing edges, and by inertial impaction due to gas jet impingement within the bed. The good mass transfer characteristics of packings can also make for efficient collection of particles by diffusion if the particles are small enough. Collection efficiency in the inertial size range (larger than 0.3 μm) rises as packing size falls. Packings are subject to plugging, but can be removed for cleaning. Temperature limitations are of special importance when plastics are used and, likewise, corrosion can have a severe effect on metallic packings. The choice of one packing material over another, or a packed tower instead of a plate column, will depend upon economics, the availability of suitable materials of construction, and design constraints such as limitation on pressure drop or tendency to become plugged.

3. **Fibrous packing.** Beds of fibre have been employed in various static or moving configurations with continuous or intermittent washing. Fibres can be made from materials such as plastic, glass and steel. Fibrous packing usually has a very large void fraction ranging from 97 to 99%. Fibres should be small in diameter for efficient operation, but strong enough to support collected particles or drops without matting together. Mixtures of large- and small-diameter fibres have been used to combine these properties. Liquid flow flushes away collected material from the fibres in concurrent, counter-current, or cross-flow arrangements similar to those for massive packings. When collection takes place by inertial impaction, accompanying the gas flow around the fibres, efficiency rises as the fibre diameter decreases and as the gas velocity increases. Diffusional collection may be very efficient for very small particles, and the efficiency of this mechanism will improve as gas velocity diminishes. Fibrous beds are very susceptible to plugging and can be impractical where scaling exists or where conditions favour deposition of suspended solids. Obviously, they will also be especially sensitive to chemical, mechanical and thermal attack.

4. **Preformed Spray.** This type of scrubber collects particles on liquid drops that have been atomized by spray nozzles. They are superior for particle collection, without danger of plugging with accumulated mud. For single-fluid atomizers,

properties of the drops are determined by the configuration of the nozzle, the liquid to be atomized and the pressure to the nozzle. With two-fluid atomizers, the flow rates and pressures of both phases (air and water or steam and water) are important. Sprays leaving the nozzle are directed into a chamber that has been shaped so as to conduct the gas past the atomized droplets. Horizontal and vertical gas flow paths have been used, as well as spray-introduced concurrent, counter-current, or cross-flow to the gas. **Ejector venturis** use high-pressure spray both to collect particles and to move the gas.

Particle collection in these units results form inertial impaction on the droplets. Efficiency is a complex function of droplet size, gas velocity, liquid-gas ratio and droplet trajectories. There is often an optimum droplet diameter that varies with fluid flow parameters. For droplets falling at their terminal settling velocity, the optimum droplet diameter for fine particle collection is around 100–500 μm; for droplets moving at high velocity within a few feet of the spray nozzle, the optimum is smaller. Spray scrubbers that take advantage of gravitational settling can achieve cut diameters around 2 μm at moderate liquid-gas ratios. High-velocity sprays can give cut diameters down to about 0.7 μm. Efficiency improves with higher spray-nozzle pressures and liquid-gas ratios.

A **Venturi** scrubber is one of several types of wet scrubber and is most commonly used for large scale control of particulates (Fig. 11). It uses the venturi effect to create water droplets in the high velocity gas stream. These droplets coalesce and collide with and scrub the particulates. Use of baffles and mist eliminators in conjunction with the velocity decrease, returns the droplets with entrained particulates to the liquid phase. These scrubbers employ gradually converging and then diverging sections to clean an incoming gaseous stream. The stream connecting the converging and diverging sections of the scrubber is called the throat. In general, the longer the throat, the higher the collection efficiency at a given pressure drop, provided the throat is not so long that frictional losses become significant. Venturi scrubbers are also available in variable throat designs that allow adjustment to the throat velocity as a means of modifying the pressure drop and efficiency.

In venturi type scrubbers **mixing** is achieved by using fans to accelerate the incoming gas stream to velocities in the range of 46 to 120 m/s in the converging venturi section. As high velocity gas enters the throat of the venturi, it encounters a stream of scrubbing liquor flowing down the walls of the chamber. The liquor is introduced by flooding, sprays, or weirs, and a spray of atomized liquid is directed into the gas stream ahead of the venturi throat. As the thin sheet of liquid reaches the venturi throat, it is sheared off by the gas stream to become a mass of droplets entrained in the atomized sprays. Typically these droplets are 25 to 100 μm in size. In the turbulence zone just beyond the entrance to the throat, these relatively massive droplets move much more slowly than dry particulates in the gas stream. As a result, collisions between droplets are frequent, so most particles are captured by the scrubbing liquor. As droplets with their burden of particulates move through the venturi, the gas/droplet mixture decelerates and the droplets collide with each other to form larger and heavier drops. When they

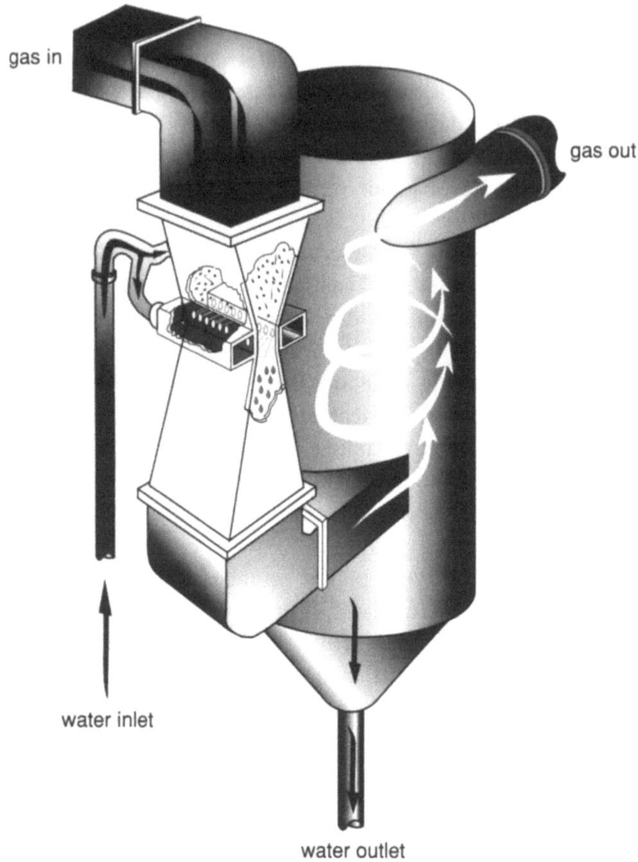

Fig. 11. Venturi wet scrubber

enter the next stage, they are separated from the gas in a cyclone (or other) separation device.

In order to collect particles coarser than 1 μm, venturi scrubbers can be operated at 2.5 kPa; for smaller particles a pressure drop of 7.5–10.0 kPa is usually required. Where appreciable amounts of particulates are finer than 0.5 μm, venturis have been operated with a pressure drop of 17.0–25.0 kPa. Venturi scrubbers produce a significant quantity of liquid waste, which may be discharged to a settling pond or piped to a water treatment plant after solids removal. The quantities of wastewater discharged are difficult to predict since systems use different L/G ratios as well as different degrees of recirculations. A venturi scrubber on a pulverized coal boiler operating at an L/G ratio of 0.9 without recirculation will discharge about 2000 l/min. Usually this discharge is pumped to a settling pond where the fly ash settles to the bottom and the liquid is either discharged, evaporated, or recycled.

Spray scrubbers are practically immune to plugging on the gas flow side but are subject to severe problems on the liquid side. The liquid-gas ratio required is high, usually running 4 to 13 l/m^3 of treated gas, depending on efficiency. The recirculating scrubber liquor can erode and corrode nozzles, pumps and piping. Nozzles can plug with pieces of scale or agglomerates of particles. A heavy loading of liquid entrainment is generated which must be collected. Gas-phase pressure drop is generally low or may even be positive, and thus enhances the flow of gas.

5. **Gas-Atomized Spray.** These devices use a moving gas stream to first atomize liquid into droplets and then accelerate the droplets. Typical of these are the venturi and the various orifice-type scrubbers. High gas velocities of 60–120 m/s raise the relative velocity between the gas and the liquid drops and promote particle collection. Liquid may be introduced in various places and in different ways without having much effect on collection efficiency as long as it results in a uniform spray distribution. These scrubbers have about the simplest and smallest configurations of all the scrubbers. While difficult to plug up, they are susceptible to erosion because of their high throat velocity.

6. **Centrifugal (Cyclonic) Scrubbers.** They impart a spinning motion to the gas passing through them. The spin may come from tangential introduction of gases into the scrubber, or from the direction of the gas stream against stationary swirl vanes. In a dry centrifugal collector (cyclone) the walls can be wetted to decrease re-entrainment of particles and to wash the deposits. Drops can be projected through the rotating gas stream to catch particles on drops by impaction. The scrubbing liquid, which contains the impurities, is removed from the bottom while the clean gas is removed from the top. The basic particle collection mechanism is **inertial impaction** caused by the rotating gas stream; the centrifugal force can be 100 times the gravity force. In the absence of spray, the efficiency will be the same as for a dry collector. A particle cut diameter of 4.0 or 5.0 μm can be obtained with a centrifugal scrubber without spray. As spray is introduced or generated inside, **impaction on drops** occurs and the efficiency approaches that of a preformed spray scrubber. The optimum drop size is 40 to 200 μm, considerably finer than that recommended for gravitational spray towers. In general, these scrubbers are very inefficient in the sub-micron range of particles. The pressure drop in a single scrubber ranges from 6.3 to 11.3 cm water gage. These scrubber types are fairly simple in form and have no small passages. They are not very susceptible to plugging, although solids can deposit on sections of the wall that are not adequately washed. Other **advantages** include low horsepower requirements, relatively low liquid rate, ease of maintenance and ease of fabrication in most materials of construction.

7. **Baffle and Secondary Flow Scrubbers.** These scrubbers cause changes in gas flow direction and velocity by means of solid surfaces (louvers, zigzag baffles, etc.). Solid collected particles may be washed intermittently from the baffle plates. Particle collection is by the centrifugal deposition caused by change in the main flow direction, or by rotating secondary flows. The cut diameter can go as low as 5 to 10 μm for continuous and discontinuous zigzags and similar arrangements.

Baffles are used as pre-cleaners and as entrainment separators. Heavy-particle or slurry loadings can cause solids deposition, which can lead to plugging and corrosion.

8. **Impingement-Entrainment Scrubbers.** Self-induced spray scrubbers force the gas to impinge on a liquid surface to reach a gas exit. Some of the liquid atomizes into drops that are entrained by the gas and collecting particles. Particle collection is generally by inertial impaction caused by impingement on the liquid surface and the atomized drops. Liquid flow is induced by the gas, so that liquid pumping requirements are mostly for make-up and purge streams. Drop size and the liquid-gas flow ratio inside the scrubber depend on scrubber geometry and gas flow rate, but are not controllable or measurable. Cut diameter ranges according to the impingement velocity. Solids deposition can pose a problem on the bottom and on portions of the wall that are not well washed. Good entrainment separation is required because of the amount of spray generated. The gas velocity through the jets is 22.5–30 m/s. This atomizes the liquid into droplets of the order of 100 μm in diameter. It is these droplets, rather than the baffles, which are the effective collecting bodies in the scrubber. The pressure drop is approximately 3.8 cm water per plate.

9. **Mechanically Aided Scrubbers.** These incorporate a motor-driven device such as fan blades, which cause gas flow, atomize the liquid and accelerate it to high velocity, so that (mainly) inertial impaction takes place between the dust particles and the liquid drops. **Disintegrator** scrubbers draw on a submerged, motor driven impeller to atomize liquid into small drops. The drops spin off the impeller across the gas stream, collecting particles on the way. Particle collection mechanisms, in probable order of importance, are: inertial impaction on the atomized liquid, inertial impaction on the motor elements and centrifugal deposition on the housing. Generally, there seems to be no power advantage for mechanically aided units over other types. High-speed impaction of liquid and slurry on the scrubber parts promotes severe abrasion and corrosion conditions. Rotating parts are also subject to vibration-induced fatigue caused by solids deposition, or wear leading to unbalancing.

10. **Moving-bed (Fluidized-bed) Scrubbers.** These devices provide a zone of mobile packing, usually plastic or glass spheres, where gas and liquid can mix intimately. Gas passes upward through the packing, while liquid is sprayed up from the bottom and/or flows down over the top of the moving bed. Gas velocities are sufficient to move the packing material around when the scrubber operates. The movement aids in making the bed turbulent and keeps the packing elements clean. When hollow or low-density spheres are used, the bed fluidizes and bed depth becomes about double that when quiescent. Particle collection takes place mainly by inertial impaction on atomized liquid and on the packing elements. Cut diameters down to about 1 μm are possibly attained. Moving bed scrubbers prove beneficial where good mass transfer characteristics are needed, as well as particle collection. The agitation cleans the packing and reduces problems with solids deposition. The hydrodynamic stability of the scrubber is limited by fluidization and surging characteristics.

11. **Combination and Enhanced Scrubbers.** Separation of particles or droplets is often facilitated by first increasing the effective size of the individual particles by causing them to agglomerate or coalesce, and then separating the enlarged particles. A number of methods are available. Thus, if the dust- or mist-laden gas is brought into contact with a super-saturated vapour, **condensation** occurs on the particles, which act as nuclei. Particles may grow to several micrometers in diameter with steam addition, enhancing scrubbing efficiency, although the cost of steam may be prohibitive. The use of **condensation effects** to improve scrubbing efficiency has also been suggested. Two different mechanisms are involved. The first involves condensation on the dust particles acting as nuclei, thereby increasing the particle size and increasing efficiency by the inertial mechanism. The second mechanism is due to the phenomenon that aerosol particles are attracted to a condensing droplet and repelled by an evaporating droplet. A knowledge of the beneficial effects of condensation should permit improvement of existing scrubbers by steam or fine-mist addition ahead of the scrub water, or by the use of colder water for scrubbing a gas, which is on the borderline of condensing conditions.

Applicability

Scrubber selection for particulate emission control depends upon the particulate loading, particle size distribution and required removal efficiency and the allowable pressure drop. Generally, venturi scrubbers are utilized in those process areas where wet collection is necessary or appears desirable, after consideration of such factors as capital investment, power requirements, slurry disposal and heat transfer or absorption duties.

The main advantages and disadvantages of wet scrubbers influencing their applicability are presented in Table 6.

Other special considerations affecting scrubber selection include: gas absorption, plugging, re-entrainment, stack-condensate fallout and freezing.

Table 6. Advantages and disadvantages of wet scrubbers

Advantages	Disadvantages
1. Absorbs gas phase emissions	1. Condensate plume suggests pollution
2. Compact size	2. Corrosion
3. Efficient through wide loading range	3. Inefficient with high-temperature gases
4. Insensitive to moisture content	4. Requires high power input
5. Low capital cost	5. Waste scrubber liquid handling required
6. Low operating and maintenance cost	
7. Re-entrainment rare	
8. Versatility for hazardous emissions	

Design-Performance

Relationships between particle collection performance and design parameters may be predicted from consideration of the unit mechanisms involved, or measured by experiment, or both. The primary design features of wet scrubbers are penetration (efficiency) and pressure drop across the device, which is a function of the liquid to gas ratio. Collection performance of venturi scrubbers is affected by the following parameters:

1. Pressure drop, caused by friction with stationary surfaces and by the acceleration of liquid. Frictional loss is very dependent upon the geometry of the scrubber and must generally be determined experimentally. Acceleration loss is fairly insensitive to scrubber geometry and is frequently the predominant cause of pressure drop.
2. Liquid-to-Gas ratio (L/G).
3. Gas velocity.
4. Particle size distribution.

The typical **range of design data** for a venturi scrubber application are the following: pressure drop 1–20 kPa, gas velocity at throat 75–100 m/s, liquid application rate 0.4–1 l/m^3, possible recirculation. A relationship between **pressure drop and L/G ratios** for venturis has been developed, assuming that all energy from the imposed pressure drop is used to accelerate the liquid droplets to the throat velocity of the gas:

$$\Delta P = 10^{-8}(U_G)^2(Q_L/Q_G) \tag{6}$$

indicating that the pressure drop across a venturi is proportional to the square of gas velocity and directly proportional to the L/G ratio. Therefore, within limits, increasing gas velocity will result in increasing pressure drop and decreasing the device cut diameter, other things being equal. Typical gas velocities employed commercially are 30–120 m/s. The lower end, i.e. 30–40 m/s, is typical of power plant applications, while the upper end of the range has been applied to lime kilns and blast furnaces.

Another empirical approach indicates, that the pressure drop for venturis is proportional to U_G^2 and $(Q_L/Q_G)^{0.78}$. Hence, for a specified gas velocity, the pressure drop across venturis is approximately proportional to the L/G ratio. The **factor *f***, which represents the ratio of relative velocity between gas and liquid droplet to gas velocity, is approximately 0.25 for fly ash and hydrophobic aerosols and significantly higher for hydrophilic (water attracted) aerosols. Care should also be given to ensure that adequate liquid is supplied to provide good gas sweeping; a minimum L/G ratio of approximately 1 is recommended.

The particle scrubber **performance characteristics** of primary importance are collection efficiency as a function of particle size and power requirement. For a given scrubber, performance will depend on gas and liquid flow rates and other factors. Particle collection may also be enhanced by increasing the particle size through agglomeration, condensation, or other particle growth mechanism. The optimum size of droplets is affected by the forces that are operating on them. For instance, for scrubbers operating with forces 100 times that of gravity, maximum target efficiencies are obtained with droplets in the range of 40 to 200 μm.

Surfactants have been used in many attempts to enhance the efficiency of scrubbers, with varying success. While some experimental studies have shown that efficiency increases as particle wettability increases, other research and large-scale tests have shown little effect of interfacial tension. The wetting characteristics of the liquid do not necessarily play a major role in the actual process of dust collection. Wetting agents, however, may serve to avoid re-entrainment of dust particles once they have been impinged on liquid droplets. In general, improvements in collection efficiency resulting from the use of wetting agents have been comparatively minor in magnitude.

Reliability

Entrainment of drops in the gas can result in liquid being carried out of the scrubber into any preheater, duct, fan and stack and then into the atmosphere. Changes in the liquid (which also includes suspended solids) can cause scaling, corrosion, erosion, plugging, fan damage, pollutant emission and other serious problems in the scrubber, unless the entrainment is separated from the gas after particle collection. The contaminant that accumulates in the scrubbing liquid must be discharged from the system. Because of this and the environmental consequences, liquid treatment and waste disposal deserve careful attention.

Maintenance is critical in venturi scrubbing systems when corrosive gases (such as sulphur oxides) are present. In addition, handling of large amounts of water containing potentially corrosive or abrasive materials causes more problems than those found in dry systems. Hence, frequent and thorough inspection of equipment is essential for reliable operation.

Comparison of Treatment Methods: Overview and Discussion

Present control methods for particulates are least efficient in the size range of 0.2 to 2.0 μm; this range is the most costly to collect. Health studies indicate that particles in this range are those which penetrate most deeply into the human respiratory system. Therefore, this is the size range for which the greatest improvements in technology are needed. Improvement in collection requires the use of a separating force which is independent of gas velocity, or the growth of fine particles to sizes which can be collected more readily. Particle growth can be accomplished through coagulation (agglomeration), chemical reaction, condensation and electrostatic attraction. Promising separating forces are the "flux forces" involving diffusion-phoresis, thermo-phoresis and electro-phoresis. Although par-

Table 7. Efficiency of dust collectors

Dust collector	Efficiency at 5 μm (%)	Efficiency at 2 μm (%)	Efficiency at 1 μm (%)
Medium-efficiency cyclone	27	14	8
High-efficiency cyclone	73	46	27
Low-pressure-drop cellular cyclone	42	21	13
Tubular cyclone	89	77	40
Irrigated cyclone	87	60	42
Electrostatic precipitator	99	95	86
Wet electrostatic precipitator	98	97	92
Fabric filter	99.8	99.5	99
Spray tower	94	87	55
Wet impingement scrubber	97	95	80
Self-induced spray deducter	93	75	40
Disintegrator	98	95	91
Venturi scrubber	99.8	99	97

Table 8. Summary of dust collector performance

Type of Equipment	Field of application	Pressure loss
Settling chambers	Removal of coarse particles, larger than about 100–150 μm	Below 50 N/m^2
Low-pressure drop cyclones	Removal of fairly coarse dusts down to about 50–60 μm	Below 250 N/m^2
High efficiency cyclones	Removal of average dusts in the range 10–100 μm	250–1000 N/m^2
Scrubbers (venturi, spray towers, etc.)	Removal of fine dusts down to about 5 μm (or down to sub-micron sizes for the high-pressure-type)	250–600 N/m^2 or more
Bag filters	Removal of fine dusts and fumes, down to about 1 μm or less	100–1000 N/m^2
Electrostatic precipitators	Removal of fine dusts and fumes down to 1 μm or less	50–250 N/m^2

Table 9. Limitations of collection equipment

Equipment	Disadvantage
Settling chamber	Size
Baffle chamber	Re-entrainment and dust buildup
Packed cross-flow	Plugging
Electrostatic precipitator	Size and maintenance
Bag-house	Bag life
Dry cyclone	Plugging with damp conditions
Wet cyclone	Plugged nozzles and misting
Packed countercurrent	Plugging
Venturi	Erosion, minimum gas temperature 1200 °F

ticle growth technique and flux-force collection can theoretically be considered independently, both phenomena are frequently at work in many practical devices.

In Tables 7, 8 and 9 a comparison between the performance of several dust collectors is presented.

Further Reading

1. American Chemical Society (1978) Cleaning our environment: A Chemical Perspective pp 138–149
2. Calvert C, Englund (1984) Handbook of air pollution technology, Wiley-Interscience
3. Cheremisinoff PN, Young RA (eds) (1975) Pollution engineering practice handbook, Ann Arbor, USA pp 61–184
4. Corbitt RA (ed.) (1989) Standard handbook of environmental engineering. McGraw-Hill, New York pp 4.1–115
5. Coulson JM, Richardson JF (in collaboration with Backhurst JR, Harker JH (1978) Chemical engineering-vol 2, 3rd edn. Pergamon Press, Oxford
6. Crocker BB, Novak DA, Scholle WA (1987) Air pollution control methods, in Encyclopedia of chemical technology, Kirk-Othmer (ed.), 3rd edn. pp 649–716
7. Environmental Protection Agency, Handbook of control technologies for hazardous air pollutants, Report EPA 625/6–91/014
8. Environmental Protection Agency (1985) Operation and maintenance manual for electrostatic precipitators, Report EPA 625/1–85/017
9. Freeman AM (1982) Air and water pollution control: a Benefit-Cost Assessment, J. Wiley, New York
10. Gordon G, Peisakhov I (1982) Dust collection and dust cleaning, Mir Publs., Moscow
11. Industrial Pollution Control Association of Japan, Industrial pollution control: general review and practice in Japan, Vol. I: Air and water, Japan (1981) pp 176–225
12. Jorgensen SE, Johnsen I (1989) Principles of environmental science and technology, Elsevier, Amsterdam, pp. .
13. Kornelius G (1991) Re-use of dust from gas cleaning operations in the ferro-alloy industry, Proceedings ACHEMA 91, Vol. on Mechan. Engng.
14. Loeffler F (1988) Abscheidung von Feinstaub aus Gasen, Chem. Ing. Tech., 60 (6), 443–452
15. Lund HF (ed.) (1971) Industrial pollution control handbook, McGraw-Hill, pp 23.1–45
16. McGlamery GG, et al. (Jan. 1975) Detailed cost estimates for advanced effluent desulphurization processes, Report NTIS PB-242541
17. Perry R, Chilton CH (eds) (1973) Chemical engineers' handbook, 5th edn. McGraw-Hill, New York
18. Reynolds TD (1982) Unit operations and processes in environmental engineering, Wadsworth, USA
19. Richter LA, Volkov EP, Pokrovsky VN (1984) Thermal power plants and environmental control, Mir Publs., Moscow
20. Salvato J (1972) Environmental engineering and sanitation, Wiley-Interscience, USA
21. Sax NI (ed.) (1974) Industrial pollution, Van Nostrand Reinhold, New York
22. Smith C (1993) Flue filtration, Chemical processing-technology international, 99–104
23. Stern AC (ed.) (1977) Air pollution-Vol. IV: Engineering control of air pollution, 2nd edn. Academic Press Inc., New York.
24. Stone R, Smallwood H (1973) Intermedia aspects of air and water pollution control, Report EPA 600/5–73–003 (NTIS PB-224.812)
25. Strauss W (1975) Industrial gas cleaning, 2nd edn. Pergamon Press, New York
26. Tomany JP (1975) Air Pollution: the emissions, the regulations and the controls, Elsevier, New York
27. Triangle Research Park, Inc. and Radian Corp (1983) Control technology appendices for pollution control: Technical Manual, Report EPA 600/8–83–009 (NTIS PB-83–214.734)
28. Vesilind PA, Peirce JJ (1982) Environmental engineering, Ann Arbor Science, Michingan
29. Erste Allgemeine Verwaltungsvorschrift zum BundesEmmissionsschutzgesetz, Technische Anleitung zur Reinhaltung der Luft - TALuft vom 27.2.1986, 95
30. Bretschneider B, Kurfurst J (1987) Air pollution control technology, Elsevier, New York

Subject Index

Accumulation mode 3
Aerosol 127, 135, 137
- collection, filters 179-186, 197
- emission 76
- removal 77
- sources 72
- transport, modelling 83-89
Aerosols, atmospheric 70-81, 92-94
-, carbonaceous 75
-, composition 58-60
-, growth 104, 109, 110
-, indoor 23
-, inorganic 37
-, organic 45, 47
-, primary/secondary 3
-, size distribution 65, 66
-, volcanic 11
Air mass back trajectory 137
Air-to-cloth ratio 307
Airborne particulate matter 234
Ammonia 45, 56, 204
Ammonium nitrate 39
Ammonium sulfate 39, 56-58
Analytical methods 241-247
- - validation 247, 248
Anderson equation 318
Atomic absorption spectrometry (AAS) 180-189, 196

Bioaerosols 22
Boltzmann charge equilibrium 151
Br/Pb ratio 135

Calorimeter 123
Carbon, elemental 24-27
Carbonaceous aerosol 75
Carrier gas 283-286
Centrifugal separators 293
Chemical potential 38-42
Chemical time bomb 140
CHN analysis 184, 189
Clean-up and fractionation 240
Cloud condensation nuclei 108-112
Coagulation 71
Coal 127-129, 138
Coarse mode 3
Collection efficiency 289, 291, 308, 332
Combustion 126, 127
Condensation nucleus counter (CNC) 266, 270, 271
Corona 311
Cottrell precipitator 313
Cunningham factor 148, 151
Cut diameters 326, 329
Cut-off size 220
Cyclones 209, 293

Deliquescence 104-108
Denuder Difference Method (DDM) 210, 211
Denuders 208, 209, 212, 214-216
Deutsch equation 318
Diesel vehicles 18, 19
Differential mobility analyzer 104
Diffusion coefficient 148, 151
Dimethylsulfide (DMS) 43, 44
Dimethylsulfone 44
Drag coefficient 291
Dry deposition 88, 132, 134
Dust holding capacity 307

Eddy current 295
Electrostatic precipitation 309
Elements, crustal 37
Emission factor 128, 130
Enrichment factor 127, 130
Epiphaniometer, particle counting 271
Eulerian model 85
Extinction coefficient 114
Extraterrestrial sources 10

Filters 207, 208
-, aerosol collection 179-186, 197
Filtration 301

Forest fires 11
Fractal dimension 149
Fugitive dust 18

Gamma spectroscopy 187
Gas-to-particle conversion 20
Gibbs free energy 38, 40, 46, 48
Gibbs-Duhem equation 105
Global cooling 112, 121
Gravity collectors 290

Hospital incinerator emissions 185
Hydrochloric acid 204
Hylocomium splendens 186, 187
Hypogymma physodes 187

Impactors 220-222, 266
Inertial collectors 290
Inlet gas losses 219
Instrumental photon activation analysis (IPAA) 192
Interception 287
Ion chromatography 179, 181, 185, 226
Ionic strength 41
Isoprene 48, 63

Junge distribution 259

Kelvin diameter 271
- equation 109
Knudsen number 53
Köhler equation 109
Koschmeider equation 121

Lagrangian model 85
Lead (Pb) 126-140

Mass conversion 126
Maxwell equation 115
Mechanical collectors 290
Mercury (Hg) 126-133, 138, 141
Methanesulfonic acid (MSA) 44
Microscopy, particle counting 262-264
Mie theory 115, 116
Millikan cell 104
Mineral dust 7
Momentum collectors 290
Monoterpenes 48, 63

Nephelometer 122
Neutron activation analysis (NAA) 175, 177-193, 196-198
Nitrate radical 44
Nitric acid 204
Nitrous acid 204
Nucleation, binary 50-52
- mode 3

Optical particle counter 269, 270
Organic acids 205
- particles 12, 25
- particulates, sampling 235-238
Oxidation rates 113

Parmelia sulcta 187
Particle counter, optical 269, 270
- diameter 258-272
- origin 72, 76-83
- removal mechanisms 287-289
- transmission 219
Particles, coagulation 289
-, coarse/fine 71
Particulates 281
- control system 283, 285
Picea abies 187, 188
PM10 145, 167
Poisson distribution 255
Polytrichum commune 187
Precipitators 316,317
Prompt gamma activation analysis (PGAA) 192

Rault's law 41
Receptor 126, 131, 134, 137
- model 83
Refractive index 113, 117
Relaxation time 151, 153, 162
Resistivity 313, 319
Reynolds number 291
Ricatti-Bessel functions 115

Sample preparation 238, 239
Sampling artefacts 207
- efficiency 154, 156
-, health related 145, 167
-, isokinetic 155
Scattering coefficient 114
- efficiency 115, 116
Scavenging ratio 91
Scrubbers 325

Scrubbing 323-325
Sea salt particles 10
Sedimentation velocity 151, 153, 160, 162
Separation factor 298
Separators, centrifugal 293
Shape factor 149, 150, 154
Sieving 289
Sink 126, 133
Sparkover 312
Sphagnum 187
Spray, preformed 326
Stokes diameter 102
Stokes law 148
Stokes number 155, 156, 159, 161, 220
Stopping distance 153, 292
Sulfate 222-227
- monitor, continuous 206, 224
- particles 119
Sulfur dioxide 42
Sulfur emissions 43
Sulfuric acid 42-44, 51, 52, 203
Surfactants 333

Terminal settling velocity 291
Tillandsia usneoides 187
Tobacco smoke 188, 189, 198
Toluene 63
Transmissionmeter 122
Transport, atmospheric, global 92-94
-, -, regional 84-89
TSP 167
Twomey effect 112

U.S. Clean Air Acts Amendments 177

V/Ni ratio 134, 135
Vaccinium myrtillus 187
Venturi 323, 327
Visibility degradation 120-123
Volcanic aerosol 11
Volcano emissions 180, 183
Vortex 293

Waste incineration 126, 127, 129, 130
Wet deposition 91, 132, 134

X-ray fluorescence (XRF) 175-178, 185, 189, 196
-, total reflection (TRXRF) 190-196

MIX
Papier aus verantwortungsvollen Quellen
Paper from responsible sources
FSC® C105338

If you have any concerns about our products,
you can contact us on
ProductSafety@springernature.com

In case Publisher is established outside the EU,
the EU authorized representative is:
Springer Nature Customer Service Center GmbH
Europaplatz 3, 69115 Heidelberg, Germany

Printed by Libri Plureos GmbH
in Hamburg, Germany